Supramolecular Science:
Where It Is and Where It Is Going

NATO ASI Series

Advanced Science Institute Series

A Series presenting the results of activities sponsored by the NATO Science Committee, which aims at the dissemination of advanced scientific and technological knowledge, with a view to strengthening links between scientific communities.

The Series is published by an international board of publishers in conjunction with the NATO Scientific Affairs Division

A Life Sciences **B Physics**	Plenum Publishing Corporation London and New York
C Mathematical and Physical Sciences **D Behavioural and Social Sciences** **E Applied Sciences**	Kluwer Academic Publishers Dordrecht, Boston and London
F Computer and Systems Sciences **G Ecological Sciences** **H Cell Biology** **I Global Environment Change**	Springer-Verlag Berlin, Heidelberg, New York, London, Paris and Tokyo

PARTNERSHIP SUB-SERIES

1. Disarmament Technologies	Kluwer Academic Publishers
2. Environment	Springer-Verlag / Kluwer Academic Publishers
3. High Technology	Kluwer Academic Publishers
4. Science and Technology Policy	Kluwer Academic Publishers
5. Computer Networking	Kluwer Academic Publishers

The Partnership Sub-Series incorporates activities undertaken in collaboration with NATO's Cooperation Partners, the countries of the CIS and Central and Eastern Europe, in Priority Areas of concern to those countries.

NATO-PCO-DATA BASE

The electronic index to the NATO ASI Series provides full bibliographical references (with keywords and/or abstracts) to about 50,000 contributions from international scientists published in all sections of the NATO ASI Series. Access to the NATO-PCO-DATA BASE is possible via a CD-ROM "NATO Science and Technology Disk" with user-friendly retrieval software in English, French, and German (©WTV GmbH and DATAWARE Technologies, Inc. 1989). The CD-ROM contains the AGARD Aerospace Database.

The CD-ROM can be ordered through any member of the Board of Publishers or through NATO-PCO, Overijse, Belgium.

Series C: Mathematical and Physical Sciences – Vol. 527

Supramolecular Science: Where It Is and Where It Is Going

edited by

Rocco Ungaro

and

Enrico Dalcanale

Dipartimento di Chimica Organica e Industriale,
Università di Parma,
Parma, Italy

Springer Science+Business Media, B.V.

Proceedings of the NATO Advanced Research Workshop on
Supramolecular Science: Where It Is and Where It Is Going
Lerici, Italy
September 4–7, 1998

A C.I.P. Catalogue record for this book is available from the Library of Congress.

ISBN 978-0-7923-5656-1 ISBN 978-94-011-4554-1 (eBook)
DOI 10.1007/978-94-011-4554-1

Printed on acid-free paper

This book contains the proceedings of a NATO Advanced Research Workshop held within the programme of activities of the NATO Special Programme on Supramolecular Chemistry as part of the activities of the NATO Science Committee.

Other books previously published as a result of the activities of the Special Programme are:

WIPFF, G. (Ed.), *Computational Approaches in Supramolecular Chemistry*. (ASIC 426) 1994.
ISBN 0-7923-2767-5

FLEISCHAKER, G.R., COLONNA, S. and LUISI, P.L. (Eds.), *Self-Production of Supramolecular Structures*. From Synthetic Structures to Models of Minimal Living Systems. (ASIC 446) 1994.
ISBN 0-7923-3163-X

FABBRIZZI, L. and POGGI, A. (Eds.), *Transition Metals in Supramolecular Chemistry*. (ASIC 448) 1994.
ISBN 0-7923-3196-6

BECHER, J. and SCHAUMBURG, K. (Eds.), *Molecular Engineering for Advanced Materials*. (ASIC 456) 1995. ISBN 0-7923-3347-0

LA MAR, G.N. (Ed.), *Nuclear Magnetic Resonance of Paramagnetic Macromolecules*. (ASIC 457) 1995.
ISBN 0-7923-3348-9

SIEGEL, JAY S. (Ed.), *Supramolecular Stereochemistry*. (ASIC 473) 1995. ISBN 0-7923-3702-6

WILCOX, C.S. and HAMILTON A.D. (Eds.), *Molecular Design and Bioorganic Catalysis*. (ASIC 478) 1996.
ISBN 0-7923-4024-8

MEUNIER, B. (Ed.), *DNA and RNA Cleavers and Chemotherapy of Cancer and Viral Diseases*. (ASIC 479) 1996. ISBN 0-7923-4025-6

KAHN, O. (Ed.), *Magnetism: A Supramolecular Function*. (ASIC 484) 1996. ISBN 0-7923-4153-8

ECHEGOYEN, L. and KAIFER ANGEL E. (Eds.), *Physical Supramolecular Chemistry*. (ASIC 485) 1996. ISBN 0-7923-4181-3

DESVERGNE J.P. and CZARNIK A.W. (Eds.), *Chemosensors of Ion and Molecule Recognition*. (ASIC 492) 1997. ISBN 0-7923-4555-X

MICHL J. (Ed.), *Modular Chemistry*. (ASIC 499) 1998. ISBN 0-7923-4730-7

ENS W., STANDING K.G. and CHERNUSHEVICH V. (Eds.), *New Methods for the Study of Biomolecular Complexes*. (ASIC 510) 1998. ISBN 0-7923-5003-0

VECIANA J. et al. (Eds.), *Supramolecular Engineering of Synthetic Metallic Materials*. (ASIC 518) 1998. ISBN 0-7923-5311-0

TSOUCARIS G. (Ed.), *Current Challenges on Large Supramolecular Assemblies*. (ASIC 519) 1999 ISBN 0-7923-5314-5

Table of Contents

PREFACE

The study of noncovalent interactions takes the roots in Chemistry, where it was soon recognised as an important area of research, with the 1987 Nobel prize awarded to the pioneers of Supramolecular Chemistry D.J. Cram, J.-M. Lehn and C. Pedersen. The initial studies were mainly devoted to crown ethers (Pedersen, 1967), cryptands (Lehn, 1969) and spherands (Cram, 1979) and on their interaction with ammonium and s block metal cations, which occurs with high efficiency and selectivity. Soon later, it was discovered that cavity containing compounds, such as cyclodextrins, cyclophanes and calixarenes were able to encapsulate also neutral molecules in a well defined structural arrangement and this widened the scope of Supramolecular Chemistry. The molecular recognition processes have been studied in great detail and in different experimental conditions (gas phase, organic media, water solution) leading to the disclosure of the important role played by weak interactions, such as hydrogen bonding, cation/π, C-H/π, *et cetera*. The success of the molecular inclusion and recognition by relatively simple hosts is also evidenced by the large use of cyclodextrins (native and semisynthetic) in pharmaceutical and food industry and by the application of molecular sensors in several fields.

The knowledge on how to master weak noncovalent interactions, which greatly benefited from theoretical and molecular modelling calculations, encouraged scientists to design more complex systems, mostly based on self-assembly, which led to the development of new materials and devices. Supramolecular conductor, semiconductor and magnetic materials have been synthesised, together with liquid crystals, sensors, wires, and grids. Supramolecular devices which are able to perform specific mechanical, photophysical, or electrochemical functions resembling logic gates, have been developed over the years. Both these topics had a profound impact also on a close discipline such as Physics.

At the same time, it has been recognised that several biological processes such as enzyme catalysis, binding to DNA, transport through membranes, cell-cell recognition, drug delivery and action, *et cetera,* are largely regulated by non-covalent interactions and this led to the penetration of supramolecular concepts inside Life Sciences.

Therefore, in the last decade Supramolecular Chemistry has developed more and more into Supramolecular Science, with a broader scope than it had at the beginning. The perception of this development led the organizers of the NATO Advanced Research Workshop entitled Supramolecular Science: where we are and where we are going, held in Lerici (La Spezia, Italy) from September 4-7, 1998, to select a large variety of topics, putting together scientists from different fields, having in common the supramolecular approach in their research.

One hundred and fourteen scientists attended the Lerici Workshop, with a large participation of young people. The beatiful setting of Villa Marigola, just in front of the "Golfo dei Poeti", helped to create a relaxed and friendly atmosphere, which greatly stimulated the exchange of information. It was a discussion oriented and forward looking meeting, with 22 lectures and seventy contributions for the poster sessions. All participants were asked to report on very new topics of their research. The following fields were covered: (i) synthetic receptors for ions, carbohydrates, peptides and small molecules; (ii) molecular recognition in chemical and biological systems, (iii) catenanes, rotaxanes, dendrimers and new materials; (iv) molecular and supramolecular devices; (v) self-assembly and nanotechnology , (vi) supramolecular catalysis, (vii) computational studies on supramolecular systems. The present volume contains most of the invited lectures presented at the Lerici NATO Workshop.

The format of the meeting was strongly encouraged by the scientific panel of NATO Special Programme on Supramolecular Chemistry. We want to thank them for having chosen the "Parma group" as the local organizers of the Workshop. A special thank to the Chairman of the NATO panel Prof. J. Fraser Stoddart and to Prof. David N. Reinhoudt, for having accepted to co-organize the meeting and for their continuous support and advise. We thank also the other members of the local organizing committee, Arturo ARDUINI, Alessandro CASNATI, Roberto CORRADINI, Andrea POCHINI, Franco UGOZZOLI and the students who enthusiastically helped to solve all the problems connected with the organization and to Ms. Paola PEDRETTI who handled all duties connected with the Workshop Secretariat in a very gentle and efficient way.

The first Advanced Research Workshop of this series was also held in Italy (Taormina, December 1991), organized by V.Balzani and L.

De Cola, to celebrate the birth of the NATO Special Programme on Supramolecular Chemistry. Since then, 23 events of this sort and one Advanced Research Institute took place within the Programme, which also provided 93 Collaborative Research Grants to scientists in at least 17 countries. The Lerici Advanced Research Workshop was the last one of the series and was also financially supported by NATO. On behalf of all scientists who benefited from the various initiatives, we want to thank the NATO Scientific Affair Division for the generous support to the field. A special thank to the Director of the NATO Special Programme on Supramolecular Chemistry Dr. Alain Jubier for his interest and for the constant help and advise given to all organisers.

Rocco Ungaro, Enrico Dalcanale
Università degli Studi di Parma
Parma, Italy
December 1998

Organizing Committee

Rocco **Ungaro** (Director)
Dipartimento di Chimica Organica e Industriale
Università di Parma
I-43100 Parma, ITALY

David N. **Reinhoudt**
Laboratory of Supramolecular Chemistry and Technology
University of Twente
7500 AE Enschede, THE NETHERLANDS

J. Fraser **Stoddart**
Department of Chemistry and Biochemistry
University of California at Los Angeles
CA 90095 Los Angeles, U.S.A.

List of Participants and Observers

Engin U. **Akkaya**
Department of Chemistry
Middle East Technical University
06531 Ankara
TURKEY
tel.: +90-312-210-5126
fax: +90-312-2101280
e-mail: akkayaeu@rorqual.cc.metu.edu.tr

Igor S. **Antipin**
Department of Organic Chemistry
Kazan State University
Kremlevskaya, 18
420008 Kazan - RUSSIA
tel.: +7-8432-315486
fax: +7-8432-752253
e-mail: iantipin@ksu.ru

Arturo **Arduini**
Dipartimento di Chimica Organica e Industriale
Università degli Studi
Viale delle Scienze
43100 Parma
ITALY
tel.: +39-0521-905555
fax: +39-0521-905472
e-mail: arduini@ipruniv.cce.unipr.it

Anja **Baehr**
ETH-Zurich
Jmfeldstr. 59
CH-8037 Zurich
SWITZERLAND
tel.: +41-1-6326010
fax: +41-1-6321109
e-mail: anja@carb.org.chem.ethz.ch

Vincenzo **Balzani**
Dipartimento di Chimica "G. Ciamician"
Università degli Studi di Bologna
Via Selmi 2
40126 Bologna
ITALY
tel.: +39-51-259560
fax: +39-51-259456
e-mail: vbalzani@ciam.unibo.it

Francesco **Barigelletti**
Istituto FRAE-CNR
Via P. Gobetti, 101
40129 BOLOGNA - ITALY
tel.: +39-51-6399819
fax: +39-51-6399844
e-mail: franz@bofra3.frae.bo.cnr.it

Jan **Becher**
Kemisk Institut - Department of Chemistry
Odense Universitet
Campusvej 55
DK-5230 ODENSE M
DENMARK
tel.: +66-15-86-00
fax: +66-15-87-80

Klaus **Bechgaard**
Department of solid State Physics
Risø National Laboratory
DK-4000 Roskilde
DENMARK
tel.: +45-46774701
fax: +45-42360115
e-mail: klaus.bechgaard@risoe.dk

Maria **Bochenska**
Department of Chemistry
Technical University of Gdansk
ul.Narutowicza 11/12
80-952 Gdansk
POLAND
tel.: +48-58-347-21-69
fax: +49-58-347-26-94
e-mail: marboch@altis.chem.pg.gda.pl

Roberta **Cacciapaglia**
Dipartimento di Chimica
Università "La Sapienza"
P.le Aldo Moro, 5 - Box N. 34 -
roma 62
00185 Roma - ITALY
tel.: +39-6-49913698
fax: +39-6-490421-31
e-mail: cacciapaglia@axrma.uniroma1.it

Sebastiano **Campagna**
Dipartimento di Chimica
Inorganica, Chimica Analitica,
Chimica Fisica - Università di
Messina
Via Sperone, 31
98166 S. Agata (Messina) - ITALY
tel.: +39-90-6765737
fax: +39-90-393756
e-mail: photochem@chem.unime.it

James W. **Canary**
Department of Chemistry
New York University
NY-10003 New York - U.S.A.
tel.: +1-212-998-8422
fax: +1-212-260-7905
e-mail: James.Canary@NYU.edu

G. W. **Canters**
Leiden University
Leiden Institute of Chemistry
Einstenweg 55
2333 CC Leiden
THE NETHERLANDS
tel.: +31-71-5274256
fax: +31-71-5274349
e-mail: CANTERS@Chem.LeidenUniv.nl

Stuart J. **Cantrill**
Dept. of Chemistry and
Biochemistry
University of California at Los
Angeles
405 Hilgard Avenue
CA 90095 Los Angeles - U.S.A.
tel.: +1-310-2067032
fax: +1-310-2061843
e-mail: cantrill@chem.ucla.edu

Alessandro **Casnati**
Dipartimento di Chim.Org. e Ind.le
Università degli Studi di Parma
Viale delle Scienze
43100 Parma-ITALY
tel.: +39-0521-905555
fax: +39-0521-905472
e-mail: casnati@ipruniv.cce.unipr.it

Stefano **Colonna**
Facoltà di Farmacia
Università di Milano
Via Golgi, 19
20133 Milano-ITALY
tel.: +39-2-70602231
fax: +39-2-70638473
e-mail: colonna@imiucca.csi.unimi.it

Roberto **Corradini**
Dipartimento di Chimica Organica e Industriale
Università degli Studi
Viale delle Scienze
43100 Parma-ITALY
tel.: +39-0521-905555
fax: +39-0521-905472
e-mail: corradini@ipruniv.cce.unipr.it

Alberto **Credi**
Dipartimento di Chimica "G. Ciamician"
Università di Bologna
Via Selmi, 2
40126 Bologna
ITALY
tel.: +39-51-259541
fax: +39-51-259456
e-mail: acredi@ciam.unibo.it

Richard M. **Crooks**
Department of Chemistry
Texas A&M University
P.O. Box 300012
TX 77842-3012 College Station
U.S.A.
tel.: +1-409-8455629
fax: +1-409-8451399
e-mail: crooks@chemvx.tamu.edu

Enrico **Dalcanale**
Dipartimento di Chimica Organica e Industriale
Università degli Studi
Viale delle Scienze
43100 Parma - ITALY
tel.: +39-0521-905555
fax: +39-0521-905472
e-mail: dalcanal@ipruniv.cce.unipr.it

Antonella **Dalla Cort**
Dipartimento di Chimica
Università "La Sapienza"
P.le Aldo Moro, 5 - Box N. 34 - roma 62
00185 Roma
ITALY
tel.: +39-6-49913386
fax: +39-6-490421 - 490631
e-mail: adallacort@axrma.uniroma1.it

Anthony P. **Davis**
Trinity College
University of Dublin - Dublin 2
IRELAND
tel.: +353-1-6082050
fax: +353-1-6712826
e-mail: adavis@tcd.ie

Silvia M. **de Brito Costa**
Centro de Quimica Estruttural/Complexo 1
Instituto Superior Técnico
Av. Rovisco Pais
P-1096 Lisbon Cocex
PORTUGAL
tel.: +351.1.316.1083/8419271
fax: +351.1.846.4455/57
e-mail: sbCista@alfa.ist.utl.pt

Javier **de Mendoza**
Universidad Autonoma de Madrid
Quimica Organica (C-1)
Facultad de Ciencias - Cantoblanco
E-28049 Madrid
SPAIN
tel.: +341-3974710
fax: +341-3973966
e-mail: iquimorg@vm1.sdi.uam.es

Amilra Prasanna **de Silva**
School of Chemistry
The Queen's University of Belfast
David Keir Building
BT9 5AG BELFAST
UNITED KINGDOM
tel.: +44-1232-274422
fax: +44-1232-382117
e-mail: a.desilva@qub.ac.uk

François **Diederich**
Laboratorium für Organische Chemie
ETH-Zentrum
Universitätstrasse 16
CH-8092 Zürich
SWITZERLAND
tel.: +41-1-6322992
fax: +41-1-6321109
e-mail: diederich@aida.ethz.ch

Abderrazzak **Douhal**
Dept. De Quimica Fisica, Sec. De Toledo
Univ. de Castilla - La Mancha
45002 TOLEDO
SPAIN
tel.: +34-25-212761
fax: +34-25-253614
e-mail: adouhal@qui-to.uclm.es

Luis **Echegoyen**
Department of Chemistry
University of Miami
Coral Gables
Florida 33124-0431 - USA
tel.: ++1-305-6612847
fax: ++1-305-2844571
e-mail: lechegoyen@umiami.ir.miami.edu

Luigi **Fabbrizzi**
Dipartimento di Chimica Generale
Università di Pavia
Viale Taramelli 12
27100 Pavia
ITALY
tel.: +39.382.507328
fax: +39.382.528544
e-mail: fabbrizz@ipv36.unipv.it

Salvatore **Failla**
Istituto Chimico - Facoltà di Ingegneria
Università di Catania
Viale A. Doria, 6
95125 Catania - ITALY
tel.: +39-095-7372452
fax: +39-095-7372450
e-mail: ch12@mail.rz.uni-sb.
desfailla@cdc.unict.it

Alida **Falai**
Istituto di Ricerca sulle Onde Elettromagnetiche "Nello Carrara"
C.N.R.
Via Panciatichi, 64
50127 Firenze - ITALY
tel.: +39-55-4235258
fax: +39-55-4379569
e-mail: falai@iroe.fi.cnr.it

Riccardo **Ferdani**
Dipartimento di Chimica Organica e Industriale
Università degli Studi
Viale delle Scienze
43100 Parma - ITALY
tel.: +39-0521-905555
fax: +39-0521-905472

Ben L. **Feringa**
Department of Chemistry
University of Groningen
9747 AG Groningen
THE NETHERLANDS
tel.: +31-50-3634278
fax: +31-50-3634296
e-mail: b.l.feringa@chem.rug.nl

Francine **Furtado Nactigal**
Coordenadoria de Aperfeicoamento do Ensino Superior - CAPES - BRASIL

Jürgen-H. **Fuhrhop**
Institut für Organische Chemie
Freie Universität Berlin - Takustr. 3
D-1000 Berlin 33 - GERMANY
tel.: +49-30-838-5394
fax: +49-30-838-55-89
e-mail: fuhrhop@chemie.fu-berlin.de

Daniel **Funeriu**
Laboratoire de Chimie Supramoléculaire
Université Louis Pasteur
4, rue B. Pascal
F-67000 Strasbourg - FRANCE
tel.: +33-3-88416159
fax: +33-3-88411020
e-mail: jml292@chimie-u-strasbg.fr

M. Reza **Ghadiri**
Department of Chemistry
The Scripps Research Institute
10666 North Torrey Pines Road
CA 92037 La Jolla - U.S.A.
tel.: +1-619-784-2700
fax: +1-619-784-2798
e-mail: ghadiri@scripps.edu

Bruce C. **Gibb**
Department of Chemistry
University of New Orleans
LA 70148 New Orleans - U.S.A.
tel.: +1-504-2803152
fax: +1-504-2806860
e-mail: BCGCM@UNO.EDU

Claudia **Giorgi**
Dipartimento di Chimica
Università di Firenze
Via Maragliano, 75/77
50127 Firenze - ITALY
tel.: +39-055-354841
fax: +39-55-354845
e-mail: giorgi@chim1.unif.it

Ernest **Giralt**
Departamento de Quimica Organica
Universitat de Barcelona
Marti I Franqués 1
E-08028 Barcelona - SPAIN
tel.: 34-3-4021269
fax: 34-3-3397878
e-mail: giralt@mafalda.qui.ub.es

Peter **Göhlitz**
Angewandte Chemie
Postfach 101161
D-69451 WEINHEIM - GERMANY
e-mail: chemistry@vchgroup.de

Silvia **Gonzales Garcìa**
Departamento Quimica Organica,
Universidad de Salamanca
Salamanca - SPAIN
tel.: +34-23-294474
fax: +34-23-294574
e-mail: silviagg@gugu.usal.es

Andrew D. **Hamilton**
Dept. of Chemistry
Yale University
CT 06520 NEW HAVEN
USA
tel.: +1-203-432-5570
fax: +1-203-432-3221
e-mail: andrew.hamilton@yale.edu

Michael John **Hannon**
Department of Chemistry
University of Warwick
CV4 7AL Coventry - U.K.
tel.: +44-1203-524107
fax: +44-1203-524112
e-mail: m.j.hannon@warwick.ac.uk

William S. **Jorgensen**
Department of Chemistry
Yale University
P.O. Box 6666
CT 06511 New Haven
U.S.A.
tel.: +1-203-432-6278
fax: +1-203-432-6299
e-mail: bill@adrik.chem.yale.edu

Alain H. **Jubier**
Division des Affaires Scientifiques
et de l'Environnement - N.A.T.O.
B-1110 BRUXELLES
BELGIUM
tel.: +32.2.7074111
fax: +32.2.7074232

Burkhard **König**
Institut fur Organische Chemie der
Technischen
Universitat Braunschweig
Hagenring 30
D-38106 Braunschweig -
GERMANY
tel.: +49-531-3915276
fax: +49-531-3915388
e-mail: b.konig@tu-bs.de

Alexander I. **Konovalov**
Arbuzov Institute of Physical &
Organic Chemsitry
Arbuzov str. 8
420083 Kazzan - RUSSIA
tel.: +7-8432-768254
fax: +7-8432-752253
e-mail: konovalov@sci.ksn.ru

Márcio **Lazzarotto**
Coordenadoria de Aperfeicoamento
do Ensino Superior
CAPES - BRASIL

Jean-Marie **Lehn**
Université Strasbourg 1
Institut Le Bel
4 rue Blaise Pascal
F-67070 Strasbourg Cedex -
FRANCE
tel.: +33-3-884160
fax: +33-3-88411020
e-mail: lehn@chimie.u-strasbg.fr

Pavel **Lhotak**
Prague Institute of Chemical
Technology
PICT Technicka 5,
16628 Prague
CZECH REPUBLIC
tel.: +420-2-2435-4280
fax: +420-2-2435-4288
e-mail: lhotakp@vscht.cz

Maurizio **Licchelli**
Dipartimento di Chimica Generale
Università di Pavia
Via Taramelli, 12
27100 Pavia - ITALY
tel.: +39-382-507329
fax: +39-382.528544
e-mail: mali@unipv.it

Aude **Livoreil**
L'OREAL
Advanced Research Center
1, Avenue E. Schueller, BP22
93600 Aulnay-Sous-Bois
FRANCE
tel.: +33-1-48689657
fax: +33-1-48689489

Linda **Lucas**
Department of Chemistry
University of Groningen
9747 AG Groningen
THE NETHERLANDS
tel.: +31-50-3634278
fax: +31-50-3634296

Wanderlei Gageiro **Machado**
Coordenadoria de Aperfeicoamento do Ensino Superior
CAPES - BRASIL

Luigi **Mandolini**
Dipartimento di Chimica
Università "La Sapienza"
Piazzale Aldo Moro, 5 - Box·N. 34 - roma 62
00185 ROMA - ITALY
tel.: +39-6-4957808
fax: +39-6-490421
e-mail: lmandolini@axrma.uniroma1.it

Irene **Mavridis**
Institute of Physical Chemistry
NCSR Demokritos
P.O. Box 60228
Aghia Paraskevi, 153 10
GREECE
tel.: +301-650-3793 or 650 3962
fax: +301-651-1766
e-mail: mavridi@cyclades.nrcps.ariadne-t.gr

Sandro **Mecozzi**
The Scripps Research Institute
MB-26
10550, North Torrey Pines Road
CA 92037 La Jolla - U.S.A.
tel.: +1-619-784-2861
fax: +1-619-784-2876
e-mail: mecozzi@scripps.edu

Pierangelo **Metrangolo**
Dipartimento di Chimica
Politecnico di Milano
Via Mancinelli, 7
20133 Milano
ITALY
tel.: +39-2-23993054
fax: +39-2-23993080
e-mail: fluorine@dept.chem.polimi.it

Josef **Michl**
Department of Chemistry and Biochemistry
University of Colorado
CO 80309-0215 Boulder - U.S.A.
tel.: +1-303-492-6519
fax: +1-303-492-0799
e-mail: michl@eefus.colorado.edu

Marco **Montalti**
Dipartimento di Chimica "G. Ciamician"
Università di Bologna
Via Selmi, 2
40126 Bologna
ITALY
tel.: +39-51-259541
fax: +39-51-259456
e-mail: mmont@ciam.unibo.it

José J. G. **Moura**
Faculdade de Ciencias e Tecnologia
Departamento de Quimica
Universidade Nova de Lisboa
P-2825 Monte de Capaica
PORTUGAL
tel.: +351-1-294-83-82
fax: +351-1-294-8550
e-mail: jose.moura@dq.fct.unl.pt

Piero **Paoletti**
Dipartimento di Chimica
Università degli Studi di Firenze
Via Maragliano, 75/77
50144 Firenze - ITALY
tel.: +39-55-354843
fax: +39-55-354845
e-mail: paoletti@chim1.unifi.it

David **Parker**
Department of Chemistry
University of Durham, South Road
DH1 3LE Durham
UNITED KINDGOM
tel.: +44-191-374-3133 (3129)
fax: +44-191-386-1127
e-mail: David.Parker@durham.ac.uk

Soledad **Penades**
Instituto de Investigationes Quimicas
Amerigo Vespucio s/n
Isla de la Cartuja
E-41092 Sevilla
SPAIN
tel.: +34-95-4489561
fax: +34-95-4460565
e-mail: penades@cica.es

Zoe **Pikramenou**
Department of Chemistry
University of Edinburgh - King's Buildings
West Mains Rd.
EH9 3JJ Edimburg
U.K.
tel.: +44-131-650-4761
fax: +44-131-650-4743
e-mail: z.pikramenou@ed.ac.uk

Roberta **Pinalli**
Dipartimento di Chimica Organica e Industriale
Università degli Studi
Viale delle Scienze
43100 Parma-ITALY
tel.: +39-521-905555
fax: +39-521-905472

Andrea **Pochini**
Dipartimento di Chimica Organica e Industriale
Università di Parma
Viale delle Scienze
43100 Parma - ITALY
tel.: +39-521-905408
fax: +39-521-905472
e-mail: pochini@ipruniv.cce.unipr.it

Miquel **Pons**
Universidad de Barcelona
Dept. de Quimica Organica
Marti I Franques 1
08028 Barcelona - SPAIN
tel.: +343-4021263
fax: +343-3397878
e-mail: miquel@guille.qo.ub.es

Silvio **Quici**
Dipartimento di Chimica Organica e Industriale
Università di Milano
Viale Golgi, 19
20133 Milano-ITALY
tel.: +39-2-2663354
fax: +39-2-2364369
e-mail: quici@iumchz.chimorg.unimi.it

Françisco Maria **Raymo**
Dept. Of Chemistry and Biochemistry
University of California at Los Angeles
405 Hilgard Avenue
CA 90095 Los Angeles - U.S.A.
tel.: +1-310-2067032
fax: +1-310-2061843
e-mail: raymo@chem.ucla.edu

David N. **Reinhoudt**
University of Twente
Department of Chemistry
P.O. Box 217
NL-7500 AE Enschede
THE NETHERLANDS
fax: +31-534894645
e-mail: C.E.vanderPloeg@ct.utwente.nl

Giuseppe **Resnati**
Dipartimento di Chimica del Politecnico
CNR - Centro di Studio Sost. Organiche Naturali
Via Mancinelli, 7
20131 Milano-ITALY
tel.: +39-2-23993032
fax: +39-2-23993080
e-mail: resnati@dept.chem.polimi.it

Helmut **Ringsdorf**
Institut für Organische Chemie
Joannes-Gutenberg Universität Mainz
J. J. Becher-Weg 18-20
D-55099 Mainz - GERMANY
tel.: +49.6131-39-2402
fax: +49.6131-39-3145
e-mail: ringsdorf@mzdmza.zdv.uni-mainz.de

John A. **Ripmeester**
Institute for Molecular Science
National Research Council of Canada
100 Sussex Drive
Ottawa - CANADA
tel.: +1-613-993-2011
fax: +1-613-998-7833
e-mail: jar@ned1.sims.nrc.ca

Kari **Rissanen**
Department of Chemistry
University of Jyvaskyla
P.O. Box 35
FIN-40351 Jyvaskyla
FINLAND
tel.: +358-14-602672
fax: +358-14-602501
e-mail: Kari.Rissanen@jyu.fi

Enrico **Rizzarelli**
Dipartimento di Scienze Chimiche
Università di Catania
Via A. Doria, 5
95025 CATANIA-ITALY
tel.: +39-95-337678
fax: +39-95-325194
e-mail: erizzarelli@dipchi.unict.it

Stefano **Roelens**
Dipartimento di Chimica Organica
C.N.R. - Centro Composti Eterociclici
Via G. Capponi, 9
50121 Firenze-ITALY
tel.: +39-55-2757658
fax: +39-55-2476964
e-mail: roelens@cesce.fi.cnr.it

Paola **Rossi**
Dipartimento di Scienze Chimiche
Università degli Studi di Trieste
Via L. Giorgieri, 1
30100 Trieste
ITALY
tel.: +39-40-6763927
fax: +39-40-6763903
e-mail: rossi@dschsun3.univ.trieste.it

Vincent **Rotello**
Department of Chemistry
University of Massachussetts
MA 01003 Amherst
U.S.A.
tel.: +1-413-545-2058
fax: +1-413-545-4490
e-mail: rotello@chem.umass.edu

Stuart J. **Rowan**
Dept. Of Chemistry and Biochemistry
University of California at Los Angeles
405 Hilgard Avenue
CA 90095 Los Angeles - U.S.A.
tel.: +1-310-2067032
fax: +1-310-2061843
e-mail: @chem.ucla.edu

Laurent **Ruhlmann**
Institut fur Organische Chemie (WE 02)
Freie Universitat Berlin - Fachbereich Chemie
Takustrasse 03
14195 Berlin - GERMANY
tel.: +49-30-838-5394
fax: +49-30-838-5589
e-mail: ruhlmann@chemie.fu-berlin.de

Jeremy K.M. **Sanders**
University Chemical Laboratory
University of Cambridge - Lensfield Road
CB2 1EW Cambridge - U.K.
tel.: +44-1223-336411
fax: +44-1223-336017
e-mail: jkms@cam.ac.uk

Francesco **Sansone**
Dipartimento di Chimica Organica e Industriale
Università degli Studi - Viale delle Scienze
43100 Parma - ITALY
tel.: +39-0521-905555
fax: +39-0521-905472
e-mail: borchi04@ipruniv.cce.unipr.it

Jean-Pierre **Sauvage**
Laboratoire de Chimie Organo-Minérale
Inst. de Chim - 1, rue Blaise Pascal, B.P. 296/R8
F-67008 Strasbourg Cedex - FRANCE
tel.: +33-3-88416130
fax: +33-3-88607312
e-mail: sauvage@chimie.u-strasbg.fr

Gian Franco **Savelli**
Dipartimento di Chimica - Università di Perugia
Via Elce di Sotto, 8
06100 PERUGIA - ITALY
tel.: +39-75-595 5538
fax: +39-75-595 5538
e-mail: savelli@unipg.it

Paolo **Scrimin**
Dipartimento di Chimica Organica - Università
Via Marzolo, 1
35131 PADOVA - ITALY
tel.: +39-49-8275276
fax: +39-49-8275239
e-mail: scrimin@mail.chor.unipd.it

Andrea **Secchi**
Dipartimento di Chimica Organica e Industriale
Viale delle Scienze
43100 Parma - ITALY
tel.: +39-0521-905555
fax: +39-0521-905472
e-mail: borchi02@ipruniv.cce.unipr.it

Ken R. **Seddon**
School of Chemistry
Queen's University
BT9 5AG Belfast - UNITED KINGDOM
tel.: +44-1232-335420
fax: +44-1232-665297
e-mail: k.seddon@qub.ac.uk

Scolastica **Serroni**
Dipartimento di Chimica Inorganica, Chimica Analitica, Chimica Fisica
Università di Messina
Via Sperone, 31
98166 S. Agata (Messina)-ITALY
tel.: +39-90-6765737
fax: +39-90-393756
e-mail: photochem@chem.unime.it

Stefano **Sforza**
Dipartimento di Chimica Organica e Industriale
Università degli Studi
Viale delle Scienze
43100 Parma - ITALY
tel.: +39-0521-905555
fax: +39-0521-905472
e-mail: cochim02@ipruniv.cce.unipr.it

Seiji **Shinkai**
Faculty of Engineering
Kyushu University
Hakazaki 6-10-1, Higashi-Ky
812-81 Fukuoka
JAPAN
tel.: +81-92-6422583
fax: +81-92-6324357
e-mail: seijitcm@mbox.nc.kyushu-u.ac.jp

Jay S. **Siegel**
Department of Chemistry
University of California - San Diego
CA 92093-0358 La Jolla
USA
tel.: +1.619.5345659
fax: +1.619.822-0386
e-mail: jss@chem.ucsd.edu

Yurii **Simonov**
Institute of Applied Physics
Moldova Academy of Science
Academii Street 5
MD 2028 Chishinau
MOLDOVA
tel.: +373-2-738-154/245-653
fax: +373-2-738-149
e-mail: xray@cc.acad.md

Ken **Standing**
Physics Department
University of Manitoba
R3T 2N2 Winnipeg MB
CANADA
tel.: +1-204-4749358
fax: +1-204-4747622
e-mail: standin@cc.UManitoba.CA

Ivan **Stibor**
Department of Organic Chemistry
Institut of Chemical Technology
University
16628 Prague
CZECH REPUBLIC
tel.: +420-2-2435 4240
fax: +420-2-2435 4288
e-mail: ivan.stibor@vscht.cz

Masayuki **Takeuchi**
Dept. Of Chemistry & Biochemistry
- Graduate School of Engineering -
Kyushu University
Hakozaki 6-10-1, Higashi-ku
Fukuoka
812-8581 Fukuoka - JAPAN
tel.: +81-92-642-3585
fax: +81-92-642-3611
e-mail: taketcm@mbox.nc.kyushu-u.ac.jp

Paolo **Tecilla**
Dipartimento di Chimica Organica
Università di Padova
Via Marzolo, 1
35131 Padova - ITALY
tel.: +39-49-8275662
fax: +39-49-8275239
e-mail: tecilla@mail.chor.unipd.it

Iris **Thondorf**
Dept. of Biochem./Biotechnology
Martin-Luther-University
Kurt-Mothes-Str. 3
D-06099 Halle - GERMANY
tel.: +49-345-5524862
fax: +49-345-5527011
e-mail: thondorf@biochemtech.uni.halle.de

Peter **Timmerman**
Dept. of Supramol. Chemistry and
Technology (SMCT)
University of Twente
P.O. Box 217 - 7500 AE Enschede
THE NETHERLANDS
tel.: +31-534892980
fax: +31-534894645
e-mail: P.Timmerman@ct.utwente.nl

Umberto **Tonellato**
Dipartimento di Chimica
Università degli Studi di Padova
Via Marzolo, 1
35131 PADOVA - ITALY
tel.: +39-49-8275269
fax: +39-49-8275239
e-mail: tonum@chor02.unipd.it

Georges **Tsoucaris**
Centre Pharmaceutique C.N.R.S.
ERS 128
Universitè Paris-Sud
92290 CHATENAY-MALABRY
FRANCE
tel.: +331-46-83-56-33
fax: +331-46-83-13-03
e-mail: georges.Tsoucaris@cep.u-psud.fr

Franco **Ugozzoli**
Dipartimento di Chimica Generale
ed Inorganica, Analitica, Chimico
Fisica dell'Università
Viale delle Scienze
43100 Parma-ITALY
tel.: +39-0521-905555
fax: +39-0521-905472
e-mail: ugoz@ipruniv.cce.unipr.it

Rocco **Ungaro**
Dipartimento di Chimica Organica e
Industriale
Università degli Studi
Viale delle Scienze
43100 Parma-ITALY
tel.: +39-0521-905555
fax: +39-0521-905472
e-mail: ungaro@ipruniv.cce.unipr.it

Alessandro **Vigato**
Istituto di Chimica e Tecnologie
Inorganiche e dei Materiali Avanzati
C.N.R.
Corso Stati Uniti, 4
35127 PADOVA-ITALY
tel.: +39-49-8295962
fax: +39-49-8702911
e-mail: vigato@nt.ictr.pd.cnr.it

Ralf **Warmuth**
Department of Chemistry
Kansas State University
111 Willard Hall - KS 66506
Manhattan
U.S.A.
tel.: +1-785-5326684
fax: +1-785-5326666
e-mail: warmuth@ksu.edu

Georges **Wipff**
Laboratoire MSM
Institut de Chimie
4, rue B. Pascal
6700 Strasbourg-FRANCE
tel.: +33-3-88416071
fax: +33-3-88-416104
e-mail: wipff@chimie.u-strasbg.fr

Rudolf **Zentel**
Fachbereich 9
Bugh Wuppertal
Gauss Strasse 20
D-42097 Wuppertal
GERMANY
tel.: +49-202-439-2493
fax: +49.202-439-2493
e-mail: zentel@uni-wuppertal.de

MOLECULAR–LEVEL DEVICES

V. BALZANI, A. CREDI, M. VENTURI
Dipartimento di Chimica "G. Ciamician"
Università di Bologna
via Selmi 2, 40126 Bologna, Italy

1. Introduction

In everyday life we make extensive use of *macroscopic devices*. A macroscopic device is an assembly of components designed to achieve a specific function. Each component of the device performs a simple act, while the entire device performs a more complex function, characteristic of the assembly. For example, the function performed by a hairdryer (production of hot wind) is the result of acts performed by a switch, a heater, and a fan, suitably connected by electric wires and assembled in an appropriate framework.

The concept of device can be extended to the molecular level [1–3]. A *molecular–level device* can be defined as an assembly of a discrete number of molecular components (that is, a *supramolecular* structure) designed to achieve a specific function. Each molecular component performs a single act, while the entire supramolecular structure performs a more complex function, which results from the cooperation of the various molecular components. The extension of the concept of device to the molecular level is of interest not only for basic research, but also for the growth of nanoscience and the development of nanotechnology [4]. Looking at supramolecular chemistry from the viewpoint of functions with references to devices of the macroscopic world is a very interesting exercise that helps the development of chemistry by introducing new concepts.

Molecular–level devices operate via electronic and/or nuclear rearrangements, and like macroscopic devices they need energy to operate and signals to communicate with the operator. Light is the most important answer to this dual requirement, as shown by Nature where photons are used as energy by the devices responsible for photosynthetic processes and as signals by the devices responsible for vision–related processes. In the same way, in artificial supramolecular systems photons can be used to cause (by photochemical reactions) and to monitor (by absorption and emission spectroscopy) the occurrence of electronic and nuclear rearrangements [5]. Electrochemistry is also very useful to cause and monitor the occurrence of electronic and nuclear rearrangements in molecular and supramolecular systems [6].

R. Ungaro and E. Dalcanale (eds.), Supramolecular Science: Where It Is and Where It Is Going, 1–22.

Two kinds of macroscopic devices play a very important role in our civilization: machines and electronic computers. One can wonder whether it is possible to construct devices of these kinds at the molecular level. In the last few years prototypes of very simple molecular–level machines [7] and molecular–level components related to the construction of molecular–based (chemical) computers [8] have been developed by several research groups. For space reasons, we will only illustrate a few of the systems investigated in our laboratory. We would like to recall, however, that an outstanding result has recently been obtained with molecular–level devices, namely the light–driven production of ATP in an artificial photosynthetic membrane [9].

2. Molecular–level Machines

2.1. INTRODUCTION

The concept of machines at the molecular level is not new. Our body can be viewed as a very complex ensemble of molecular–level machines that power our motions, repair damage, and orchestrate our inner worlds of sense, emotion, and thought [10]. The 1997 Nobel Prize in Chemistry went to three researchers (P. Boyer, B.J. Walker, and J.C. Skou) for their pioneering work on ATP synthase, one of the most important machines of our body [11].

The problem of the construction of artificial molecular–level machines was posed for the first time by Richard P. Feynman, Nobel Laureate in Physics, in his famous address *There Is Plenty of Room at the Bottom* to the American Physical Society in 1959 [12]: "*...What are the possibilities of small but movable machines? ...An internal combustion engine of molecular size is impossible. Other chemical reactions, liberating energy when cold, can be used instead. ...Lubrication might not be necessary; bearings could run dry; they would not run hot because heat escapes from such a small device very rapidly...*".

Very clever examples of molecular–level machines (e.g., a phototweezer) were reported by Shinkai, et al. [13] several years ago, but a substantial growth in this research field has only occurred in the last few years. In collaboration with the group of Professor J. F. Stoddart, we have investigated a number of systems based on pseudorotaxanes, rotaxanes, and catenanes where mechanical movements can be induced by (i) chemical energy, (ii) electrical energy, or (iii) light. For space reasons, only a few examples will be discussed.

2.2. PSEUDOROTAXANES, ROTAXANES, AND CATENANES

Rationale and efficient synthetic approaches for the preparation of complicated supramolecular systems like pseudorotaxanes, rotaxanes and catenanes have been devised only recently [14]. The strategies chosen by Stoddart and coworkers are based on (i) charge–transfer (CT) interactions between electron–acceptor (e.g., bipyridinium–type) and electron–donor (e.g., dimethoxybenzene–type) units [15], or

(ii) hydrogen–bonding interactions between secondary ammonium functions and crown ethers [16]. Figure 1 schematises one of the possible routes that can be followed for obtaining rotaxanes and catenanes based on CT interactions through the self–assembly of a pseudorotaxane between an electron–donor component (such as macrocycle **1**) and an electron–acceptor species (like a bipyridinium derivative). Figure 2 shows the formation of a pseudorotaxane based on hydrogen–bonding interactions between an ammonium centre and a crown ether; rotaxanes and catenanes can then be obtained by stoppering or cyclization reactions similar to those schematised in Figure 1.

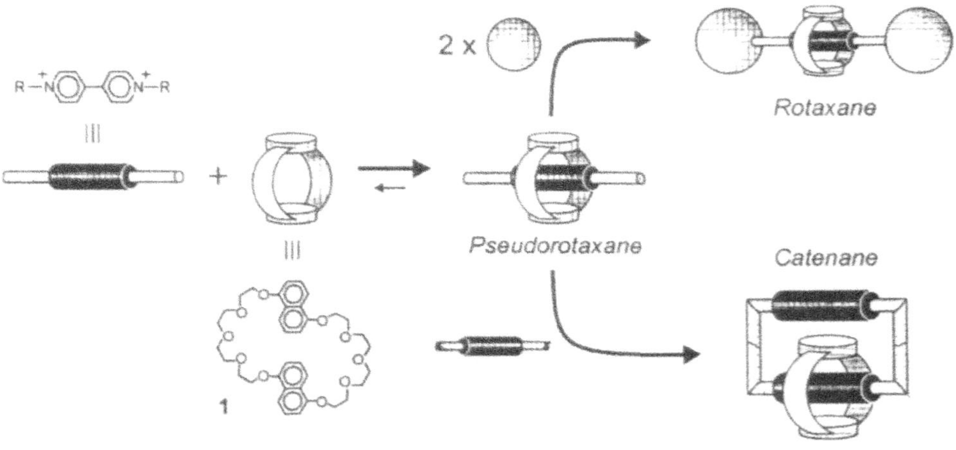

Figure 1. Self–assembly of a pseudorotaxane based on charge–transfer interactions, and synthesis of rotaxanes and catenanes.

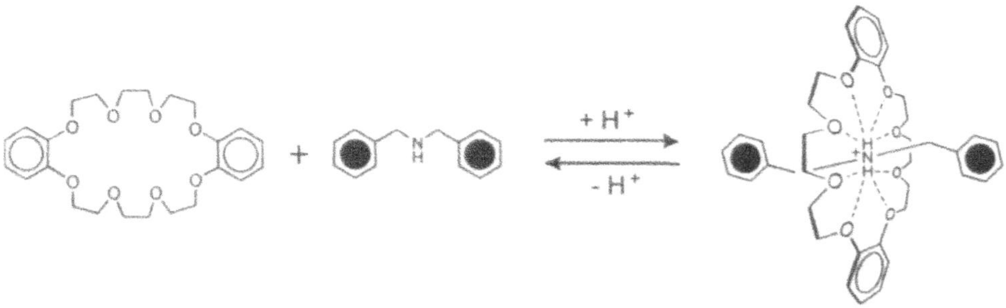

Figure 2. Self–assembly of a pseudorotaxane based on hydrogen–bonding interactions.

2.2.1. *Systems Based on CT Interactions*

The charge–transfer interaction between electron–acceptor and electron–donor units has several important consequences from the spectroscopic and electrochemical viewpoints and plays a fundamental role as far as the machine–like behaviour of these supramolecular species is concerned. The donor/acceptor interaction introduces low energy CT excited states which are responsible not only for the colour of these supramolecular species (due to the presence of broad and weak absorption bands in the visible region), but also for the quenching of the potentially luminescent excited states localised on the molecular components (Figure 3) [15].

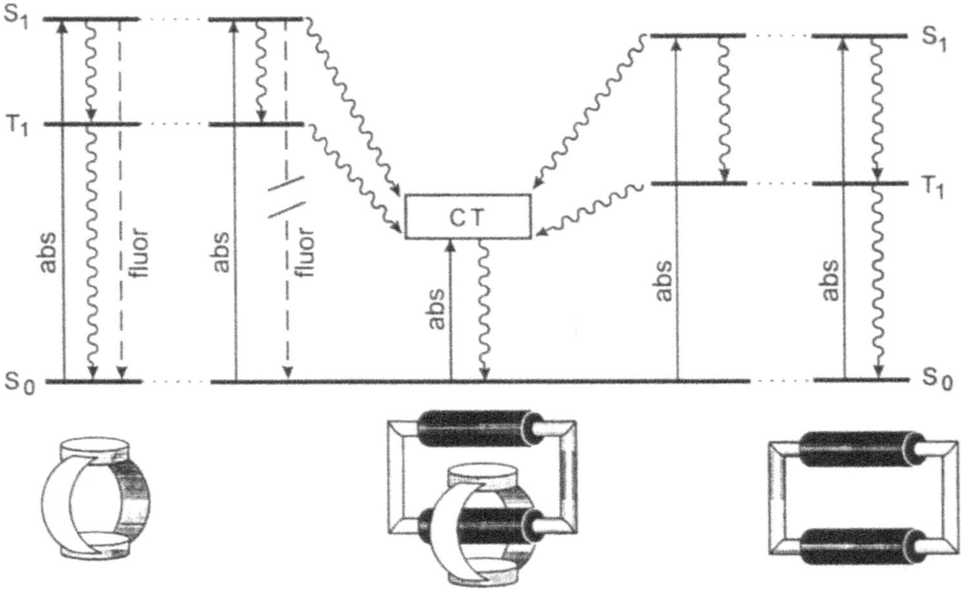

Figure 3. Schematic energy–level diagram for a catenane based on charge–transfer interactions. The potentially luminescent excited states localised on the molecular components are quenched through low lying CT excited states.

As far as the electrochemical behaviour is concerned, it should be noted that, when engaged in CT interactions, the electron–acceptor and electron–donor units become more difficult to reduce and to oxidise, respectively. Furthermore, units which are topologically equivalent in an isolated component may not be so when the component is engaged in non–symmetric interactions with another component. Consider, for example, the macrocyclic component **2** shown in Figure 4 [17]. Such species exhibits a two–electron reduction process which corresponds to the simultaneous first reduction of the two equivalent bipyridinium–type units and, at a more negative potential, another two–electron process which corresponds to the second reduction of such units. When macrocycle **2** is catenated with macrocycle **1**, which contains two dimethoxynaphthalene–type electron–donor units, its electrochemical behaviour changes drastically: (i) all the reduction processes take place at more negative potentials, as expected because the bipyridinium–type units are engaged in CT interactions with the electron–donor units of the crown ether; and (ii) the two electron–acceptor units are no longer equivalent because the unit that resides inside the macrocycle **1** experiences a stronger CT interaction than the unit which resides alongside. Therefore, four distinct one–electron reduction processes are observed (Figure 4) which are assigned, starting from less negative potential values, to the first reduction of the alongside and inside bipyridinium–type units (first and second process), and to the second reduction of the alongside and inside units (third and fourth process).

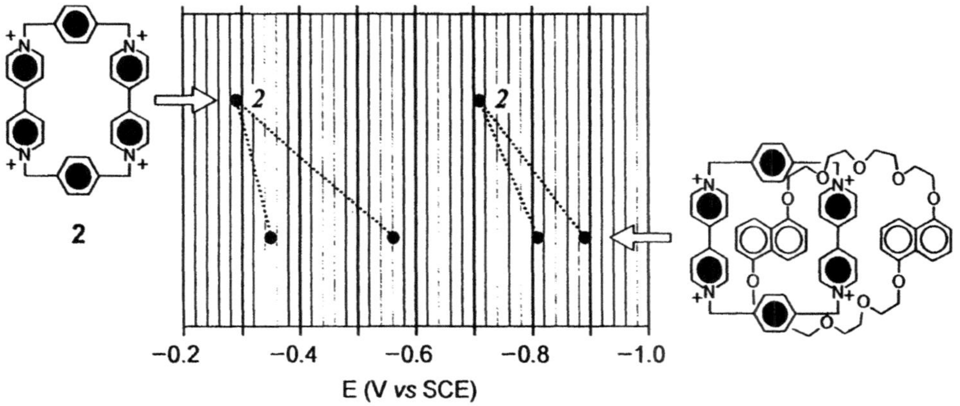

Figure 4. Comparison between the reduction potentials of the free macrocycle **2** and the same macrocycle involved in a catenane structure [17].

In this kind of pseudorotaxanes, rotaxanes, and catenanes, the stability of a specific supramolecular structure is largely due to the CT interaction. In order to cause mechanical movements, such a CT interaction has to be destroyed. This can be done by reduction of the electron–acceptor units or by oxidation of the electron–donor units through chemical, electrochemical, or photochemical redox processes. In most cases, the CT interaction can be restored by an opposite redox process, which thus promotes a reverse mechanical movement leading back to the original structure.

2.2.2. *Systems Based on Hydrogen–bonding Interactions*

Contrary to what happens in the case of CT interactions, hydrogen–bonding interactions (Figure 2) do not introduce low lying energy levels. Therefore, even if the absorption bands of the molecular components of pseudorotaxanes, rotaxanes and catenanes based on this kind of interaction are often perturbed compared with the absorption bands of the isolated molecular components, no new band is present in the visible region. As far as luminescence is concerned, in the supramolecular architecture each component maintains its potentially luminescent levels, but intercomponent energy–transfer processes may often cause quenching and sensitization processes (*vide infra*). In principle, intercomponent electron transfer can also occur.

As far as the electrochemical behaviour is concerned, the properties of the separated components are more or less modified when the components are assembled.

In these compounds, mechanical movements can be caused by destroying the hydrogen–bonding interaction which is responsible for assembly and spatial organization. This can be easily done by addition of a suitable base able to deprotonate the ammonium centre. The movement can be reversed by addition of an acid able to reprotonate the amine function.

2.3. PHOTOCHEMICALLY DRIVEN PYSTON/CYLINDER SYSTEMS

Dethreading/rethreading of the wire and ring components of a pseudorotaxane reminds the movement of a piston in a cylinder. We have shown that, in suitable pseudorotaxanes, the movement of such a rudimentary molecular machine can be feeded by chemical energy, electrical energy, or light.

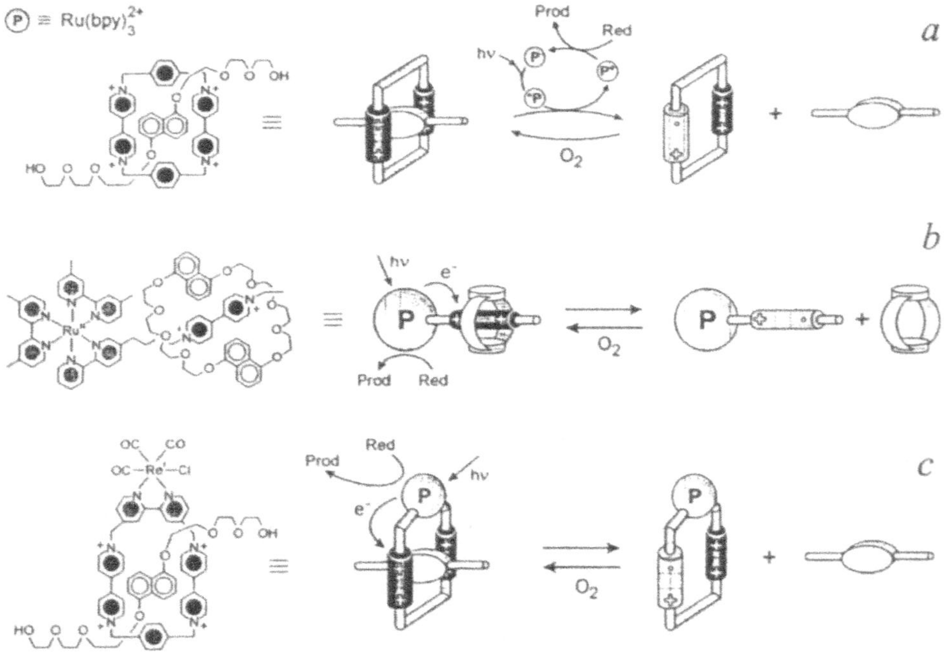

Figure 5. Light–driven dethreading of: (a) a pseudorotaxane by excitation of an external photosensitiser [18]; (b) a pseudorotaxane incorporating a photosensitiser as a stopper in the wire–type component [19]; (c) a pseudorotaxane incorporating a photosensitiser in the macrocyclic component [20].

Our first attempt at designing [18] a photochemically driven molecular–level machine of a pseudorotaxane type was based (Figure 5a) on the use of an external electron–transfer photosensitiser. As a result of the above described CT interactions, the electron–acceptor ring **2** and a naphtho–based electron–donor wire self–assemble in aqueous solution. Irradiation with visible light of the external electron–transfer photosensitiser **P** (e.g., $Ru(bpy)_3^{2+}$) causes reduction of one of the bipyridinium–type units of the ring (the back electron–transfer reaction is prevented by the presênce of a sacrificial reductant like triethanolamine). Once the ring has received an electron, the donor/acceptor interaction responsible for the self–assembly is partly destroyed and therefore the wire dethreads from the reduced ring. If oxygen is allowed to enter the solution, oxidation of the reduced bipyridinium–type unit restores the donor/acceptor interaction and causes rethreading. The threading, dethreading, and rethreading processes can be easily followed by absorption and fluorescence spectroscopy.

Second generation photochemically driven machines have then been designed where the piston/cylinder pseudorotaxane structure incorporates the "light–fueled" motor (i.e., the photosensitiser) in the wire (Figure 5b) [19] or in the macrocyclic ring (Figure 5c) [20]. We have shown that in both cases excitation of the photosensitiser with visible light in the presence of a sacrificial donor causes reduction of the electron–acceptor unit and, as a consequence, dethreading. Rethreading can again be obtained by allowing oxygen to enter the solution.

2.4. ELECTROCHEMICALLY DRIVEN ROTATION OF A RING IN CATENANES

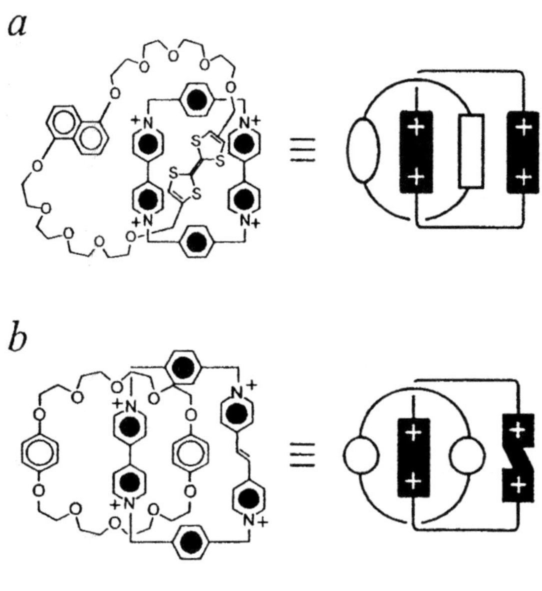

Figure 6. (a) A catenane with a non–symmetric ring containing two different electron–donor units [21]; (b) a catenane with a non–symmetric ring containing two different electron–acceptor units [17].

In a catenane, structural changes caused by rotation of one ring with respect to the other can be evidenced only when one of the two rings contains two non–equivalent units. We discuss here two examples. In the first one (Figure 6a), the ring containing the electron–acceptor units is symmetric, whereas the other ring is non–symmetric since it contains two different electron–donor units. In the second example (Figure 6b), the non–symmetric ring is that containing the electron–acceptor units. In a catenane structure, the inside electron–donor (acceptor) unit experiences the effect of two electron–acceptor (donor) units, whereas the alongside electron–acceptor (donor) unit experiences the effect of only one electron donor (acceptor). Therefore, at the beginning the better electron–acceptor (donor) unit of the non–symmetric ring occupies the inside position and the worse one stays alongside. It may happen that on electrochemical removal or addition of electron(s), the feature of better electron donor or acceptor in the non–symmetric ring moves from a unit to the other, with a consequent change in their reciprocal (inside or alongside) position in the catenane structure.

2.4.1. *A Catenane With a Ring Bearing Two Different Electron Donor Units*

In the catenane shown in Figure 6a [21], one ring is made of two different electron donors, namely a tetrathiafulvalene (TTF) and a 1,5–dioxynaphthalene (DMN) unit. The better electron donor (i.e., TTF) enters the ring and the worse one (i.e., DMN) remains alongside (Figure 7, top left), as shown by a variety of techniques, including X–ray crystallography. On electrochemical oxidation, the TTF unit is oxidised first, thereby losing its electron–donating properties. The disruption of the CT interaction and the electrostatic repulsion between TTF^+ and the tetracationic electron–acceptor ring cause circumrotation of the macrocyclic polyether to yield the 'rotational' isomer with the DMN moiety positioned inside the acceptor ring (Figure 7, bottom right). Upon reduction of TTF^+, the switching is fully reversible. The oxidation/reduction cycle, which is accompanied by a clearly detectable colour change, can be followed by ^1H NMR spectroscopy, UV/Vis spectroscopy, and cyclic voltammetry.

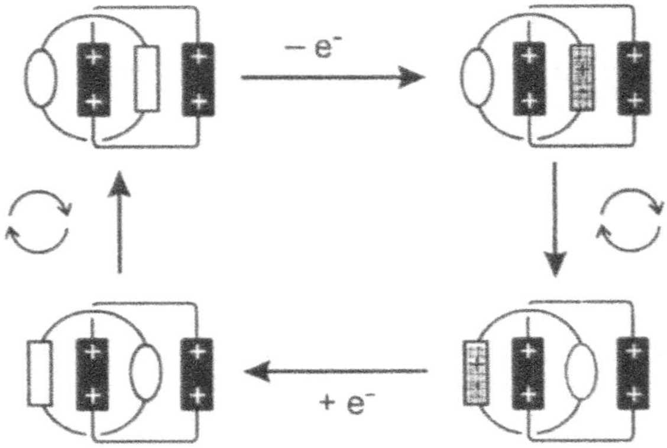

Figure 7. The electrochemically controlled movements of the ring components upon one–electron oxidation/reduction in the non–symmetric catenane shown in Figure 6a [21].

2.4.2. *A Catenane With a Ring Bearing Two Different Electron Acceptor Units*

In the catenane shown in Figure 6b [17], one ring is made of two different electron acceptors, namely a bipyridinium–type and a bis(pyridinium)ethylene–type unit. The better electron acceptor (i.e., the bipyridinium–type unit) enters the ring and the worse one [i.e., the bis(pyridinium)ethylene–type unit] remains alongside (Figure 8, top left). A detailed examination of the electrochemical and spectroelectrochemical results obtained for the catenane and simpler model compounds shows that the first reduction process (process I in Figure 8) concerns the (inside) bipyridinium–type unit. As soon as this unit is monoreduced, it becomes a worse electron acceptor than the non–reduced bis(pyridinium)ethylene–type unit. Therefore, the most stable structure (i.e., the most stable 'rotational' isomer) for the one–electron reduced catenane is that with the bis(pyridinium)ethylene–type unit inside. This leads to a mechanical movement which interchanges the positions of the two units. If the bipyridinium–type unit is reoxidised, it becomes again the better electron acceptor and goes inside. On further reduction of the one–electron reduced catenane (process II in Figure 8), the second electron enters the bis(pyridinium)ethylene–type unit which is inside. At this stage, both units are monoreduced. The second reduction potentials of the isolated components shows that the monoreduced bis(pyridinium)ethylene–type unit is a better electron acceptor than the monoreduced bipyridinium–type unit. Therefore, the second reduction should not

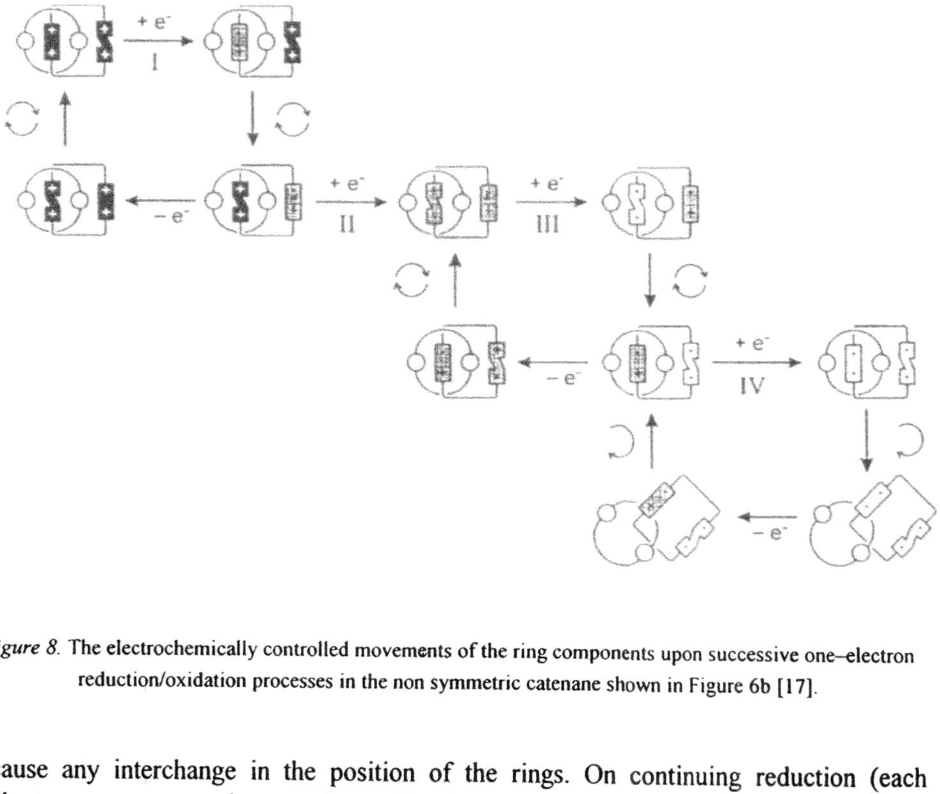

Figure 8. The electrochemically controlled movements of the ring components upon successive one–electron reduction/oxidation processes in the non symmetric catenane shown in Figure 6b [17].

cause any interchange in the position of the rings. On continuing reduction (each electron–acceptor unit can be reduced twice), the third electron enters the bis(pyridinium)ethylene–type site (process III in Figure 8), which definitely becomes a worse electron acceptor than the monoreduced bipyridinium–type unit. Therefore, at this stage the most stable structure should be that with the monoreduced bipyridinium–type unit inside. Finally, upon further reduction (process IV in Figure 8), also the bipyridinium–type unit loses completely its electron–acceptor properties, so that there should be no driving force favouring a specific structure. In conclusion, successive addition of electrons in this system is expected to cause the relative stabilization of different structures and, more generally, releases brakes to ring rotation.

3. Signal Processing at the Molecular Level

3.1. INTRODUCTION

Current computers are based on sets of components (wires, switches, transistors, etc.) constructed by the "top down" approach. This approach, however, has arrived close to its intrinsic limitations. A necessary condition for further miniaturization to increase the power of information processing and computation is the "bottom up" construction of molecular–level components capable to perform the functions needed (chemical computer) [8]. The possibility that future computer may be based on solutions, instead of on solid state systems, should also be taken into consideration [22]. Apart from such futuristic applications, the design and construction of *a molecular–level electronic set* (i.e., a set of molecular–level systems capable to play functions that mimick those performed by macroscopic components in electronic devices) is of great scientific interest since it introduces new concepts in the field of chemistry and stimulates the ingenuity of research workers engaged in the emerging field of nanotechnology. In the last few years, a number of such molecular–level components (e.g., wires, antennas, on/off switches, plug/socket devices, two–pole switches, two coupled molecular switches, three–pole switches, memories, logic gates) have been reported. For space reasons, only a few examples investigated in our laboratory will be discussed. It is worthwhile noticing that, at the molecular level, delivery of electrons (electron transfer) and electronic energy (energy transfer) can be discussed on the basis of similar theoretical treatments. Furthermore, both electron– and energy–transfer processes can be photoinduced [1b].

3.2. WIRES AND ANTENNAS

An important function for the construction of molecular–level devices is photoinduced energy and electron transfer over long distances and/or along predetermined directions. In this regard, the spacer linking the starting photoactive unit to the end unit plays a fundamental role. When the spacer is not rigid (i. e., $-(CH_2)_n-$ chains), the geometry of the system is not defined. The role played by a spacer, however, is not only structural, since its chemical nature controls the electronic communication between the selected terminal units. Spacers that are easy to oxidise or reduce and/or possess low energy electronic levels are unsuitable because they can play the role of quenchers of the excited state. Another important feature for a spacer is its modular composition, which allows changing distance without changing the electronic nature of the connection.

For wire–type compounds, very attractive candidates to play the role of spacers are polyenenes, polyalkynylenes, oligothienylene, oligophenylenes. In our laboratory we have recently prepared [23a] the rod–like compound **N–(ph)$_5$–A** (overall length 3.1 nm, Figure 9a), where a 1–naphthyl (**N**–) and a 9–anthryl (–**A**) chromophoric units are connected by a 1,4–quinquephenylene–type spacer [–**(ph)$_5$**–] bearing alkyl chains for solubility reasons. We have found that in **N–(ph)$_5$–A** the practically complete quenching of the fluorescent excited states of the naphthalene– and quinquephenylene–based components is accompanied by quantitative energy transfer to the lower lying fluorescent level of the anthracene–based unit, indicating a very efficient energy transfer. The wire–like **$Ru(bpy)_3{}^{2+}$–$(ph)_n$–$Os(bpy)_3{}^{2+}$** compounds (metal–to–metal distance, 4.2 nm for n=7, Figure 9b) have also been prepared and the photoinduced energy– and electron– (upon oxidation of the Os–based moiety) transfer processes are under investigation [23b].

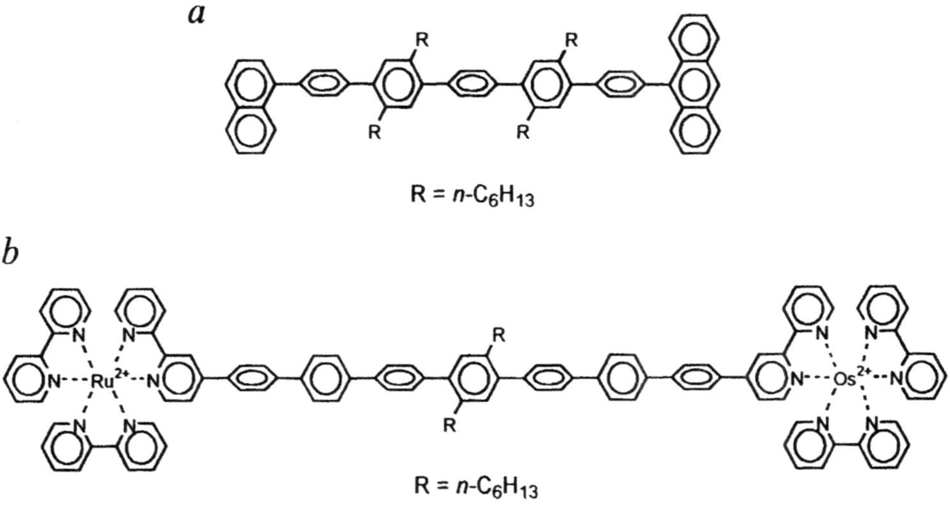

Figure 9. Molecular wires for electronic energy transfer based on oligophenylene spacers [23].

Ru^{II} and Os^{II} polypyridine complexes have also been extensively used to construct polynuclear compounds that can play the role of antennas for light harvesting. For more details on these systems, see Ref. 24.

3.3. PLUG/SOCKET SYSTEMS

A macroscopic plug/socket system is characterised by two features: (i) possibility to connect/disconnect the two components in a reversible way; (ii) electron flow from the socket to the plug when the two components are connected. We have recently designed and constructed supramolecular systems that may be considered as molecular–level plug/socket devices: "plug in/plug off" is reversibly controlled by acid/base reactions, and photoinduced flow of electronic energy (or electrons) takes place in the "plug in" state.

It has long been known that crown ethers can form adducts with organic ammonium ions [25]. As mentoned above (Section 2.2.2), it has been found that suitable "wire–type" ammonium ions can thread through crown ethers to give pseudorotaxane structures [16,26a]. Accordingly, we have recently verified that the "wire–type" (9–anthracenyl)benzylammonium ion, AH^+, and the DB24C8 crown ether give rise to such an adduct (Figure 10a) [26a]. The association process can be quantitatively reversed by addition of a suitable base like tributylamine. In the pseudorotaxane supramolecular structure, the quenching of the fluorescence of the dimethoxybenzene moieties is accompanied by a parallel sensitization of the fluorescence of the anthracene unit, indicating the occurrence of a very efficient flow of electronic energy in the "plug in" state.

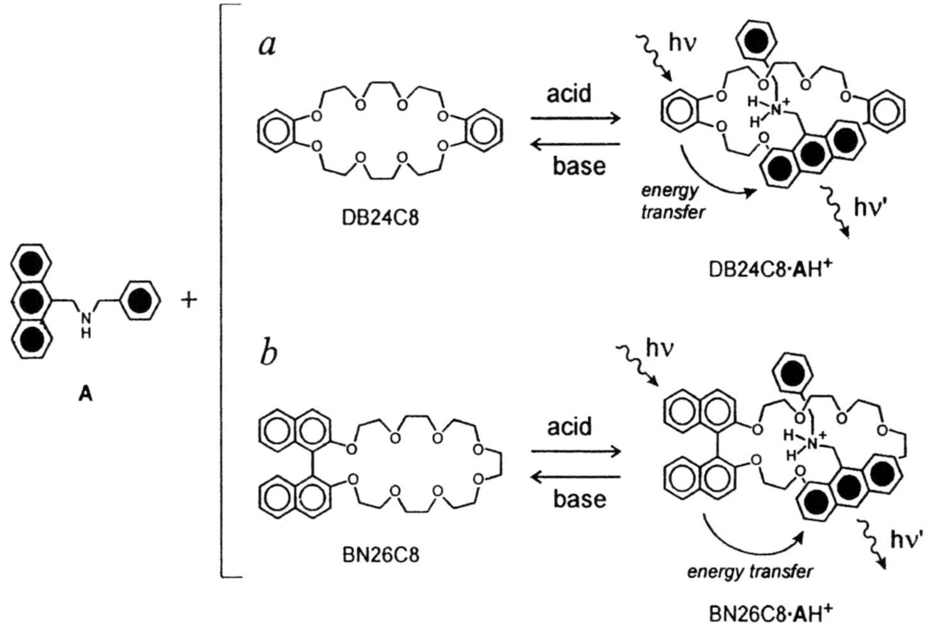

Figure 10. Acid–base controlled "plug in/plug off" of (9–anthracenyl)benzylammonium ion with the (a) DB24C8 [26a] and (b) BN26C8 [26b] crown ethers. The occurrence of photoinduced energy transfer in the "plug in" state is schematised.

An even more interesting case is that based on AH^+ and the (±) binaphthocrown ether BN26C8 [26b]. Molecular modelling shows that the minimum–energy conformation corresponds to a pseudorotaxane–like structure (Figure 10b). The quenching of the binaphthyl–type fluorescence is accompanied by the sensitization of the fluorescence of the anthracenyl unit of the ammonium ion. The rate constant for electronic energy transfer from the binaphthyl units of the crown to the anthracenyl unit of the wire is higher than 4×10^9 s^{-1}. Addition of a stoichiometric amount of base to the pseudorotaxane structure causes the revival of the binaphthyl fluorescence and the disappearance of the anthracenyl fluorescence upon excitation in the binaphthyl bands, showing that "plug off" has occurred.

The above results are interesting since the plug/socket molecular–level concept can be straightforwardly extended to suitably selected systems where (a) light excitation induces an electron flow, and (b) the "plug in/plug off" function is stereoselective since the enantiomeric recognition of chiral ammonium ions by crown ethers containing binaphthyl is a well known property.

3.4. THREE–POLE SWITCHES

Extending the concept of two–pole electrochemical switching in supramolecular species [6,17,21,27], we have developed a system that can be reversibly switched in three different states through electrochemical control of the guest properties of one component (Figure 11) [28].

It is well known [29] that tetrathiafulvalene is stable in three different oxidation states, TTF(0), TTF^{+}, and TTF^{2+}. It has already been established that TTF(0) gives a 1:1 charge transfer complex with the cyclobis(paraquat–*p*–phenylene) macrocycle **2**, playing the role of electron donor [27fg]. Reversible decomplexation/recomplexation can be performed by cyclic oxidation/reduction of the TTF moiety. Since oxidation decreases the electron–donor properties of a molecule with a concomitant increase in its electron–acceptor properties, we thought that while TTF(0) plays the role of an electron–donor guest, TTF^{2+} could play the role of an electron–acceptor guest. According to this expectation, we found that TTF^{2+} gives rise to a charge–transfer complex with the electron donor macrocycle **1**. The above results opened the way to the use of TTF as a switchable electron–donor/electron–acceptor guest and to control its association with electron–acceptor/electron–donor hosts by electrochemical techniques.

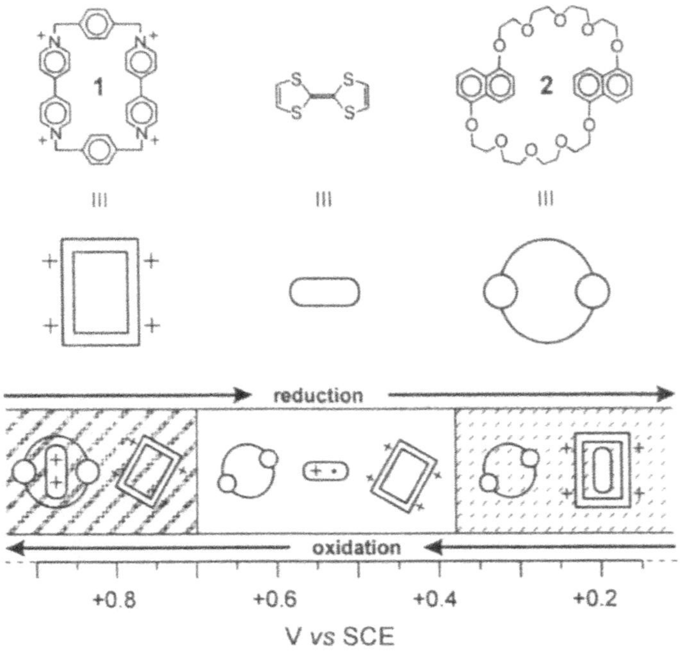

Figure 11. Components of the three–pole system and schematic representation of the ranges of electrochemical stability of the three states available to the system [28].

The investigated system (Figure 11) consists of acetonitrile solutions of TTF and the two previously mentioned macrocycles **1** and **2**. Under such conditions, the potential of the TTF^{+}/TTF(0) couple moves to *more* positive values (70 mV), whereas that of the TTF^{2+}/TTF^{+} couple moves to *less* positive values (14 mV), exactly as observed for the [**2**·TTF(0)] and [**1**·TTF^{2+}] complexes, respectively. These results show that (i) the TTF guest interchanges between the two hosts on changing its oxidation state from 0 to +2, and (ii) the TTF^{+} species does not interact with either host. In other words, depending on the potential range (Figure 11), TTF can be (i) free in the TTF^{+} state, (ii) complexed with the electron–acceptor host in the TTF(0) state, or (iii) complexed with the electron–donor host in the TTF^{2+} state. The reversibility of the electrochemical processes shows that the complexation/decomplexation reactions and, as a consequence, the exchange of the guest between the two hosts, are fast processes compared to the time scale of the electrochemical experiments. In such a three–state system, "writing" can be performed electrochemically and "reading" can be done by absorption, emission, and NMR spectroscopy, and electrochemically.

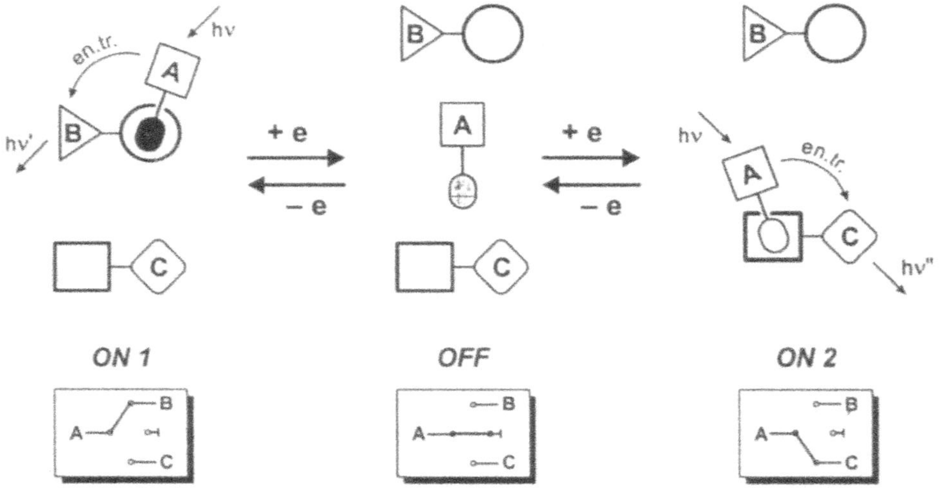

Figure 12. Schematic representation of a three–pole supramolecular switching of energy transfer.

The mechanical movements taking place in the described supramolecular system, where a "free" molecule can be electrochemically driven to associate with either of two different receptors, open the way to futuristic applications in the field of molecular–level signal processors [8]. One can conceive second generation systems where the electrochemically driven movements can control the occurrence *and* the selection of the partner in energy– or electron–transfer processes. Consider, e.g., a system (Figure 12) where the potential guest is appended to a chromophoric group **A** and the potential hosts are appended to the chromophoric groups **B** and **C**, respectively, whose lowest excited state is lower than that of **A**. In such a system, light excitation of **A** could lead to no energy transfer (OFF), energy transfer to **B** (ON 1) or energy transfer to **C** (ON 2), depending on the potential value selected by the operator. Admittedly, the donor/acceptor–based molecular–level connection poses severe limitations as to the choice of the chromophoric groups to be used if energy transfer has to proceed through the CT connections. However, it does not seem unlikely that systems can be designed where molecular association (i) plays only the role of bringing the two chromophoric groups at a suitable distance for through–space energy transfer, or (ii) relies on a different kind of interaction (e.g., hydrogen bonding) [16,25,26]. Similar switching of electron–transfer processes could also be performed and more complex energy– and/or electron–transfer patterns are conceivable.

3.5. LOGIC GATES

3.5.1. *Introduction*

Computers are based on semiconductor logic gates which perform binary algebraic operations [30]. Logic gates are switches whose output state (*0* or *1*) depends on the input conditions (*0* or *1*). Several examples of supramolecular systems capable to behave as logic gates have recently been reported [31,32]. In general, they are constituted by a fluorophore connected with one or more receptors able to interact selectively with external species; the excited state behaviour of the fluorophore depends on whether or not the receptors are interacting with the external species. It can be noticed that a logic gate needs energy to operate. In the logic gates described below energy comes from the light excitation of the fluorophore.

YES and NOT single–input gates are the simplest logic devices [30]. A YES gate passes the input bits to the output without changes (input: *1*, output: *1*; input: *0*, output: *0*), while a NOT gate inverts any input data (input: *1*, output: *0*; input: *0*, output: *1*). Molecular systems that can perform such simple logic operations are very common, and fluorescence is a particularly useful signal to monitor such operations [5b,31]. A molecule that fluoresces (output) only in acidic media (i.e., in the presence of a proton input) acts as a YES gate. Conversely, a fluorescent molecule whose emission (output) disappears under the action of an input of protons can be regarded as a NOT logic device. In order to perform more complex logic operations, however, carefully designed multi–component chemical systems are needed. Figure 13 shows schematically the changes which have to occur in a chemical system in order to perform the AND, OR, and XOR (eXclusive OR) logic operations under the action of two chemical inputs (X and Y). For illustration purposes, the equivalent (from a logic viewpoint) electric circuits are also shown. It should be pointed out, however, that the comparison between the characteristics of logic gates as different as chemical systems and electric circuits is only formal and should be made with great care. Under some aspects, such a comparison could also be inappropriate or misleading.

3.5.2. *AND Gates*

One of the fundamental logic gates used to perform additions is the AND one. This gate does the logic product between the inputs: the output signal is *1* provided that all the inputs are *1*; it is sufficient to have just one "off" input bit to make the output "off". The equivalent circuit of the AND gate has switches connected in series (Figure 13a). The scheme for a molecular–level two–input AND gate is shown in Figure 14 [32a,c].

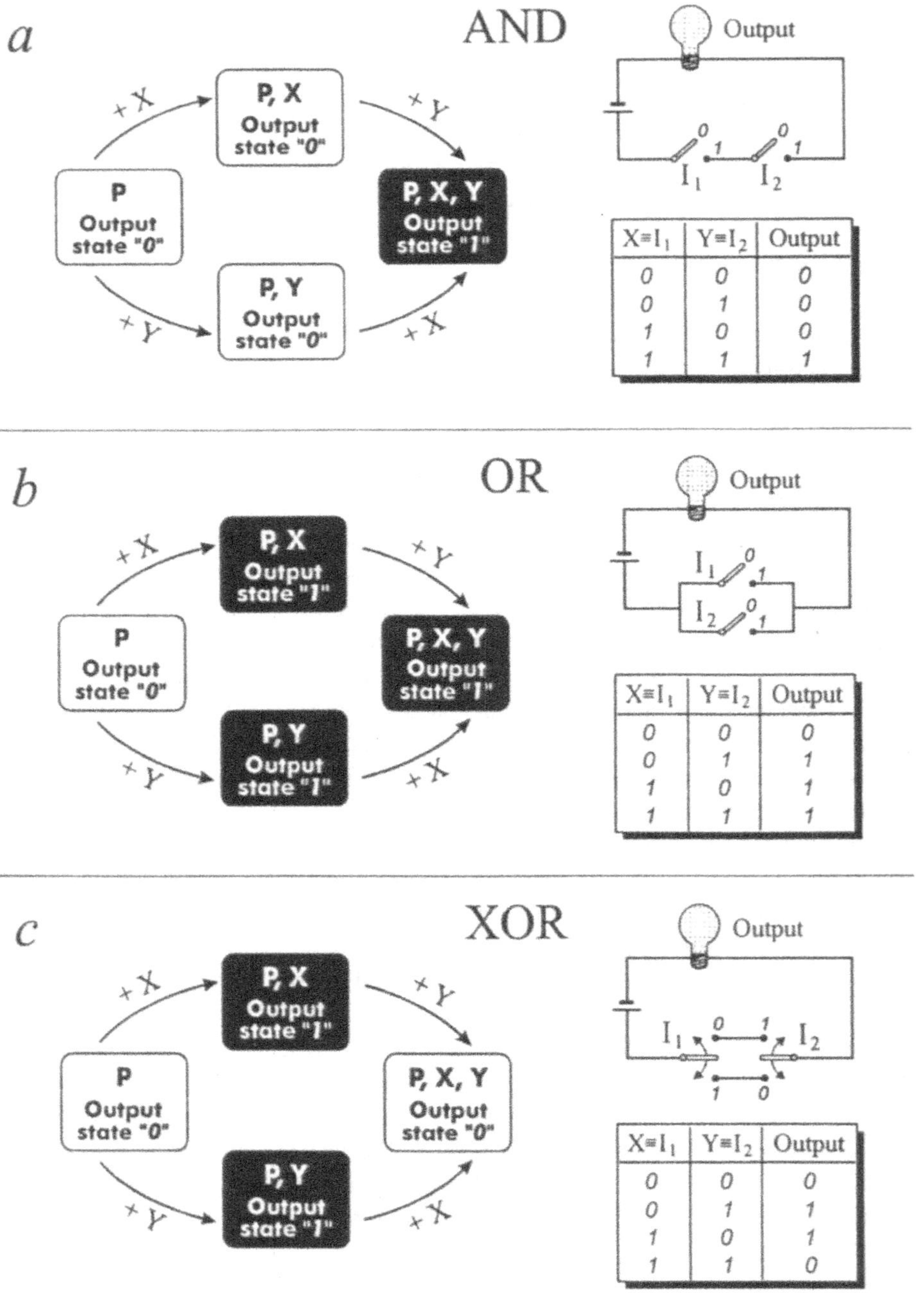

Figure 13. Schematic representation of a chemical system (P) which performs the AND (a), OR (b), and XOR (c) logic operations under the action of two chemical inputs (X and Y). The truth tables of such operations are also shown, along with their representations based on electric circuit schemes.

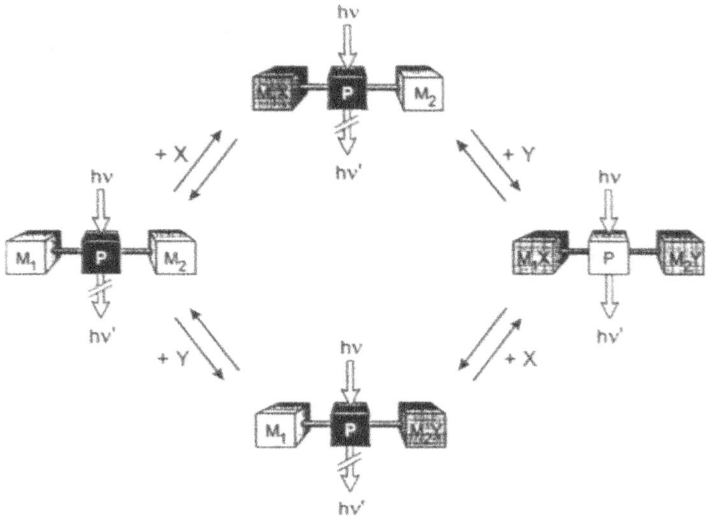

Figure 14. Scheme for a molecular–level two–input AND gate.

Such a device is made by a fluorophore (P) and two covalently linked receptors (M_1 and M_2) able to quench the luminescence of the fluorophore via, for example, photoinduced electron transfer. When one of the receptor hosts a suitable external species, it can no longer quench the luminescence of the emitting unit; however, the luminescence is quenched by the other "free" receptor. The output signal (i.e., the luminescence of P) can only be observed when both receptors are engaged by suitable external species (the input signals X and Y). Examples of such AND gates have been reported by de Silva, et al. [32a,c]. In the compound *a* shown in Figure 15 [32c] the anthracene luminescence is independently quenched by electron transfer from the amino group and the benzocrown ether unit. The electron–transfer process from the amino group and from the benzocrown ether can be prevented by adding protons and sodium ions, respectively. The anthracene moiety shows fluorescence quantum yield 0.24 when both ions are added (i.e., when both inputs signals are 1), whereas essentially no emission is seen when only one or none of these ions is present.

3.5.3. *OR Gates*

An OR gate performs the logic sum between boolean variables; the output is *1* if at least one of the inputs is *1* [30]. The equivalent circuit of an OR gate has switches connected in parallel (Figure 13b). The chemical model for a two–input OR gate is represented in Figure

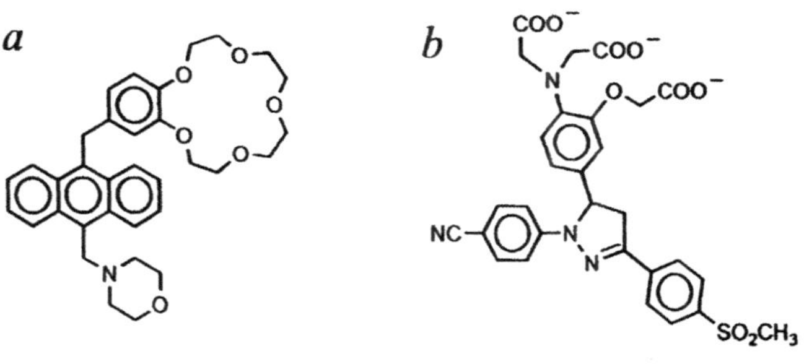

Figure 15. Supramolecular species capable to perform as an AND [32c] (a) and OR [32a] (b) gates.

16. The luminophore (P) is quenched by an attached receptor (M), which can host two different species (X and Y) with poor chemoselectivity. In such a case it is sufficient to add only one of the two external species to the solution in order to engage the receptor and recover the output luminescence. Some interesting examples of this type of gate have recently been developed [32a,b]. In compound *b* shown in Figure 15 [32a] the luminescence of the pyrazoline emitting moiety is quenched by electron transfer from the 1–amino–2–oxybenzene group. However, when the receptor binds magnesium or calcium ions, its oxidation potential increases so that the electron–transfer process is no longer allowed and the luminescence is switched on. With 10^{-3} M Ca^{2+} only, 0.5 M Mg^{2+} only, and both 10^{-3} M Ca^{2+} and 0.5 M Mg^{2+} in neutral water, the fluorescence enhancement factors are 67, 57 and 67, respectively. Therefore the behaviour of this system corresponds satisfactorily to the truth table of the OR operation.

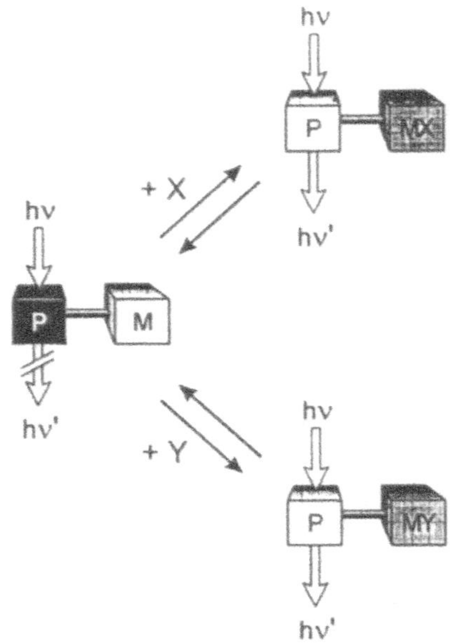

Figure 16. Scheme for a molecular–level two–input OR gate.

3.5.4. *XOR Gates*

An XOR gate is a much more complex device, as one can understand from the fact that its equivalent circuit contains two bipolar switches (Figure 13c). The familiar possibility to switch on/off the same bulb by either of two distinct switches of a room is based on this type of circuit. The truth table of the XOR operation is the same as that of the OR operation except that the output is *0* if *both* the inputs are *1*. In today's processors, addition is performed with an AND gate which gives the carry digit and an XOR gate which gives the sum digit. It is also important to notice that the XOR gate is actually a comparator because it can establish whether the two inputs have the same value.

The pseudorotaxane **3** schematised in Figure 17 results from the self–assembly of the electron–accepting 2,7–dibenzyldiazapyrenium dication **4** with the crown ether **5** which contains two 2,3–dioxynaphthalene electron–donating units. In the pseudorotaxane structure, the electron–deficient diazapyrenium unit is sandwiched between the electron–rich 2,3–dioxynaphthalene units of **5**. Because of the electron donor/acceptor interaction, a low energy CT excited state is formed which is responsible for (i) the presence of a weak and broad absorption band with λ_{max} around 400 nm, partially hidden by the bands of **4**, and (ii) the disappearance of the strong fluorescence (Figure 18) exhibited by the two separated components (λ_{max}=432 nm for **4**, λ_{max}=343 nm for **5**) [33]. In a CH_2Cl_2/CH_3CN 9:1 *v/v* solution containing 2×10^{-5} M **4** and 3×10^{-5} M **5**, about 85% of **4** and 60% of **5** are threaded in the pseudorotaxane **3**

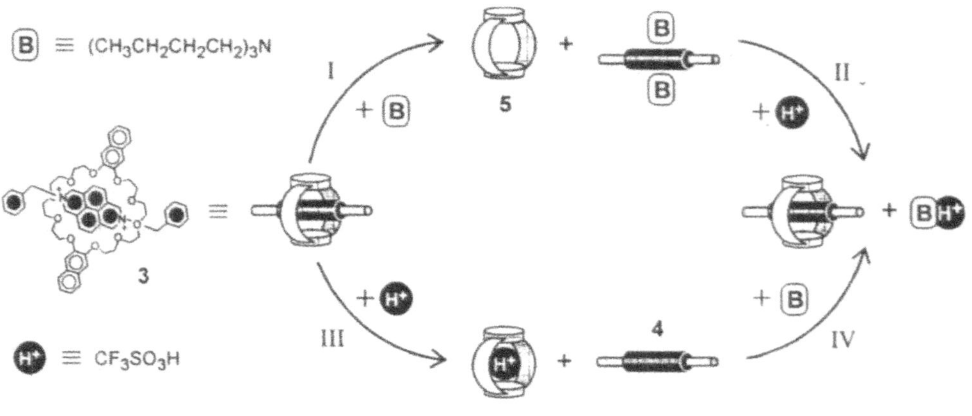

Figure 17. Schematic representation of the unthreading/rethreading pattern of pseudorotaxane **3**, which corresponds to an XOR logic function [33].

at room temperature. Upon addition of 6×10^{-4} M tributylamine (**B**, Figure 17), a 1:2 adduct (**4**·$\mathbf{B}_2$) is formed between the 2,7–dibenzyldiazapyrenium dication **4** and the amine, with the consequent unthreading of **3** (process I in Figure 17). This process causes large spectral changes, namely (i) the recovery of the emission band of **5** (Figure 18), and (ii) the appearance of a broad absorption (λ_{max}=575 nm) and emission (λ_{max}=670 nm, Figure 18) band characteristic of **4**·$\mathbf{B}_2$. From the recovery of the fluorescence of free crown ether **5**, one can estimate that 95% of the pseudorotaxane **3** undergoes unthreading. Subsequent addition of a stoichiometric amount (relative to the previously added amine) of trifluoromethanesulfonic acid (H^+, Figure 17) unlocks **4** from the **4**·$\mathbf{B}_2$ adduct and allows rethreading between **4** and **5** to give back almost quantitatively the pseudorotaxane **3** (process II in Figure 17). This process is of course accompanied by spectral changes opposite to those observed upon addition of amine. Processes I and II can be repeated on the same solution by repeating the addition of amine and acid.

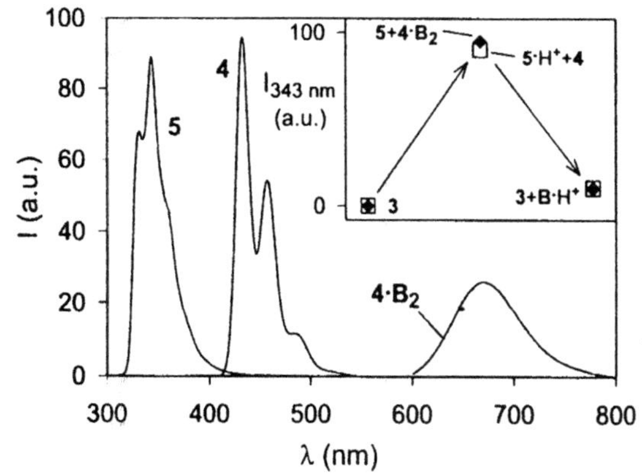

Figure 18. Emission spectra for **4**, **5**, and **4**·$\mathbf{B}_2$ in CH_2Cl_2/CH_3CN 9:1 *v/v*. The emission spectrum of **5**·H^+ is almost identical to that of **5**. Inset shows the net changes in the emission intensity at 343 nm associated with processes I–IV shown in Figure 17 [33].

As shown in Figure 17, the unthreading/rethreading cycle can also be performed by reverting the order of the two inputs. Both processes I and III cause a strong increase of emission intensity at 343 nm, which is cancelled by processes II and IV, respectively. Therefore we can conclude that the chemical system described above shows the input/output relationships indicated by the truth table of the XOR logic gate (Figure 13c) [33]. The strong fluorescent signal at 343 nm is present (output: *1*) only when either amine or H^+ (inputs X and Y in the truth table) are added (i.e, X: *1* and Y: *0*, or vice versa); conversely, the fluorescent signal is absent (output: *0*) when none or both of the inputs are present (i.e., X=Y: *0* or X=Y: *1*).

4. Conclusions

As we have seen above, in 1959 in his address to the American Physical Society R.P. Feynman said [12]: *"An internal combustion engine of molecular size is impossible. Other chemical reactions, liberating energy when cold, can be used instead"*. The described examples of molecular machines driven by redox or acid–base "cold" chemical reactions fulfil Feynman's prediction. Furthermore, we have shown that the primary energy source to drive such mechanical movements can be electricity or, even more interesting, light. In the same address, Feynman concluded his reflection on the construction of molecular–scale machines as follows [12]: *"What would be the utility of such machines? Who knows? I cannot see exactly what would happen, but I can hardly doubt that when we have some control of the rearrangement of things on a molecular scale we will get an enormously greater range of possible properties that substances can have, and of different things we can do"*. We believe that such a consideration can apply, and is the most appropriate final comment, to all the molecular–level devices described in this paper.

Acknowledgements. This work was supported by MURST (Supramolecular Devices project), University of Bologna (Funds for Selected Research Topics) and the EU (contracts FMRX–CT96–0076 and FMRX–CT96–0031).

References

[1] (a) Balzani, V., Moggi, L. and Scandola, F. (1987), in V. Balzani (ed.), *Supramolecular Photochemistry*, Reidel, Dordrecht, p. 1. (b) Balzani, V. and Scandola, F. (1991) *Supramolecular Photochemistry*, Horwood, Chichester.

[2] (a) Carter, F. L., Siatkowsky, R. E. and Woltjien, H. (eds.) (1988) *Molecular Electronic Devices*, North Holland, Amsterdam. (b) Lehn, J.–M. (1995) *Supramolecular Chemistry*, VCH, Weinheim. (c) Jortner, J. and Ratner, M. (eds.) (1998) *Molecular Electronics*, Wiley–VCH, Weinheim.

[3] (a) Verdaguer, M. (1996) *Science* **272**, 698. (b) Astruc, D. (1997) *Acc. Chem. Res.* **30**, 383.

[4] (a) Drexler, K. E. (1992) *Nanosystems: Molecular Machinery, Manufacturing, and Computation*, Wiley, New York. (b) Göpel, W. and Ziegler, Ch. (eds.) (1992) *Nanostructures Based on Molecular Materials*, VCH, Weinheim. (c) Bard, A. J. (1994) *Integrated Chemical Systems*, Wiley, New York.

[5] (a) Balzani, V. and Scandola, F. (1996), in *Comprehensive Supramolecular Chemistry*, J. L. Atwood, J. E. D. Davies, D. D. Macnicol and F. Vögtle (eds.), Pergamon, Oxford, Vol. 10, p. 1. (b) Balzani, V., Credi, A. and Scandola, F. (1997) *Chem. Ind. (Milan)* **79**, 751.

[6] (a) Boulas, P. L., Gomez–Kaifer, M. and Echegoyen, L. (1998) *Angew. Chem. Int. Ed.* **37**, 216. (b) Kaifer, A. E. and Mendoza, S. (1996), in *Comprehensive Supramolecular Chemistry*, J. L. Atwood, J. E. D. Davies, D. D. Macnicol and F. Vögtle (eds.), Pergamon, Oxford, Vol. 1, p. 701.

[7] (a) Balzani, V., Gómez–López, M. and Stoddart, J. F. (1998) *Acc. Chem. Res.* **31**, 405. (b) Chambron, J.–C. and Sauvage, J.–P. (1998) *Chem. Eur. J.* **4**, 1362.

[8] Rouvray, D. (1998) *Chem. Br.* **34**, 26.

[9] Steinberg–Yfrach, G., Rigaud, J.–L., Durantini, E. N., Moore, A. L., Gust, D. and Moore, T. A. (1998) *Nature* **392**, 479.

[10] Goodsell, D. S. (1996) *Our Molecular Nature: The body's Motors, Machines, and Messages*, Copernicus, New York.

[11] Service, R. F. (1997) *Science* **278**, 579.

[12] Feynman, R. P. (1960) *Eng. Sci.* **23**, 22.

[13] Shinkai, S. and Manabe, O. (1984) *Top. Curr. Chem.* **121**, 67.

[14] (a) Amabilino, D. B. and Stoddart, J. F. (1995) *Chem. Rev.* **95**, 2725. (b) Dietrich–Buchecker, C. O. and Sauvage, J.–P. (1991) *Bioorg. Chem. Frontiers* **2**, 195.

[15] Anelli, P. L., Ashton, P. R., Ballardini, R., Balzani, V., Delgado, M., Gandolfi, M. T., Goodnow, T. T., Kaifer, A. E., Philp, D., Pietraszkiewicz, M., Prodi, L., Reddington, M. V., Slawin, A. M. Z., Spencer, N., Stoddart, J. F., Vicent, C. and Williams, D. J. (1992) *J. Am. Chem. Soc.* **114**, 193.

[16] Ashton, P. R., Ballardini, R., Balzani, V., Gómez–López, M., Lawrence, S. E., Martínez–Díaz, M. V., Montalti, M., Piersanti, A., Prodi, L., Stoddart, J. F. and Williams, D. J. (1997) *J. Am. Chem. Soc.* **119**, 10641.

[17] Ashton, P. R., Ballardini, R., Balzani, V., Credi, A., Gandolfi, M. T., Marquis, D. J.–F., Menzer, S., Pérez–García, L., Prodi, L., Stoddart, J. F., Venturi, M., White, A. J. P. and Williams, D. J. (1995) *J. Am. Chem. Soc.* **117**, 11171.

[18] Ballardini, R., Balzani, V., Gandolfi, M. T., Prodi, L., Venturi, M., Philp, D., Ricketts, H. G. and Stoddart, J. F. (1993) *Angew. Chem. Int. Ed. Engl.* **32**, 1301.

[19] Ashton, P. R., Ballardini, R., Balzani, V., Constable, E. C., Credi, A., Kocian, O., Langford, S. J., Preece, J. A., Prodi, L., Schofield, E. R., Spencer, N., Stoddart, J. F. and Wenger, S. (in press) *Chem. Eur. J.*

[20] Ashton, P. R., Balzani, V., Kocian, O., Prodi, L., Spencer, N. and Stoddart, J. F. (in press) *J. Am. Chem. Soc.*.

[21] Asakawa, M., Ashton, P. R., Balzani, V., Credi, A., Hamers, C., Mattersteig, G., Montalti, M., Shipway, A. N., Spencer, N., Stoddart, J. F., Tolley, M. S., Venturi, M., White, A. J. P. and Williams, D. J. (1998) *Angew. Chem. Int. Ed. Engl.* **37**, 333.

[22] Ball, P. and Garwin, L. (1992) *Nature* **355**, 761.

[23] (a) Schlicke, B., De Cola, L. and Balzani, V. (submitted). (b) Schlicke, B., De Cola, L., Sabbioni, E. and Balzani, V. (work in progress).

[24] (a) Balzani, V., Juris, A., Venturi, M., Campagna, S. and Serroni, S. (1996) *Chem. Rev.* **96**, 759. (b) Balzani, V., Campagna, S., Denti, G., Juris, A., Serroni, S. and Venturi, M. (1998) *Acc. Chem. Res.* **31**, 26.

[25] (a) Ashton, P. R., Campbell, P. J., Chrystal, E. J. T., Glink, P. T., Menzer, S., Philp, D., Spencer, N., Stoddart, J. F., Tasker, P. A. and Williams, D. J. (1995) *Angew. Chem. Int. Ed. Engl.* **34**, 1865. (b) Kolchinski, A. G., Busch, D. H. and Alcock, N. W. (1995) *J. Chem. Soc. Chem. Commun.* 1289.

[26] (a) Montalti, M., Ballardini, R., Prodi, L. and Balzani, V. (1996) *Chem. Commun.* 2011. (b) Ishow, E., Credi, A., Balzani, V., Spadola, F. and Mandolini, L. (submitted).

[27] (a) Deans, R., Niemz, A., Breinlinger, E. C. and Rotello, V. M. (1997) *J. Am. Chem. Soc.* **119**, 10863. (b) Mirzoian, A. and Kaifer, A. E. (1997) *Chem. Eur. J.* **3**, 1052. (c) Wang, Y., Mendoza, S. and Kaifer, A. E. (1998) *Inorg. Chem.* **37**, 317. (d) Collin, J.–P., Gaviña, P. and Sauvage, J.–P. (1996) *Chem. Commun.* 2005. (e) Ashton, P. R., Ballardini, R., Balzani, V., Boyd, S. E., Credi, A., Gandolfi, M. T., Gómez–López, M., Iqbal, S., Philp, D., Preece, J. A., Prodi, L., Ricketts, H. G., Stoddart, J. F., Tolley, M. S., Venturi, M., White, A. J. P. and Williams, D. J. (1997) *Chem. Eur. J.* **3**, 152. (f) Asakawa, M., Ashton, P. R., Balzani, V., Credi, A., Mattersteig, G., Matthews, O. A., Montalti, M., Spencer, N., Stoddart, J. F. and Venturi, M. (1997) *Chem. Eur. J.* **3**, 1992. (g) Devonport, W., Blower, M. A., Bryce, M. R. and Goldenberg, L. M. (1997) *J. Org. Chem.* **62**, 885. (h) Bissell, R. A., Cordóva, E., Kaifer, A. E. and Stoddart, J. F. (1994) *Nature* **369**, 133. (i) Livoreil, A., Sauvage, J.–P., Armaroli, N., Balzani, V., Flamigni, L. and Ventura, B. (1997) *J. Am. Chem. Soc.* **119**, 12114. (j) Credi, A., Montalti, M., Balzani, V., Raymo, F. M., Langford, S. J. and Stoddart, J. F. (in press) *New J. Chem.*

[28] Balzani, V., Stoddart, J. F., *et al.* (work in progress).

[29] (a) Hünig, S., Kießlich, G., Quast, H. and Scheutzow, D. (1973) *Liebigs Ann. Chem.* 310. (b) Schukat, G. and Fanghänel, E. (1985) *J. Prakt. Chem.* **327**(5), 767. (d) Bryce, M. R., Devonport, W. and Moore, A. J. (1994) *Angew. Chem. Int. Ed. Engl.* **33**, 1761. (c) Jørgensen, T., Hansen, T. K. and Becher, J. (1994) *Chem. Soc. Rev.* **23**, 41.

[30] (a) Millman, J. and Grabel, A. (1988) *Microelectronics*; McGraw–Hill, London, Ch. 6. (b) Barker, J. (1991) *Chem. Br.* **27**, 728.

[31] (a) de Silva, A. P. and McCoy, C. P. (1994) *Chem. Ind.* 992. (b) de Silva, A. P., Gunaratne, H. Q. N., Gunnlaugsson, T., Huxley, A. J. M., McCoy, C. P., Rademacher, J. T. and Rice, T. E. (1997) *Chem. Rev.* **97**, 1515. (c) Ghosh, P., Bharadwaj, P. K., Roy, J. and Ghosh, S. (1997) *J. Am. Chem. Soc.* **119**, 11903. (d) Inouye, M., Akamatsu, K. and Nakazumi, H. (1997) *J. Am. Chem. Soc.* **119**, 9160.

[32] (a) de Silva, A. P., Gunaratne, H. Q. N. and McCoy, C. P. (1993) *Nature* **364**, 42. (b) de Silva, A. P., Gunaratne, H. Q. N. and Maguire, G. E. M. (1994) *J. Chem. Soc., Chem. Commun.* 1213. (c) de Silva, A. P., Gunaratne, H. Q. N. and McCoy, C. P. (1997) *J. Am. Chem. Soc.* **119**, 7891.

[33] Credi, A., Balzani, V., Langford, S. J. and Stoddart, J. F. (1997) *J. Am. Chem. Soc.* **119**, 2679.

TRANSITION METAL-CONTAINING CATENANES AND ROTAXANES : CONTROL OF ELECTRONIC AND MOLECULAR MOTIONS

J.-C. CHAMBRON, J.-P. SAUVAGE, J.-P. COLLIN, P. GAVIÑA, V. HEITZ, M. LINKE, A. LIVOREIL

Laboratoire de Chimie Organo-Minérale, CNRS-UMR 7513, Université Louis Pasteur, Institut Le Bel, 4, rue Blaise Pascal, 67070 Strasbourg cedex, France

ABSTRACT. Catenanes and rotaxanes are molecules based on mechanical bonds. Catenanes are made up with interlocking rings, whereas rotaxanes are molecules consisting of a cyclic component threaded onto a dumbbell-shaped linear component. Although first synthesized in the late sixties, they were considered as curiosities until the template methods (taking advantage of hydrophobic interactions, or using a transition metal or donor-acceptor interactions between aromatic stacks) made them relatively easily available. This allowed for the development of functional catenanes and rotaxanes, that is molecules responding to external stimuli, like injection or removal of electrons, light irradiation, and so on. Examples are rotaxanes and catenanes displaying electrochemically-triggered intramolecular motions such as translation of the ring along the dumbbell axle, rotation of one ring within the other, or photochemically-induced electron transfer from Zn(II)-porphyrin stoppers (electron donors in the excited state) to a Au(III) porphyrin electron acceptor appended to the ring component.

R. Ungaro and E. Dalcanale (eds.), Supramolecular Science: Where It Is and Where It Is Going, 23–38.

1. Introduction

Catenanes, rotaxanes and, to a lesser extent, molecular knots have recently been popularized through many publications devoted to their synthesis and to the study of their chemical or physical properties [1]. Their prototypes are represented in Figure 1, and the drawings should be sufficient to apprehend their structural specificities. Notice simply that, whereas catenanes and knots are composed of rings only, rotaxanes are made up with at least a ring and a dumbbell, the latter component bearing stoppers large enough to prevent unthreading of the ring.

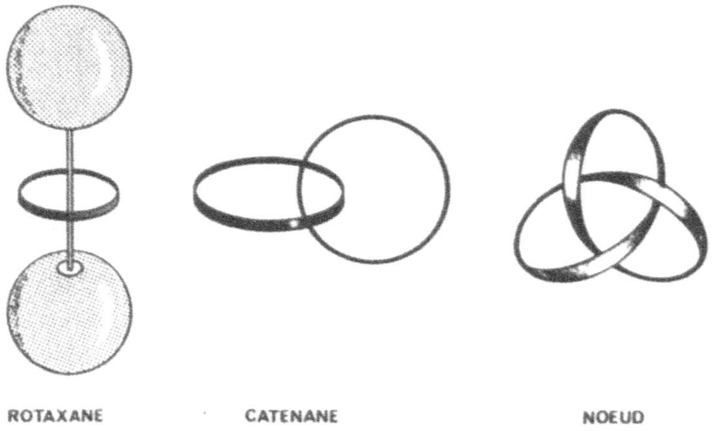

Figure 1 : Cartoons of the simplest rotaxane, catenane, and knot.

From a purely topological viewpoint, rotaxanes have never been considered as noble as interlocking ring systems and knots, although they have been associated to these species in the chemist's mind since the very first published discussions and experimental works on catenanes [1a, 2]. The reason for this is the following : whereas in both systems the components are held together by a mechanical bond, in the case of catenanes, the mechanical bond is also a topological bond, since the interlocked rings cannot be separated without cleavage of one of them. In the case of rotaxanes, only deformations of one of the components are required to dissociate the system.

Rotaxanes can be described as threaded species stabilized by steric interactions. The first threaded species in the chemical literature was obtained by Lüttringhaus, Cramer, Prinzbach, and Henglein as early as 1958 [3], and was a host-guest complex (**1**) between a cyclodextrin and a benzene nucleus bearing long chain dithiols, the latter protruding from the rims of the cyclodextrin (Figure 2). The target of these authors was to obtain a catenane species by formation of a disulfide bond. In fact, mechanically-bound species based on cyclodextrins as ring components were to be synthesized only in 1981, for the rotaxanes, by Ogino, as discussed below [4], and for the catenanes, more than a decade later, in 1993, by Stoddart and coworkers [5]. This article will focus mainly on molecular machines based on rotaxanes, but a catenane-based primitive molecular motor will also be described.

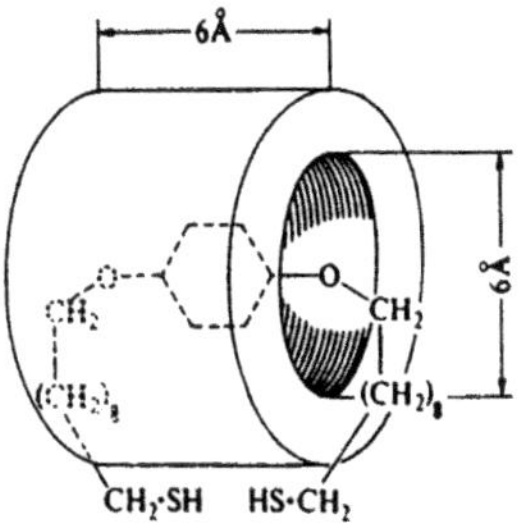

1

Figure 2 : Reproduction from ref. 3 of the first threaded molecular complex : α-cyclodextrin is threaded onto a benzene nucleus bearing long chain dithiols.

2. Overview of the synthetic methods for making rotaxanes

The introduction of new synthetic methodologies for making rotaxanes, based on templates, has certainly changed our view on them [1c-f]. In fact, they no more appear as the exotic species which they used to be before those efficient synthetic strategies were proposed. Nevertheless, rotaxanes are known for several decades thanks to the pioneering work of Schill et al. [6] (directed strategy) and Harrison (threading [7a] or slippage [7b-c]). Examples of these historical rotaxanes (**2** & **3**) are shown in Figure 3.

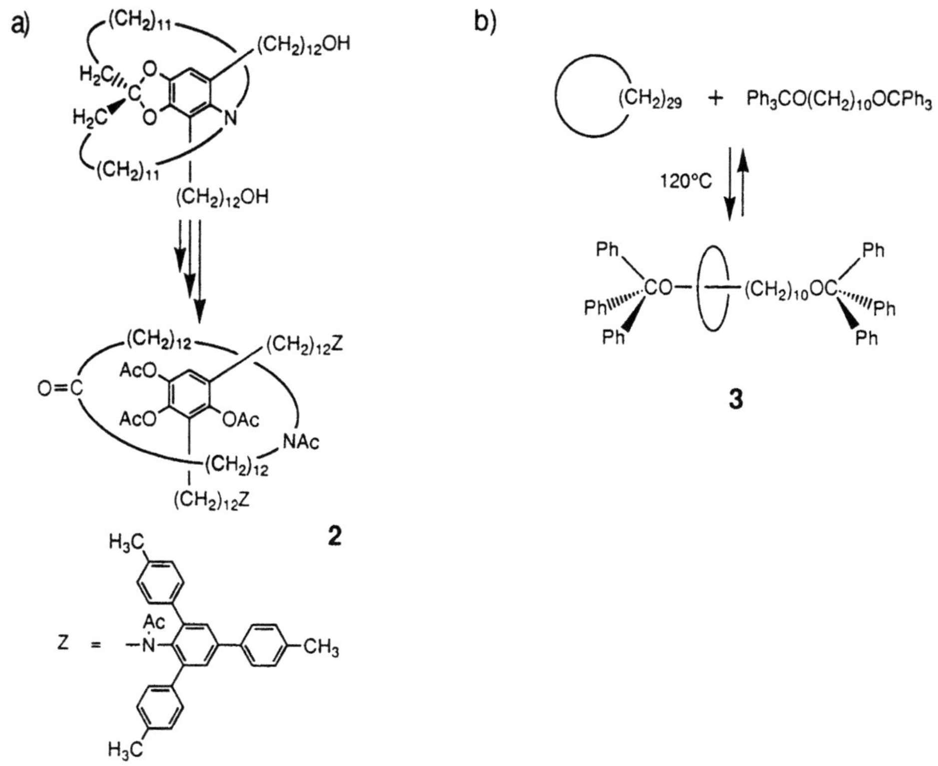

Figure 3 : (a) The first rotaxane (2) prepared by directed synthesis. Also shown is the key precursor, in which a ketal group maintains the plane of the ring orthogonal to the mean plane containing the thread precursor [6a]. (b) The first rotaxane (3) produced by statistical slippage : it is obtained by heating a mixture of dumbbell and cyclic components. In order to force the threading process, the solvent used was the macrocyclic component itself [7b].

Chemists being suddenly more successful at the making of threaded or interlocked molecular systems, they also rapidly became more ambitious in terms of functionality and complexity of the chemical groups to be introduced. This is especially true for rotaxanes. Ogino [4] and others [8] used transition metal complexes as stoppers, and this approach is potentially very rich as far as chemical and physical properties are concerned. Figure 4 shows the first rotaxane of this kind (**4**) [4]. It is also the first cyclodextrin-based rotaxane. The cyclodextrin ring is threaded onto an alkyl chain ended by amino groups, the latter being able to coordinate to cobalt(III) centres chelated with ethylenediamine ancillary ligands.

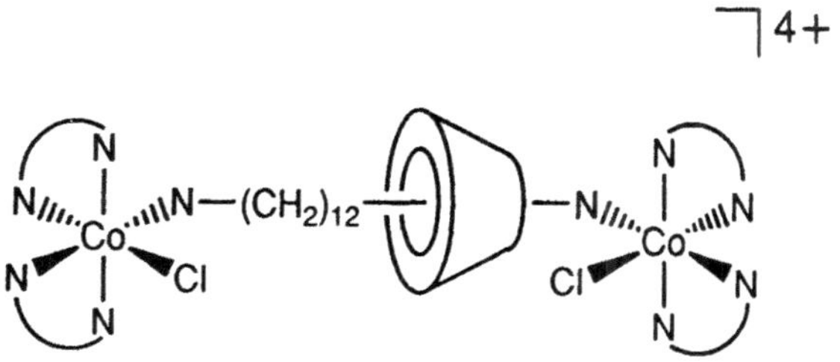

4

Figure 4 : The first rotaxane (4) bearing transition metal complex fragments as stoppers. It is also the first cyclodextrin-based rotaxane : the α-cyclodextrin ring is threaded onto an alkyl chain ended by amino groups, which bind cobalt(III) centres coordinated to ethylenediamine chelates [4a].

3. Functional rotaxanes and catenanes

The use of electro- and/or photoactive components incorporated in the central part of the linear fragment of the rotaxane or used as stoppers is of particular interest. The compounds will now respond to certain external stimuli (chemical, electrical or photonic signal) in a way dictated by their special structure. The threaded nature of the molecule will govern various processes following the triggering signal, such as molecular motions, electron or energy transfer, and so on. The "molecular shuttle" (**5**) described by Kaifer, Stoddart and their coworkers [9] is an interesting example of a molecule whose shape and internal motions can be externally controlled by electrochemical means. The threaded ring is a tetracationic electron-deficient cyclophane. The dumbbell component is made up with two "stations" able to interact with the ring, but differring in their electron-donating properties : a benzidine group, which is the strongest electron donor, and a biphenol-based moiety. In the initial state, *ca.* 84% of the macrocycle rests at the benzidine station. Upon electrochemical oxidation of the latter, the macrocycle is shifted towards the biphenol station. The process can be reversibly activated by reducing the benzidine station back to its initial state.

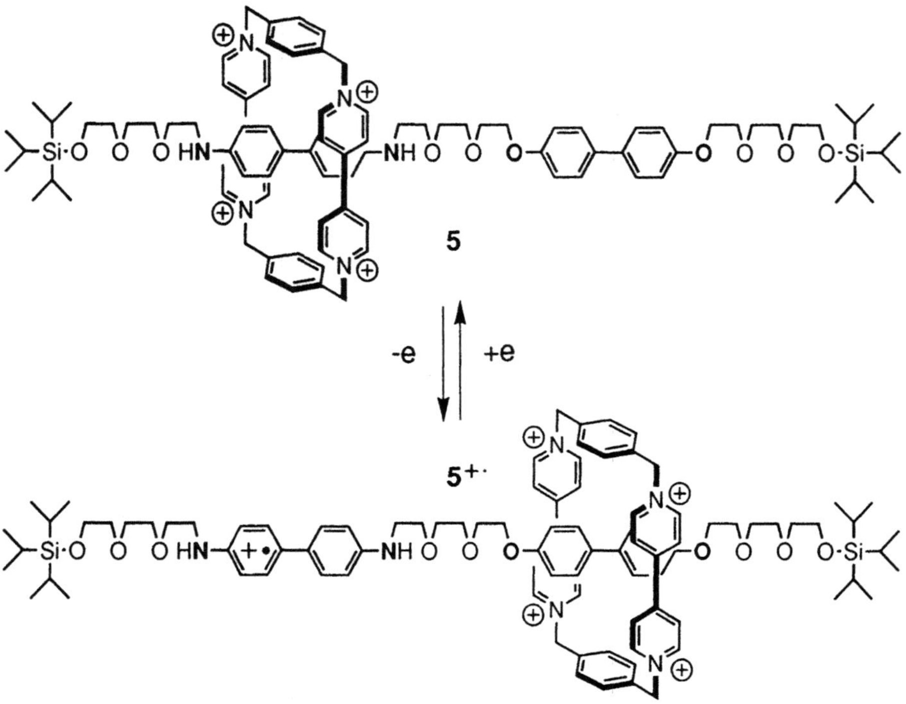

Figure 5 : A molecular shuttle (5) [9]. An electron-deficient cyclophane is threaded onto a linear component incorporating two sites differing by their electron-donor properties. In the resting state, ca. 84% of the macrocycle is bound to the more powerful electron-donor, the benzidine moiety. Electrochemical oxidation of the latter triggers the translation of the macrocycle to the biphenol moiety. The process is reversible.

3.1. ELECTRON TRANSFER WITHIN ROTAXANES

The system of Figure 6 is the C_{60}-stoppered rotaxane **6** [10]. It contains a central copper(I) complex, based on 2,9-diphenyl-1,10-phenanthroline (dpp) chelates, whose metal-to-ligand charge transfer (MLCT) excited state, generated by visible light irradiation, is a powerful electron donor. To explain the quenching of this luminescent MLCT state, it was suggested that electron transfer from the copper(I) central complex of the rotaxane to one of the C_{60} stoppers acting as electron acceptor, took place, under light irradiation [10b].

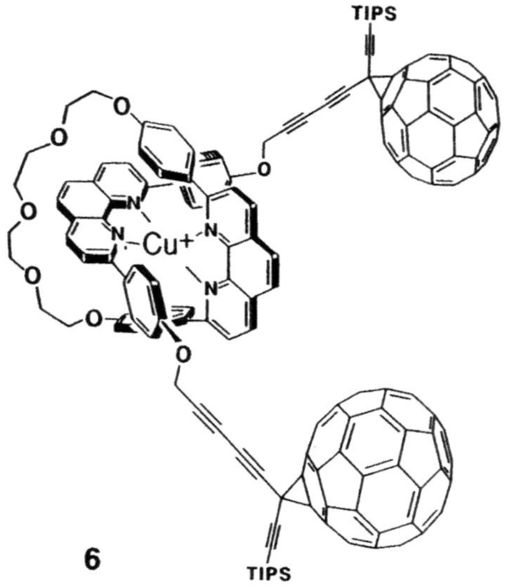

6

Figure 6 : A C_{60}-stoppered rotaxane (6) [10], based on a copper(I) bis-2,9-diphenyl-1,10-phenanthroline (dpp) complex. Thanks to the electron-donor character of the MLCT excited state of the complex fragment and the electron-acceptor properties of the C_{60} moiety, electron transfer takes place between these two components.

Rotaxanes displaying photoinduced intramolecular electron transfer were mostly based on porphyrinic stoppers. Porphyrins are ideal stoppers for this purpose, since they are at the same time voluminous and electro- and photoactive [11]. Our interest for photosynthesis has prompted us to make porphyrin-containing rotaxanes as models of precise fragments of the photosynthetic Reaction Centre [12]. The two representative examples of Figure 7 undergo fast photoinduced intramolecular electron transfer between the zinc (II) porphyrin (singlet excited state) and the gold (III) porphyrin (which is an electron acceptor in its ground state). Interestingly, the electron transfer rate can be, to a large extent, governed by the nature of the central metal or by demetalation of the central coordination site [13]. In the case of the uncomplexed rotaxane **7**, photoinduced electron transfer from the zinc to the gold porphyrin took place at a rate of $(36\ ps)^{-1}$, whereas for the Cu(I)-complexed rotaxane **8**, the same event was accelerated by a factor of *ca* 20. The rate enhancement was explained in terms of a superexchange mechanism, involving the virtual electronic states of the central copper(I) complex fragment.

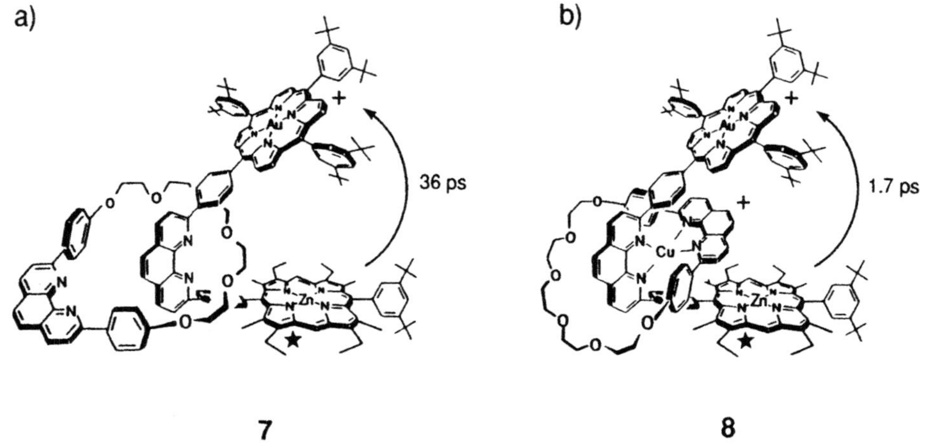

Figure 7 : (a) A rotaxane (7) containing a zinc(II) porphyrin and a gold(III) porphyrin as stoppers [12, 13]. Photoinduced electron transfer takes place between the Zn-porphyrin component in its singlet excited state (*) and the Au-porphyrin component in its ground state. (b) The same rotaxane complexing a Cu(I) cation (**8**) with the chelate fragments incorporated in the dumbbell and the macrocyclic components. The rate of electron transfer was found to be increased by a factor of 20 by comparison with the free rotaxane **7**.

In the previous examples (rotaxanes **6-8**) photoinduced electron transfer takes place between covalently linked photo- and electroactive components. In order to make impossible the identification of an electron transfer pathway involving a bond sequence, catenane [14] and rotaxane [15] systems in which the donor and acceptor components are maintained in the same molecule by mechanical bonds only seem to be ideal targets (Figure 8). In particular, this problem was addressed with rotaxane **9** composed of (i) a macrocycle incorporating a dpp chelate and bearing a pendent gold(III) porphyrin, and (ii) a dumbbell component containing the same dpp chelate and stoppered by two identical bulky zinc(II) porphyrins (Figure 9). It was obtained by demetallation of the copper(I)-complexed rotaxane **10**. Structural studies in solution suggested that demetallation leading to the template-free rotaxane, induced the translation motion of the dumbbell component towards the gold(III) porphyrin of the macrocycle. Preliminary photophysical studies showed that the fluorescence of the zinc porphyrins was strongly quenched in both cases ; with respect to the uncomplexed rotaxane, this clearly indicated that electron transfer from the zinc(II) porphyrin (singlet excited state) to the gold(III) porphyrin component occurred. Therefore electron transfer between mechanically-bound species has been achieved in this molecule.

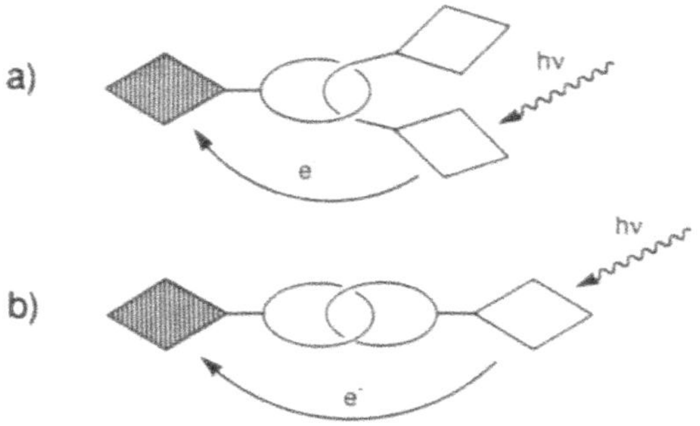

Figure 8 : Donor-acceptor multiporphyrin conjugates. The donor porphyrin (empty diamond) and the acceptor porphyrin (hatched diamond) are mechanically linked in either catenane (a) or rotaxane (b) structure. The arrows show the direction of photoinduced electron transfer.

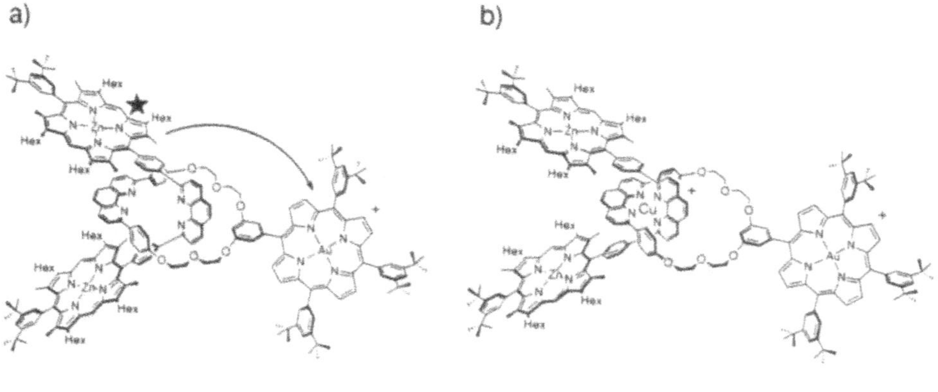

9 10

Figure 9 : (a) Through-space electron transfer (represented by an arrow) in a [2]-rotaxane (9) from a Zn-porphyrin stopper of the dumbbell in its singlet excited state (*) to the Au porphyrin covalently attached to the macrocycle. (b) The same rotaxane complexing a Cu(I) cation (10) [15].

3.2. MOLECULAR MOTIONS WITHIN CATENANES AND ROTAXANES

The first molecular motor elaborated and studied in our group was a catenane containing two different interlocking rings. Its principle is explained in Figure 10 [16]. The actual system and the full square-scheme are indicated in Figure 11. The starting copper(I) complex $\mathbf{11}_{(4)}{}^{+}$ is a 4-coordinate species, whose high redox potential (+ 0.63 V vs. SCE in CH_3CN) clearly indicates that the geometry of the system (tetrahedral or distorted tetrahedral) is well adapted to copper(I). This redox state being very stable with the environment provided by $\mathbf{11}_{(4)}{}^{+}$, a relatively high redox potential will have to be applied for the monovalent copper center to be oxidized to the divalent state. Interestingly, the 4-coordinate Cu(II) complex $\mathbf{11}_{(4)}{}^{2+}$ is an intense green species, with a d-d absorption band at 670 nm (ε = 800 $M^{-1}.cm^{-1}$) in CH_3CN. This compound can be generated either by chemical (Br_2 or $NOBF_4$) or electrochemical oxidation.

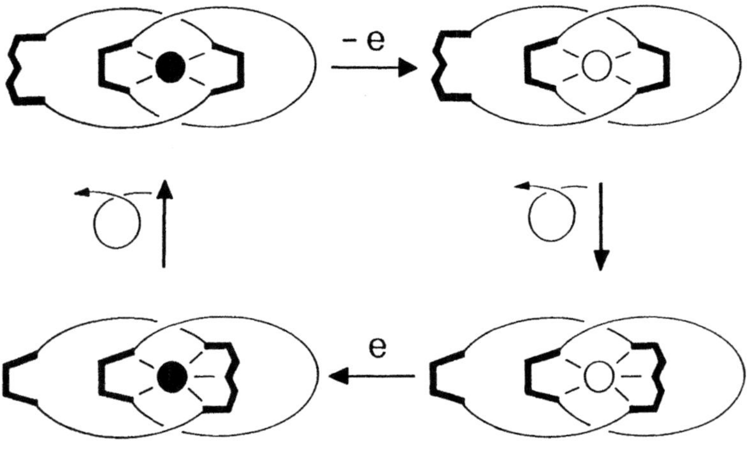

Figure 10 : Electrochemically triggerred rearrangement of a [2]-catenate containing two different rings. The stable 4-coordinate monovalent complex [top left, the black circle represents Cu(I)] is oxidized to an intermediate tetrahedral divalent species [top right, the white circle represents Cu(II)]. This compound undergoes a complete reorganization process to afford the stable 5-coordinate Cu(II) complex [bottom right]. Upon reduction, the 5-coordinate monovalent state is formed as a transient [bottom left]. Finally, the latter undergoes the conformational change which regenerates the starting complex.

The changeover reaction converting $\mathbf{11}_{(4)}^{2+}$ to the stable 5-coordinate species $\mathbf{11}_{(5)}^{2+}$ is quantitative. It is easily monitored by visible absorption spectroscopy since the product of the rearrangement reaction is only slightly colored (pale olive green ; $\lambda_{max} = 640$ nm ; $\varepsilon = 125\ M^{-1}.cm^{-1}$). Both copper(II) complexes $\mathbf{11}_{(4)}^{2+}$ and $\mathbf{11}_{(5)}^{2+}$ have electronic spectra typical for 4- and 5-coordinate species respectively, in accordance with previously reported complexes having analogous ligand sets [17].

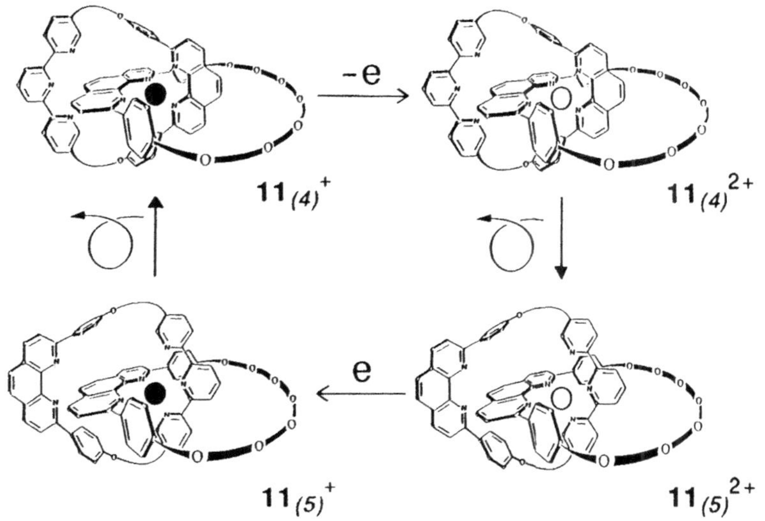

Figure 11 : Electrochemically induced molecular rearrangements undergone by the copper catenate $\mathbf{11}^{2+/+}$. In the text, the subscript *4* or *5* indicates the number of nitrogen atoms coordinated to the metal. This number is explicited on the figure [16].

The same conversion process can also be monitored by EPR [18]. It was demonstrated in an unambiguous fashion that $\mathbf{11}_{(4)}^{2+}$ is a distorted tetrahedral complex and that the product of the changeover, $\mathbf{11}_{(5)}^{2+}$, is a square pyramidal compound.

An interesting question is dealing with the rate of the ring gliding motion which transforms $\mathbf{11}_{(4)}^{2+}$ into $\mathbf{11}_{(5)}^{2+}$ or, after reduction of the latter, $\mathbf{11}_{(5)}^{+}$ into $\mathbf{11}_{(4)}^{+}$. It was observed that this last process, involving Cu(I), is fast (a few seconds at room temperature, regardless of the solvent) whereas the copper(II) complex rearrangement $\mathbf{11}_{(4)}^{2+}$ into $\mathbf{11}_{(5)}^{2+}$ is slow.

As shown above, *rotational* motions of a ring within the other were observed for the copper catenate $11^{2+/+}$. The rotaxane **12** represented in Figure 12 was designed and synthesized in order to observe electrochemically-triggered *translational* motions between two different sites of the axle [19]. The ring incorporates a dpp fragment and the molecular string consists of a terdentate ligand (terpy = 2,2',6',2"-terpyridine) linked to a bidentate coordinating unit (2,9-disubstituted-1,10-phenanthroline). Trityl-derived stoppers are covalently attached to the extremities of the string. The principle is similar to that used in the case of the catenane, and will not be detailed here. Initially the ring threaded onto the molecular string coordinates copper(I) together with the phenanthroline

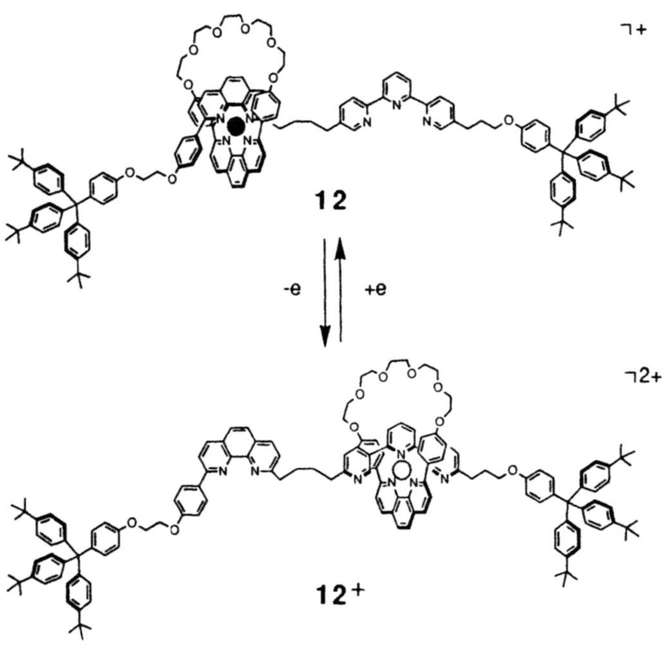

Figure 12 : A rotaxane in motion (**12**) based on the redox state dependence of the stereoelectronic requirements of a transition metal (Cu) [19]. Cu(I) and Cu(II) are represented by black and white circles respectively. The linear component incorporates a bidentate and a separated terdentate chelating site. The macrocycle coordinates Cu(I) together with the bidentate site of the linear component in a tetrahedral fashion. Electrochemical oxidation of Cu(I) to Cu(II) triggers the transfer of the Cu(II)-complexed ring to the terdentate chelate of the thread. The timescale of this process is *ca.* 1-2h. Interestingly, the back motion is faster : it takes place in minutes.

site of the string in a tetrahedral fashion. Oxidation of Cu(I) to Cu(II) triggers a modification of the coordination sphere of the copper atom, which results in the net transfer of the copper(II)-complexed ring to the terpy site of the string. This translation motion is slow (*ca.* 1-2 h) but complete under the experimental conditions used (CH_3CN, room temperature). The reversal to the starting state is initiated by reduction of Cu(II) to Cu(I) and is somewhat faster (minutes) than the forward motion. Noteworthy, the presence of the stoppers, that is, the rotaxane structure, is essential : when the string has open ends, motions of the ring along the string are accompanied by partial dethreading.

4. Conclusion

In summary, the rotaxane and catenane frameworks can be used with advantage for making molecules in which *controlled* motions occur : translational motions of a molecular ring threaded onto a molecular rod, rotation of a molecular ring within another, through-bond or through-space electron transfer from a donor to an acceptor electroactive components. These molecular or electronic motions are driven either by a redox process (by electrochemical means for example) or by a photochemical event (triggered by light). The mechanical bond of rotaxanes and catenanes comes into play in different manners : (i) stabilizing heteroleptic transition metal complexes (two differently substituted dpp fragments or a dpp and a terpy unit around a copper centre) which could not be obtained otherwise because of fast ligand exchange ("scrambling"), if the coordinating fragments were independent and free to diffuse in solution ; (ii) allowing non dissociative large amplitude motions, like translation of a ring threaded onto a rod, or rotation of a ring within the other, which mimic, at the molecular level, the motions observed in different kinds of motors ; (iii) providing mechanical bonds for studying through-space intramolecular electron-or energy-transfer processes.

Molecular machines constitute an emerging field of research, at the interface with many areas of molecular sciences. For instance, the concepts discussed in the present paper can certainly be generalized to organized assemblies of molecules (liquid crystals) or to molecular components attached to an electrode surface. The overall properties of the system (now, a "device") such as liquid crystal character, ability to carry electrons or transport molecules, etc... could be switched by just applying a given signal. The variety of impulses to be used is immense : redox (as in the present examples), photonic, heat or pressure change, magnetic field variation, pH-change, chemical signal (recognition of a molecule), electric field, etc... The effect obtained is also multifarious and could lead to dramatic modifications of the bulk properties of the system. Since a simple signal can change the shape and the volume of a compound and of its assemblies, fascinating features related to mechanics (contraction or stretching) could be imagined, reminiscent of biological systems such as muscles or other biological molecular machines. A particularly challenging biological system to be mimicked by artificial molecular assemblies is the ubiquitous enzyme ATP synthase [20]. This protein represents today the most perfect rotary motor.

References

1. a) G. Schill, *Catenanes, Rotaxanes and Knots*, Academic Press, New York, 1971; b) C.O. Dietrich-Buchecker, J.-P. Sauvage, *Chem. Rev.* **1987**, *87*, 795-810; c) C.O. Dietrich-Buchecker, J.-P. Sauvage, *Bioorganic Chemistry Frontiers* **1991**, *2*, 195-248; d) H.W. Gibson, M.C. Bheda, P.T. Engen, *Progr. Polym. Sci.* **1994**, *19*, 843; e) D.B. Amabilino, J.F. Stoddart, *Chem. Rev.* **1995**, *95*, 2725-2828; f) R. Jäger, F. Vögtle, *Angew. Chem. Int. Ed. Engl.* **1997**, *36*, 930-944.
2. a) H.L. Frisch, E. Wasserman, *J. Am. Chem. Soc.* **1961**, *83*, 3789-3795; b) V.I. Sokolov, *Russian Chem. Rev. (Engl. Transl.)* **1972**, *41*, 452-463.
3. A. Lüttringhaus, F. Cramer, H. Prinzbach, F.M. Henglein, *Liebigs Ann. Chem.* **1958**, *613*, 185-198.
4. a) H. Ogino, *J. Am. Chem. Soc.* **1981**, *103*, 1303-1304; b) H. Ogino, K. Ohata, *Inorg. Chem.* **1984**, *23*, 3312-3316.

5. D. Armspach, P.R. Ashton, C.P. Moore, N. Spencer, J.F. Stoddart, T.J. Wear, D.J. Williams, *Angew. Chem. Int. Ed. Engl.* **1993**, *32*, 854-858.

6. a) G. Schill, H. Zollenkopf, *Liebigs Ann. Chem.* **1969**, *721*, 53-74; b) G. Schill, R. Henschel, *Liebigs Ann. Chem.* **1970**, *731*, 113-119; c) G. Schill, C. Zürcher, W. Vetter, *Chem. Ber.* **1973**, *106*, 228-235.

7. a) I.T. Harrison, S. Harrison, *J. Am. Chem. Soc.* **1967**, *89*, 5723-5724; b) I.T. Harrison, *J. Chem. Soc., Chem. Commun.* **1972**, 231-232; c) I.T. Harrison, *J. Chem. Soc. Perkin Trans. I* **1974**, 301-304.

8. a) K. Yamanari, Y. Shimura, *Bull. Chem. Soc. Jpn.* **1983**, *56*, 2283-2289; b) R. Isnin, A.E. Kaifer, *J. Am. Chem. Soc.* **1991**, *113*, 8188-8190; c) R.S. Wylie, D.H. Macartney, *J. Am. Chem. Soc.* **1992**, *114*, 3136-3138; d) D.J. Cárdenas, P. Gaviña, J.-P. Sauvage, *Chem. Commun.* **1996**, 1915-1916; e) D.J. Cárdenas, P. Gaviña, J.-P. Sauvage, *J. Am. Chem. Soc.* **1997**, *119*, 2656-2664.

9. R. A. Bissell, E. Cordova, A.E. Kaifer, J.F. Stoddart, *Nature* **1994**, *369*, 133-137.

10. a) F. Diederich, C. Dietrich-Buchecker, J.-F. Nierengarten, J.-P. Sauvage, *J. Chem. Soc., Chem. Commun.* **1995**, 781-782; b) N. Armaroli, F. Diederich, C.O. Dietrich-Buchecker, L. Flamigni, G. Marconi, J.-F. Nierengarten, J.-P. Sauvage, *Chem. Eur. J.* **1998**, *4*, 406-416.

11. a) P.R. Ashton, M.R. Johnston, J.F. Stoddart, M.S. Tolley, J.W. Wheeler, *J. Chem. Soc., Chem. Commun.* **1992**, 1128-1131; b) F. Vögtle, F. Ahuis, S. Baumann, J.L. Sessler, *Liebigs Ann.* **1996**, 921-926.

12. a) J.-C. Chambron, V. Heitz, J.-P. Sauvage, *J. Chem. Soc., Chem. Commun.* **1992**, 1131-1133, b) J.-C. Chambron, V. Heitz, J.-P. Sauvage, *J. Am. Chem. Soc.* **1993**, *115*, 12378-12384; c) J.-C. Chambron, S. Chardon-Noblat, A. Harriman, V. Heitz, J.-P. Sauvage, *Pure & Appl. Chem.*, **1993**, *65*, 2343-2349.

13. a) J.-C. Chambron, A. Harriman, V. Heitz, J.-P. Sauvage, *J. Am. Chem. Soc.* **1993**, *115*, 6109-6114; b) J.-C. Chambron, A. Harriman, V. Heitz, J.-P. Sauvage, *J. Am. Chem. Soc.* **1993**, *115*, 7419-7425.

14. a) D.B. Amabilino, J.-P. Sauvage, *Chem. Commun.* **1996**, 2441-2442; b) D.B. Amabilino, J.-P. Sauvage, *New J. Chem.* **1998**, *22*, 395-409.

15. M. Linke, J.-C. Chambron, V. Heitz, J.-P. Sauvage, *J. Am. Chem. Soc.* **1997**, *119*, 11329-11330.

16. A. Livoreil, C.O. Dietrich-Buchecker, J.-P. Sauvage, *J. Am. Chem. Soc.* **1994**, *116*, 9399-9400.

17. For previous studies on 5-coordinate Cu(I) and Cu(II) complexes with imine-type ligands, see : a) J.A. Goodwin, D.M. Stanbury, L.J. Wilson, C.W. Eigenbrot, W.R. Scheidt, *J. Am. Chem. Soc.* **1987**, *109*, 2979-2991; b) J.A. Goodwin, G.A. Bodager, L.J. Wilson, D.M. Stanbury, W.R. Scheidt, *Inorg. Chem.* **1989**, *28*, 35-42; c) J.A. Goodwin, L.J. Wilson, D.M. Stanbury, R.A. Scott, *Inorg. Chem.* **1989** *28*, 42-50; d) C.M. Harris, T.N. Lockyer, *Aust. J. Chem.* **1970**, *23*, 673-682; e) G. Arena, R.P. Bonomo, S. Musumeci, R. Purello, E. Rizzarelli, S. Sammartano, *J. Chem. Soc. Dalton Trans.* **1983**, 1279-1283.

18. a) F. Baumann, A. Livoreil, W. Kaim, J.-P. Sauvage, *Chem. Commun.* **1997**, 35-36; for the EPR properties of copper(II) complexes having similar coordination spheres as in $\mathbf{11}_{(4)}{}^{2+}$ or $\mathbf{11}_{(5)}{}^{2+}$ see : a) B.A. Goodman, J.B. Raynor, *Adv. Inorg. Chem. Radiochem.* **1970**, *13*, 135-362; b) M. Geoffroy, M. Wermeille, C.O. Dietrich-Buchecker, J.-P. Sauvage, G. Bernardinelli, *Inorg. Chim. Acta* **1990**, *167*, 157-164; c) S.B. Sanni, H.J. Behm, P.T. Beurskens, G.A. Van Albada, J. Reedijk, A.T.H. Lenstra, A.W. Addison, M. Palaniandavar, *J. Chem. Soc., Dalton Trans.* **1988**, 1429-1435; d) W. Kaim, *Coord. Chem. Rev.* **1987**, *76*, 187-235.

19. a) P. Gaviña, J.-P. Sauvage, *Tetrahedron Lett.* **1997**, *38*, 3521-3524; b) J.-P. Collin, P. Gaviña, J.-P. Sauvage, *New J. Chem.* **1997**, *21*, 525-528.

20. T. Elston, H. Wang, G. Oster, *Nature* **1998**, *391*, 510-513.

ELECTROCHEMICAL TRANSFORMATIONS OF METHANOFULLERENES

L. ECHEGOYEN
Department of Chemistry
University of Miami
Coral Gables, FL 33124
USA

J.-P. BOURGEOIS, F. CARDULLO, J. CRASSOUS, F. DIEDERICH, M. GÓMEZ-LÓPEZ, A. HERRMANN, R. KESSINGER, M. RÜTTIMANN,
Laboratorium für Organische Chemie, ETH-Zentrum,
Universitätstrasse 16. CH-8092 Zürich
Switzerland

1. Introduction

The synthesis of fullerene derivatives started almost as soon as the parent compounds were first isolated in large quantities in 1990, after extraction from the soot produced by arching graphite rods under a helium atmosphere [1]. A wide variety of reactions have been developed to prepare fullerene derivatives, mainly from C_{60} and C_{70}, since these are the most abundant carbon allotropes produced in the soot. Given the spherical symmetry of C_{60}, along with its unique and interesting electronic properties, it has been used extensively as the core from which well-defined three-dimensional architectures containing specific functional groups have been built [2,3]. Single as well as multiple adducts can be easily attached to C_{60} using a variety of protocols, but the persistent challenge is to prepare the multiple adducts in a regioselective way [4].

Of the many derivatization reactions involving C_{60}, the so-called *Bingel* reaction, first reported in 1993, is undoubtedly one of the most versatile and consequently one of the most used for the preparation of fullerene derivatives [5]. It works under mild conditions and it provides relatively high yields of the corresponding products, which are called methanofullerenes. The resulting cyclopropane rings attached at [6,6] junctions on the surface of C_{60} (see structure **1**) are very stable, even under relatively extreme thermal conditions. Further synthetic elaboration of the ester-functionalized methano bridges offers a convenient and versatile way to prepare a wide variety of functional methanofullerenes.

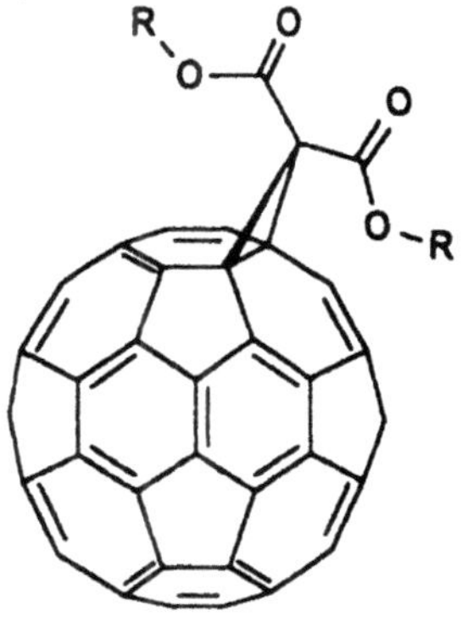

1

The *Bingel* reaction has also been used to prepare bis-adducts of C_{60}, and the relative distribution of the regioisomers that are obtained has been analyzed in detail by Hirsch and his coworkers, and partially rationalized using statistical and theoretical arguments [6]. All of the

R. Ungaro and E. Dalcanale (eds.), Supramolecular Science: Where It Is and Where It Is Going, 39–52.

regioisomers obtained using the *Bingel* reaction exhibit methano-adducts connected exclusively at [6,6] junction positions on the C_{60} surface. Of the eight theoretically possible regioisomers, seven were detected and characterized [6]. The nomenclature introduced by Hirsch to define the possible bis-adduct regioisomers is presented in scheme 1, where the cyclopropane at the top represents the first methano-adduct and the highlighted bonds are the relative positions where the second adduct can be found. No *cis*-1 bis-adduct was isolated, presumably due to the steric compression that would result from the close proximity of the two adducts in this isomer. The *trans*-1 derivative is almost always the least abundant due to kinetic and statistical reasons, and is isolated in 0.8-2% yields [6].

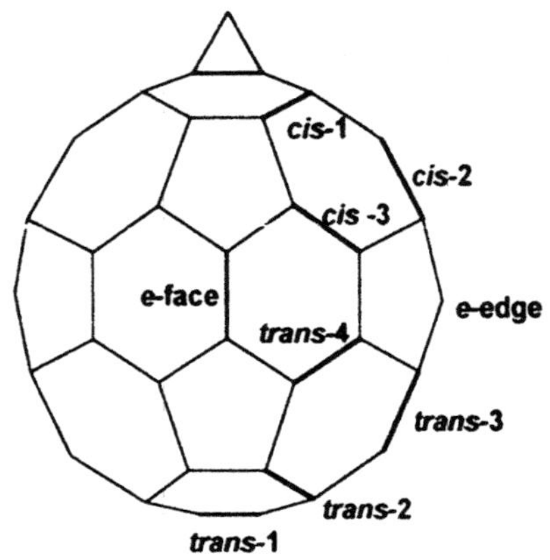

Scheme1. Nomenclature for Bis-Adducts of C_{60}

In order to have regioisomeric control of bis- and tris-adducts formation, some of us introduced the concept of tether-directed remote functionalization [7]. In the case of a double Bingel reaction [4b] a tether separates two malonate groups and by its length and rigidity enforces a particular regioisomeric addition pattern on the surface of the fullerene. Using this approach a variety of bis(cyclopropanated) [60]fullerene derivatives have been prepared, both regio- and stereoselectively by macrocyclization of the carbon sphere *via* double *Bingel* addition [10b]. This reaction has provided ready access to bis(methano)fullerenes with *cis*-2, *cis*-3, *e*, *trans*-4, and *trans*-3 addition patterns [6, 8]. More recently, using a dibenzo-18-crown-6 tether, a *trans*-1 bis-adduct was also prepared in relatively high yields, structure **2** [9]. It is thus possible, using cleverly designed tethered bis-malonate precursors, to prepare regioiselective bis-adducts using the *Bingel* macrocyclization reaction.

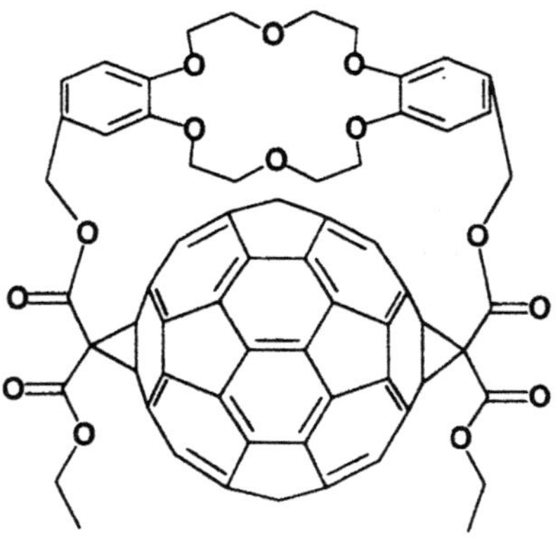

2

The electrochemical properties of all of these bis-adducts have been studied in reasonable detail, using either cyclic voltammetry (CV) or steady-state voltammetry (SSV) at a rotating disk electrode [10]. While many interesting observations have been made about their electrochemical behavior, the most relevant to the present work concerns the *cis*-2 isomer, **3** [11]. The bis(methano)fullerenes with a *cis*-2 addition pattern exhibit a chemically irreversible second one-electron reduction, as determined by using CV and SSV on a rotating disk electrode. Of all the isomers studied, the *cis*-2 is the only one that exhibits such an irreversible second reduction step on the voltammetric time scales. This observation led to the initiation of a project that eventually culminated in the observations that are presented in this chapter. Most of the work reported here was conducted while one of the authors (LE) was on sabbatical leave at the ETH in Zürich, between July, 1997 and July, 1998.

2. The Methano-adducts Can Move on the Surface: "The Shuffle"

The starting point of this project, as briefly mentioned in the introductory section, was the analysis of the behavior of the *cis*-2 bis-methanofullerene, **3**, under conditions of controlled potential electrolysis (CPE). The experiment was conducted in CH_2Cl_2 containing 0.1 M Bu_4NPF_6, in a home-built H-cell with a glass frit #4 separating the two compartments [12]. The electrolyte and the compound (approximately 6 mg) were degassed and evacuated to 10^{-6} mmHg, before the solvent was vapor transferred directly into the cell from CaH_2, after being independently degassed by 4-5 freeze-pump-thaw cycles. The cell compartment that contained the compound was fitted with a glassy carbon (GC) working electrode (3 mm diameter) and a Pt mesh electrode for reductive electrolysis. A silver wire, separated from the solution by a Vycor tip, served as the reference electrode. The other cell compartment, which contained only the electrolyte (no **3** present) was also fitted with a Pt mesh electrode that was used at the counter electrode during CPE. The complete apparatus was kept under vacuum, or more precisely, under the vapor pressure of CH_2Cl_2 during all of the electrochemical manipulations until the end, when the products were removed for analysis.

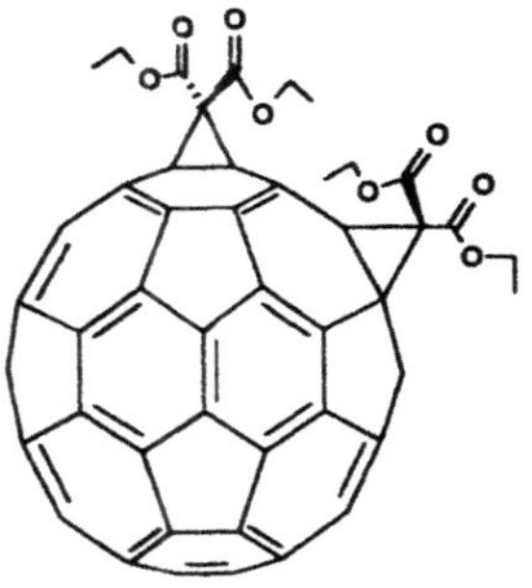

3

The initial CV of compound **3**, obtained using a Princeton Applied Research Model 263 potentiostat/galvanostat, is shown in figure 1a. The irreversibility of the second reduction process is evident from a simple inspection of figure 1a. After recording this CV, the solution was submitted to CPE at −1.2 V *vs* the Ag wire pseudo-reference electrode, some 100 mV negative of the first reduction peak potential. The electrolysis behaved well, and the current decreased to the background level in approximately 20 min. The integrated current yielded a value of charge transferred that was equivalent to 1e⁻ per molecule. There were no apparent surprises during this first electrolytic process, as anticipated from the CV and SSV behavior. The voltammetric

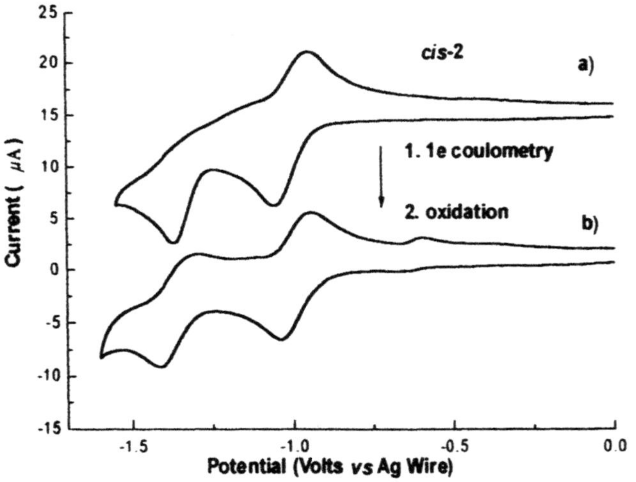

Figure 1. CV of *cis*-2 isomer before (a) and after (b) 1e-electrolysis.

results indicated that the second electron reduction would be the one leading to an irreversible chemical process. However, recording the CV of the 1e$^-$ electrolyzed solution revealed something that was quite unexpected. The second reduction appeared to be more chemically reversible than in the original CV, a very dramatic change, see figure 1b. Therefore, rather than continue with the next electrolysis to prepare the dianion. the solution was electrolyzed back to the neutral state of the fullerene, at 0 V *vs* Ag. 80% of the charge was recovered upon oxidation, and the solution was submitted to column chromatography (SiO_2, CH_2Cl_2:hexane 3:1), to yield a mixture of bis-adducts (44%), which was analyzed by HPLC. Analytical HPLC was performed on a Macherey-Nagel AG Nucleosil 100-7 (7 μm) 250 mm x 4 mm column using a toluene mobile phase at a flow rate of 2 ml/min. The percentages of the individual bis-adducts in the mixture were determined from the integrated HPLC peak areas with optical detection at 310 nm and corrected for the difference in UV-vis absorption at this wavelength.

Interestingly, the analytical chromatogram showed that the *cis*-2 isomer had completely disappeared and the *e* bis-adduct had become the predominant isomer in the newly formed product mixture (*e*: 57%; *trans*-3: 31%; *trans*-4: 8%; *cis*-3: 4%). These unexpected results, recently reported [13], led us to conduct similar experiments starting with some of the other bis-adduct isomers. Thus the pure *cis*-3, *e*, *trans*-4, and *trans*-3 bis-methanofullerenes were submitted to the exact same protocol: 1e$^-$ CPE around −1.2 V *vs* Ag, followed by oxidative CPE at 0 V and product analysis using chromatographic techniques. These four regioisomers were recovered in 75-90% yields and did not show any isomerization. Therefore, isomeric conversion upon 1e$^-$ CPE occurs only for the *cis*-2 bis-methanofullerene.

These observations prompted us to analyze the behavior of the bis-methanofullerenes upon 2e$^-$ CPE and these experiments led to several interesting observations. The CPE experiments were conducted around −1.55 V *vs* Ag, but in every case they had to be manually interrupted before the current had decreased to background levels. This is a very important point, which actually leads to other observations that are described in detail in the next section, under the title of "*retro-Bingel*". Suffice it to say at this point that CPE at the potential corresponding to the second voltammetric reduction peaks were interrupted after determining, via coulometry, that the amount of charge passed corresponded to 2e$^-$ per molecule.

TABLE 1. Relative Distribution of Bis-methanofullerene isomers obtained after 2e$^-$ coulometry followed by oxidation at 0 V *vs* Ag for different starting pure isomers

	cis -(1-3)	*e*	*trans* -4	*trans* -3	*trans* -2	*trans* -1	mono	Yield
cis -2	----	23	5	8	52	12	29	49
cis -3	----	22	6	10	52	11	28	53
e	----	26	9	12	44	9	10	44
trans -4	----	23	10	12	47	8	14	61
trans -3	----	23	9	12	46	10	23	43
trans -2	----	22	6	9	53	10	19	39

In most cases the currents that were observed at the time the coulometries were stopped were quite substantial, indicating that the CPE process was still not over [13]. After interrupting the 2e$^-$ reductive CPE's, the solutions were submitted to oxidative CPE at 0 V, to return to the neutral states of the compounds. The products were again analyzed using the same chromatographic protocol already described. The results are summarized in table 1 [13]. It is interesting to note that the isomer distribution observed after this reduction-oxidation procedure is essentially the same no matter which pure isomer is initially electrolyzed. Thus, the methano-adducts appear to be migrating around the fullerene sphere, but not in a random manner. Also notice that no *cis*-2 nor *cis*-3 isomers are detected at all in the product mixtures. The major isomer that forms is always the *trans*-2 derivative (40-50%). It is important to point out that approximately 10% of the *trans*-1 bis-adduct is produced in every case and that the *trans*-1 and *trans*-2 isomers, therefore, appear roughly in their expected 1:4 statistical ratio. This result is especially noteworthy since, as already mentioned, when diethyl 1,2-methano[60]fullerene-61,61-dicarboxylate undergoes a *Bingel* reaction with one equivalent of diethyl 2-bromomalonate, only 0.8 - 2% of the *trans*-1 regioisomer can be isolated [6]. After the CPE procedure the *trans*-1 and *trans*-2 derivatives predominate over the *trans*-3 and *trans*-4 ones, while the *e*-isomer is also formed in relatively high yield (approximately 25%). The preferred formation of both *e* and *trans* isomers could be, at least partially, a consequence of their inherently higher stability, as compared to the *cis* derivatives, which had been predicted by theoretical calculations [6]. The fact that the same product distribution is observed after coulometric reduction of the different isomers with two electrons suggests that the isomerization occurs under thermodynamic control.

In addition to these observations, notice that in every case there is a substantial amount of the mono-methanofullerene adduct formed, anywhere between 10 and almost 30%. Therefore, electrolysis with 2e$^-$ per molecule has the ability to "shuffle" the adducts until an equilibrium distribution of isomers is attained and it can also lead to the removal of one of the adducts. In some cases small amounts of C_{60} were also detected after completing the electrolytic redox cycle. It is thus obvious that the reduction electrons can "move" and eventually "remove" the bis-ester methano-adducts from C_{60}. The removal process is essentially the reverse of the *Bingel* reaction, and these results are described in more detail in the next section. The detailed mechanisms involved in these processes are currently not known, but some experiments have been performed to answer some of the more immediate questions.

One important concern is whether the "shuffle" occurs intra- or intermolecularly. We designed a series of crossover experiments to address this question by preparing bis(diethoxycarbonylmethano) and bis(dipropoxycarbonylmethano) fullerene derivatives [13]. These were prepared via the *Bingel* reaction to yield the typical bis-adduct isomer distributions, as already described [6], seven isomers of $(EtOOC)_2C{<}C_{60}{>}C(EtOOC)_2$ and seven of $(PrOOC)_2C{<}C_{60}{>}C(PrOOC)_2$. A 1:1 mixture of these isomeric products was submitted to the 2e$^-$ reduction-followed-by-oxidation protocol in an effort to detect any crossover products that contained both ethyl- as well as propyl-ester methano-adducts on the same fullerene. Analysis of the final product mixtures by HPLC as well as by MALDI-TOF mass spectrometry failed to detect any crossover products. The experiment was also done backwards, by preparing an isomer mixture of $(EtOOC)_2C{<}C_{60}{>}C(PrOOC)_2$ via sequential *Bingel* addition of diethyl- and then dipropyl-malonates to C_{60}. Electrolysis with 2e$^-$ followed by oxidation and analysis with

HPLC and MALDI-TOF failed to detect any of the pure crossover products, $(EtOOC)_2C{<}C_{60}{>}C(EtOOC)_2$ or $(PrOOC)_2C{<}C_{60}{>}C(PrOOC)_2$. We thus feel fairly confident that the isomerization of the two adducts is an intramolecular process initiated by the reduction electrons [13]. Although we have no experimental evidence to support a particular mechanism at he present time, it seems likely that the isomerization proceeds initially by a heterolytic C-C bond breakage of one of the bridgehead cyclopropane bonds, to form an ester-stabilized carbanion and a C_{60}-based radical. The radical then delocalizes over the sphere and thus the cyclopropane ring re-forms (over a [5,6] junction) and the process is alternately repeated as the methano bridge "walks" on the surface from a [6,6] to a [5,6] junction, one "leg" at a time. The fact that such a predominance of the *trans*-2 and *trans*-1 isomers is observed in every case in the product mixture could be the result, at least partially, of a coulombic repulsion effect. Further experimental work is currently underway to try to elucidate this interesting process. We are also trying to use the "shuffle" reaction to see if it is possible to maximize the formation of the *trans*-1 and *trans*-2 isomers, which are synthetically very difficult to prepare. We are also investigating the generality of this isomerization by studying methano-adducts derived from C_{70} and some of the other higher fullerenes.

3. Removal of the Methano-adducts: The *Retro-Bingel Reaction*

As already alluded to under the previous headings, reductive electrolysis of ester-containing methanofullerenes can also result in the removal of the adducts. The manner in which we found this out is somewhat interesting and thus we return to a description of how these experiments are conducted and what the observations are.

In order to simplify the description, let's describe the electrolysis behavior of the simple mono-adduct, **1**, where R=ethyl. The CV of this compound is reversible, at least through the second reduction process, see figure 2. If CPE is conducted at approximately −1.0 V *vs* Ag, a reversible one-electron process is observed, as described in the previous section for the bis-adducts. If electrolysis is then conducted at −1.4 V, in an attempt to form the dianion, the current does not decrease exponentially, as during the first electrolysis. If allowed to proceed to completion, the current does decrease to background level, indicating that the process(es) end by themselves under controlled-potential conditions. However, the total number of electrons transferred is 4, counting from the neutral state. When the resulting solution is reoxidized at 0 V, only 1.5-2 e^- are recovered, and the solution exhibits the typical magenta color of C_{60}, along with a considerable amount of a black precipitate. Analysis of the product showed that the material was indeed C_{60}, recovered in 82% yield in a relatively pure state [14]. This reaction, which we have called the *retro-Bingel* reaction for obvious reasons, thus proceeds in relatively high yields, at least in the removal of a single methano-adduct from C_{60}. Prior to our work, Haddon, Wudl, and co-workers [15] reported that the mono-anion of 1,2-methano[60]fullerene-61,61-dinitrile, was unstable under CV conditions and, based upon voltammetric observations, was reported to revert back to C_{60}. Reversal to C_{60} upon CV studies had also been reported for $C_{60}H_2$ [16], but reference 15 was the only report of the electrochemical transformation of a methanofullerene to C_{60}.

In order to test the generality of this electrolytic reaction, we also investigated the exhaustive electrolysis of the bis-methanofullerene adducts at the second reduction potential. Recall that in the previous section we interrupted the second electrolysis of the bis-adducts, and did not allow them to go to completion, since the current levels were high after 2e⁻ electrolysis. If electrolysis of any of the bis-adducts is conducted around –1.5 V *vs* Ag, and allowed to proceed to completion (to background current levels), a total

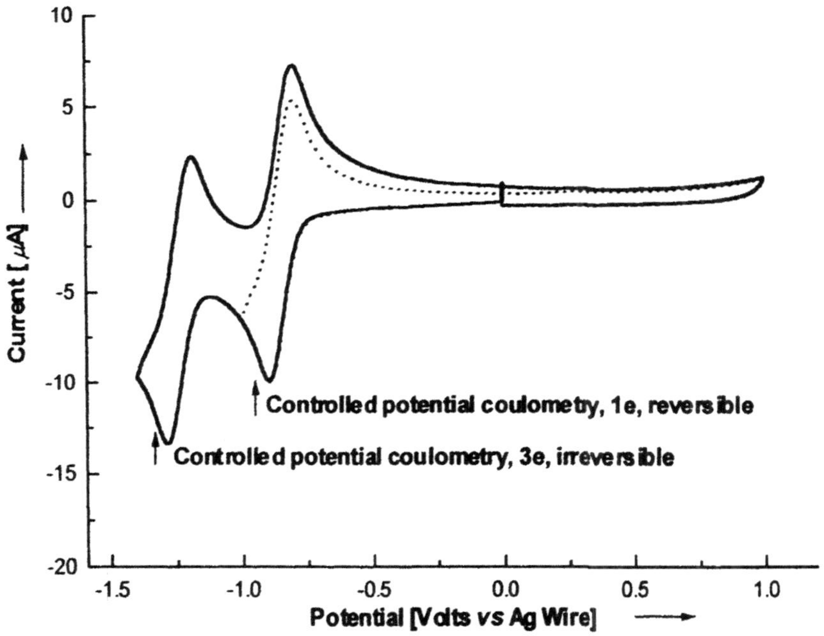

Figure 2. CV of Compound **1**, showing one-electron and two-electron waves

transfer of charge of approximately 6e⁻ per molecule is observed. Reoxidation of the resulting solutions at 0 V produces mainly C_{60}, in over 75% yield. It is thus possible to remove two methano-adducts electrolytically and with very respectable yields. As expected, the overall yields decrease as the number of methano-adducts that are removed increases. We have recently converted a tetra-adduct to C_{60} with an overall yield of 15% using this procedure.

Further generality of the *retro-Bingel* reaction was obtained by submitting compound **4**, a C_{70} methano-adduct, to the same electrolysis protocol. Purification of the solution after coulometric reduction (which took 4e⁻ per molecule) followed by oxidation (1.5e⁻ recovered) resulted in a 70% yield of isolated C_{70}. It thus appears as if the *retro-Bingel* reaction is a general procedure to remove ester-substituted methano-adducts from fullerenes.

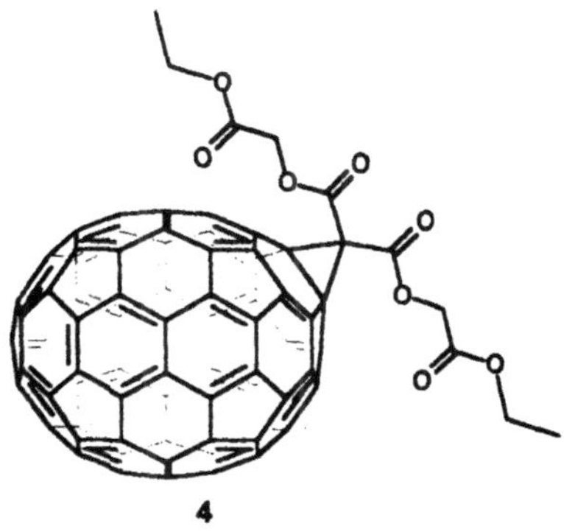

4

With these preliminary results in-hand, we decided to exploit this reaction in order to obtain compounds that were either not available by other routes, or which were very difficult to prepare or isolate, to show the usefulness of the reaction in preparative fullerene chemistry.

3.1. OPTICALLY-PURE C_{76} ENANTIOMERS USING THE *RETRO-BINGEL* REACTION

Prior to our work, Hawkins and Meyer had performed the kinetic optical resolution of the inherently chiral fullerene D_2-C_{76} by employing the asymmetric *Sharpless* osmylation reaction [17]. A comparison of the circular dichroism (CD) spectra reported by Hawkins and Meyer for the C_{76} enantiomers that they obtained [17] to those of a variety of optically active, covalent derivatives of C_{76}, prepared by us [18] revealed a large, unexpected difference in the magnitude of the Cotton effects. Whereas our covalent C_{76} derivatives displayed bands reaching $\Delta\varepsilon$ values up to 250 $M^{-1}cm^{-1}$, the enantiomers of the pure fullerenes reported by Hawkins and Meyer displayed bands with $\Delta\varepsilon$ values up to only 32 $M^{-1}cm^{-1}$ [17, 19]. In order to reinvestigate the chiroptical

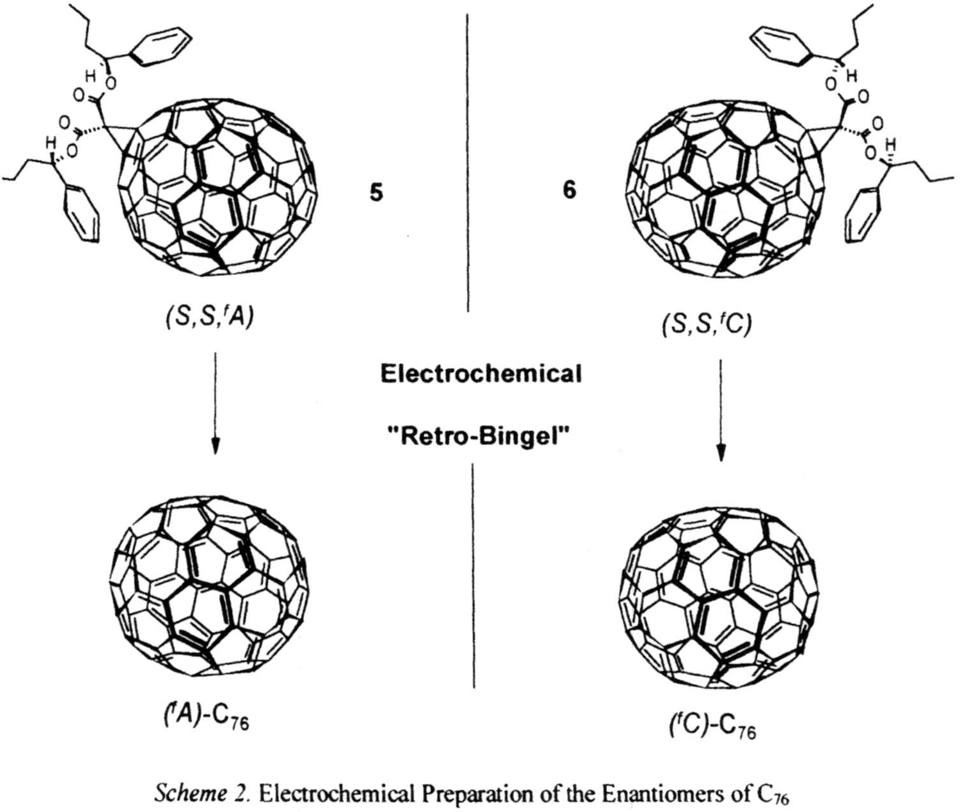

Scheme 2. Electrochemical Preparation of the Enantiomers of C_{76}

properties of C_{76} and to test the power and generality of the *retro-Bingel* reaction, we decided to prepare the enantiomerically pure isomers starting with the diastereoisomerically pure mono-adducts **5** and **6**, which were previously reported [18].

Derivatives **5** and **6**, with enantiomeric carbon cores [18], were independently submitted to CPE and to subsequent purification by HPLC, see scheme 2. Preparative HPLC was conducted on a Regis Buckyclutcher I Trident-Tri-DNP (10 μm), 500 x 21.1 mm id column with a *n*-hexane/toluene 60/40 mobile phase at 8 ml/min. The two enantiomers of C_{76} were obtained in low yields (between 5-10 %). So far, it has been impossible to optimize these yields due to the limited availability of the starting diastereoisomers.

The CD spectra of the enantiomers of C_{76} are shown in figure 3. They exhibit the expected mirror image shapes with band positions that are in full agreement with those reported by Hawkins and Meyer [17, 19]. In contrast with their results, however, the Cotton effects observed in our spectra are almost a full magnitude larger than those previously reported. The $\Delta\varepsilon$ values reach up to 210 $M^{-1}cm^{-1}$ and are therefore much more in agreement with the values measured for the optically pure C_{76} derivatives such as **5** and **6** [18]. Two possible reasons may account for the noticeably low values reported by Hawkins and Meyer: either the enantiomeric purity of their samples was not very high or their concentrations were not accurately determined. We feel confident that our results correspond to enantiomerically pure samples of the C_{76} fullerene.

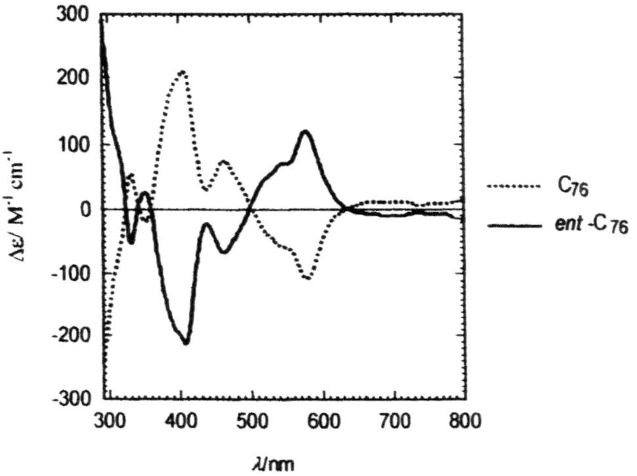

Figure 3. CD spectra for the enantiomers of C_{76}

3.2 ISOLATION OF PURE C_{2V}-C_{78} AND A NEW C_2-SYMMETRIC BIS-ADDUCT

A similar procedure to that described in the preceding section was used to prepare pure C_{2V}-C_{78}, starting from a previously isolated tris-methanoadduct, compound **7** [20]. This compound was isolated using HPLC and the symmetry of the fullerene core was determined to be C_{2V} by analysis of the ^{13}C NMR spectrum [20]. It was thus an excellent precursor to prepare C_{2V}-C_{78} *via* the *retro-Bingel* removal of the three adducts. This turned out to be the case, as described below and, in addition, a bis-methanoadduct was also isolated after removal of the first adduct via the *retro-Bingel* reaction [21]. More interestingly, this bis-adduct, with C_2-symmetry, was not formed during the *Bingel* addition process to the C_{78} isomers. Some of the experimental details that led to these observations are presented next [21].

A total of 3 mg of (±)-7 (0.002 mmol) was submitted to CPE at −1.3 V for 30 min, during which 3e⁻ per molecule were transferred. Although we anticipated that approximately 8 e⁻ would be required for complete removal of the three adducts, we analyzed the CV after this initial transfer of 3e⁻. The observed changes in the CV were quite dramatic, as can be seen in figure 4. The CV of the original tris-adduct shows considerable irreversibility and the first reduction wave occurs at a very negative potential. For the tris-adduct, only the first reduction is reversible, while three reversible couples are observed for the product bis-adduct, figure 4. The bis-adduct was isolated after reoxidizing the solution at 0 V, in 55% yield, and was found to be fairly pure. It was fully characterized using ^{13}C NMR and MS [21]. It should be noted that the first reduction of the bis-adduct is approximately 500 mV anodically shifted (easier to reduce) than the corresponding process for the tris-adduct. In addition, electrolyses of other bis-adducts of C_{78}, prepared *via* the *Bingel* reaction, lead to CV's that are very similar to the one shown at the top of figure 4. Even during the electrolysis of D_3-C_{78} adducts, dramatic changes of the CV's were almost invariably observed, and the intermediate CV's were always similar to the one in figure 4. We believe that this is the result of "shuffling" the adducts on the surface of the fullerene, to bis-adducts with very similar electronic properties. At the time of writing there is no substantive proof of methano-adduct "shuffling" on the surface of C_{78} so this is only a hypothesis that needs to be studied in more detail in the future.

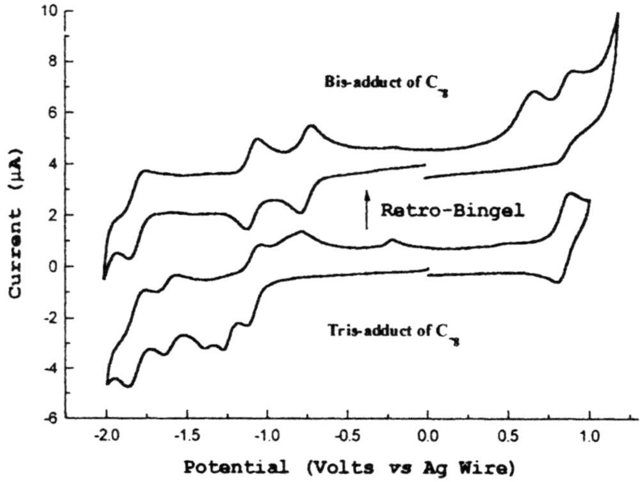

Figure 4. CV's of tris- and bis-adducts of C_{78}, before and after the *retro-Bingel* reaction

If the electrolysis of (±)-7 is continued beyond the first 3e⁻, at −1.8 V *vs* Ag, and then it is followed by reoxidation at 0 V, the three adducts are eventually removed and pure C_{2V}-C_{78} can be isolated in 31% yield. This is after removal of the electrolyte by passing the mixture through a short column (SiO_2, toluene), followed by HPLC (Buckyclutcher I Trident-Tri-DNP (10 µm), 500 x 21.1 mm id; *n*-hexane/toluene 1:1). The other isomer, D_3-C_{78}, was obtained by HPLC of a mixture of C_{2V} - and D_3-C_{78} as previously described [22]. Having these pure isomers allowed us, for the first time, to examine their electrochemical behavior independently [21].

One of us had previously reported the Osteryoung Square Wave Voltammetry (OSWV) of an isomer mixture of C_{78} [23]. At that time we observed what appeared to be two sets of two oxidation peaks each, one set being approximately four times as intense as the other one. We proposed then that the first oxidation observed (at +0.70 V *vs* the ferrocene/ferricinium couple (Fc/Fc^+) in 1,1,2,2-tetrachloroethane (TCE) [23]), which belonged to the smaller set of peaks, arose from D_3-C_{78}, which is obtained in a 1:4 ratio relative to the C_{2V}-C_{78} isomer. The proposed first oxidation of the C_{2V}-C_{78} was reported at +0.95 V, an amazing 250 mV more positive than the other isomer. Since the pure isomers were not available, these assignments were made on a tentative basis. We thus decided to take a look at the voltammetric behavior of the pure isomers.

Figure 5 shows the CV's of D_3- and C_{2V}-C_{78}, in CH_2Cl_2. Although the potential window available in this solvent:electrolyte system was not as large as the one obtained in TCE [23], it can be clearly appreciated that the D_3 isomer exhibits an oxidation at +0.74 V *vs* Fc/Fc^+, while no oxidation process is observed for the corresponding C_{2V} isomer. More recent electrochemical results (not shown) with these samples in TCE confirmed that the first oxidation of the C_{2V} isomer occurs 250 mV more positive, at +0.99 V.

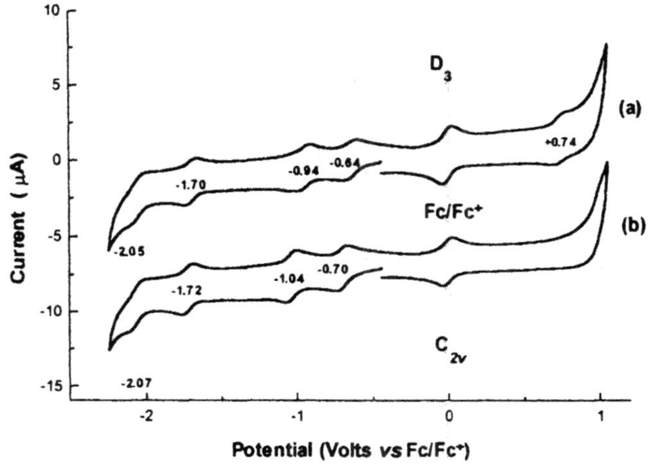

Figure 5. CV's of the pure isomers of C_{78}: (a) D_3. (b) C_{2v}

The results presented in this section are summarized in scheme 3. Initial electrolysis of the tris-adduct 7 with 3e⁻ results in the formation of a C_2-symmetric bis-methanoadduct in 55% yield. This bis-adduct exhibits an electrochemical response that is very different from that of the precursor tris-adduct and quite similar to that of the parent C_{2V}-C_{78}. Continued electrolysis of the solution results in the formation of the parent fullerene, C_{2V}-C_{78}, in 31% yield based on the initial amount of 7.

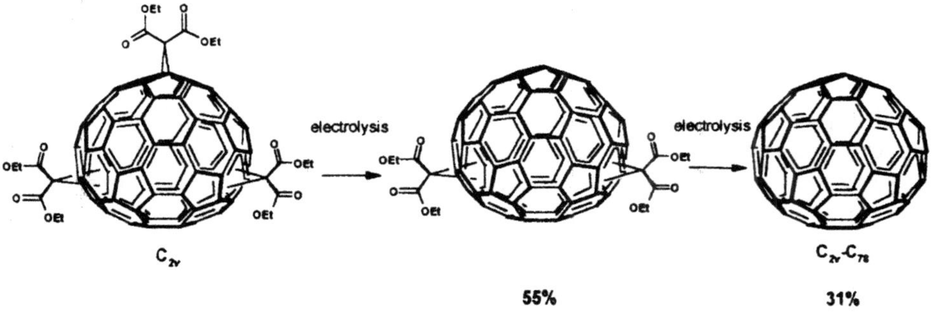

7

Scheme 3. Electrochemical transformation of tris-adduct 7 into a bis-adduct, and into C_{2v}-C_{78}.

4. Summary, Conclusions and Future Work

We have shown that fullerenes containing bis-ester methano-adducts undergo unusual transformations when reduced electrolytically. Mono- as well as multiple-adducts can easily be removed if enough electrons are supplied to the corresponding methanofullerenes, in a reaction that we have called the *retro-Bingel*. The removal of one adduct from C_{60} occurs in high yield, about 80%, while removal from C_{70} occurs in 70% yield. Removal of two adducts from C_{60} occurs in over 75% yield, and removal of four adducts in 15%.

The generality and practical utility of the *retro-Bingel* reaction was tested by electrolyzing diastereoisomers of C_{76} that contain a chiral methano-adduct. In this manner, it was possible to isolate the pure enantiomers of C_{76}, as judged by their CD spectra and the correspondingly high values of their Cotton effects. The *retro-Bingel* reaction was also applied to a tris-methano adduct of C_{2v}-C_{78}. Along the electrolysis, the tris-adduct was converted cleanly and in 55% yield to a bis-adduct with C_2-symmetry. Interestingly, this bis-adduct was different from any of the C_{78} bis-adducts that were isolated from the *Bingel* reaction. Therefore, the *retro-Bingel* is able to afford methanofullerene compounds which are otherwise not accessible. Continued electrolysis of the C_{78} derivative resulted in the complete removal of the three adducts and the consequent formation of pure C_{2v}-C_{78} in 31% yield.

Perhaps more interesting than the *retro-Bingel* reaction was the electrochemically-induced isomeric transformations that we observed for bis-methano adducts of C_{60}, which we have termed the "shuffle". No matter which one of the pure bis-methano adducts was reduced with $2e^-$, the same bis-adduct isomer distribution was observed in every case. This distribution shows no *cis* isomers (even when the starting bis-methanofullerene is *cis*-2 or *cis*-3) and a predominance of the *trans*-2 and (if corrected for statistical factors) the *trans*-1 isomer. Preliminary mechanistic studies have shown that the isomerization is an intra- and not an inter-molecular process. Thus the adducts are electrochemically-induced to "walk" on the surface of the fullerene sphere, and this leads to a final product mixture that is under thermodynamic control. Although the potential synthetic utility of the "shuffle" appears to be less than that of the *retro-Bingel*, it is mechanistically more interesting and will be the subject of continued investigation in our laboratories.

There are several important questions to answer and avenues to pursue that derive from the preliminary work presented in this chapter. The first one, and perhaps the most evident, is related to the true general synthetic utility of the *retro-Bingel* reaction as a protection-deprotection tool in fullerene derivative synthesis. So far we have shown specific applications that work well, but these are relatively unique and thus lack general applicability. The key question is how selectively can methano-adducts be removed *via* the *retro-Bingel* reaction without affecting other types of adducts on the fullerenes. So far we have only conducted preliminary studies to address this question. One example was the *retro-Bingel* reaction of compound **8**, shown in scheme 4. Although the yields appear to be somewhat low, UV-Vis evidence supports the formation of **9**, indicating that it is possible to remove a methano-adduct while leaving the cyclohexene ring unperturbed. It is necessary to repeat this experiment and to systematically investigate the *retro-Bingel* reaction in the presence of other functional groups on the fullerene core. If generally selective (and we believe and hope that it will be) it could become a useful tool for the elaboration of exotic derivatives with controlled regiochemistry.

Another important question that needs to be addressed concerns the generality of the "shuffle" reaction. In particular, it is necessary to find out if it occurs for other fullerene cores, such as C_{70} and the higher fullerenes. Thus mono- as well as multiple-adducts of C_{70} should be electrolyzed and the products analyzed. While a mono-methanoadduct "shuffle" on C_{60} is impossible to detect due to the spherical symmetry of the core, it should be easy to observe if it occurs on C_{70}. It will also be interesting to see how the bis-adducts of C_{70} behave and, if they isomerize, to analyze the final isomer distributions.

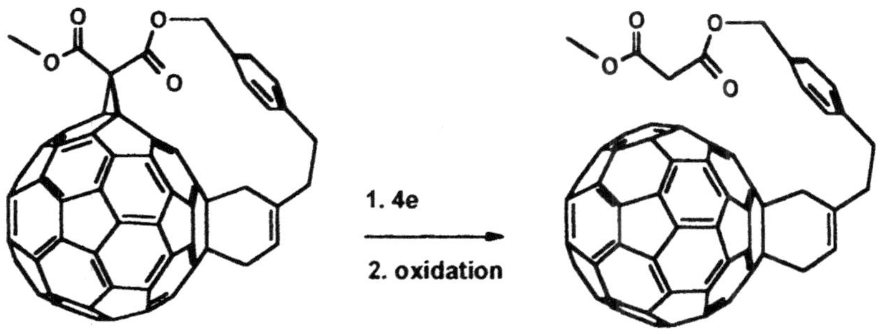

8 9

Scheme 4. Retro-Bingel transformation of **8** to **9**.

5. Acknowledgements

L. E. wishes to express his gratitude to the Chemistry Division of the National Science Foundation (CHE-9313018) for generous financial support. He especially acknowledges the NIH Fogarty Center for a Senior International Fellowship (IF06 TW-02231-01) that allowed him to spend a sabbatical year (1997-8) at the ETH in Zürich, Switzerland, during which most of the work reported here was conducted. The work was also generously supported by the Swiss National Science Foundation and by Hoechst AG.

6. References

1. Krätschmer, W., Lamb, L.D., Fostiropoulos, K., and Huffman, D.R. (1990) *Nature* **347**, 354-358.
2. a) Wooley, K.L., Hawker, C.J., Fréchet, J.M.J., Wudl, F, Srdanov, G., Shi, S., Li, C., and Kao, M. (1993) *J. Am. Chem. Soc.* **115**, 9836-9837; b) Diederich, F., and Thilgen, C. (1996) *Science (Washington)* **271**, 317-323. c) Hirsch, A. (1994) *The Chemistry of Fullerenes*, Thieme, Stuttgart; d) Okamura, H., Ide, N., Minoda, M., Komatsu, K., and Fukuda, T. (1998) *Macromolecules* **31**, 1859-1865.
3. Fagan, P.L., Calabrese, J.C., and Malone, B. (1992) *Acc. Chem. Res.* **25**, 134-142.

4. a) Isaacs, L., Haldimann, R.F., and Diederich, F. (1994) Angew. Chem. Int. Ed. Engl. **33**, 2339-2342; b) Nierengarten, J.-F., Gramlich, V., Cardullo, F., and Diederich, F. (1996) *Angew. Chem. Int. Ed. Engl.* **35**, 2242-2244.
5. Bingel, C. (1993), *Chem. Ber.* **126**, 1957-1959.
6. Hirsch, A., Lamparth, I., and Karfunkel, H. R. (1994), *Angew. Chem. Int. Ed. Engl.* **33**, 437-438.
7. a) Isaacs, L., Seiler, P., and Diederich, F. (1995) *Angew. Chem. Int. Ed. Engl.* **34**, 1466-1469; c) Isaacs, L., Diederich, F., and Haldimann, R.F. (1997) *Helv. Chim. Acta* **80**, 317-342.
8. Q. Lu, D. I. Schuster, and S. R. Wilson, (1996) *J. Org. Chem.* **61**, 4764-4768.
9. Bourgeois J.-P., Echegoyen, L., Fibbioli, M., Pretsch, E., and Diederich, F. (1998) *Angew. Chem. Int. Ed.* **37**, 2203-2207.
10. (a) Boudon, C., Gisselbrecht, J.-P., Gross, M., Isaacs, L., Anderson, H. L., Faust, R., and Diederich, F. (1995) *Helv. Chim. Acta* **78**, 1334-1344. (b) Nierengarten, J.-F., Habicher, T., Kessinger, R., Cardullo, F., Diederich, F., Gramlich, V., Gisselbrecht, J.-P., Boudon, C., and Gross, M. (1997) *Helv. Chim. Acta* **80**, 2238-2276.
11. Cardullo, F., Seiler, P., Isaacs, L., Nierengarten, J.-F., Haldimann, R. F., Diederich, F., Mordasini-Denti, T., Thiel, W., Boudon, C., Gisselbrecht, J.-P., and Gross, M. (1997) *Helv. Chim. Acta* **80**, 343-371.
12. Boulas, P. L., Zuo, Y., and Echegoyen, L. (1996) *Chem. Commun.*, 1547-1548.
13. Kessinger, R., Gómez-López, M., Boudon, C., Gisselbrecht, J.-P., Gross, M., Echegoyen, L., and Diederich, F. (1998) *J. Am. Chem. Soc.*, in press.
14. Kessinger, R., Crassous, J., Herrmann, A., Rüttimann, M., Echegoyen, L., and Diederich, F. (1998) *Angew. Chem. Int. Ed.* **37**, 1919-1922.
15. Keshavarz-K. M., Knight, B., Haddon, R.C., and Wudl, F. (1996) *Tetrahedron* **52**, 5149-5159.
16. a) Boulas, P., D'Souza, F., Henderson, C.C., Cahill, P.A., Jones, M.T., and Kadish, K.M. (1993) *J. Phys. Chem.* **97**, 13435-13437; b) Guarr, T.F., Meier, M.S., Vance, V.K., and Clayton, M. (1993) *J. Am. Chem. Soc.* **115**, 9862-9863; c) Lerke, S.A., Parkinson, B.A., Evans, D.H., Lerke, P.J., and Fagan, P.J. (1995) *J. Electroan. Chem.* **383**, 127-132; d) Niyazymbetov, M.E., Evans, D.H., Lerke, S.A., Cahill, P.A., and Henderson, C.C. (1994) *J. Phys. Chem.* **98**, 13093-13098.
17. Hawkins, J.M., and Meyer, A. (1993) *Science* **260**, 1918-1920.
18. Herrmann, A., and Diederich, F. (1996) *Helv. Chim. Acta* **79**, 1741-1756.
19. Hawkins, J.M., Nambu, M., and Meyer, A. (1994) *J. Am.Chem.Soc.* **116**, 7642-7645.
20. Herrmann, A., and Diederich, F. (1997) *J. Chem. Soc., Perkin Trans. 2*, 1679-1684.
21. Boudon, C., Gisselbrecht, J.-P., Gross, M., Herrmann, A., Rüttimann, M., Crassous, J., Cardullo, F., Echegoyen, L., and Diederich, F. (1998) *J. Am. Chem. Soc.*, in press.
22. Diederich, F., Whetten, R. L., Thilgen, C., Ettl, R., Chao, I., and Alvarez, M. M. (1991) *Science* **254**, 1768-1770.
23. Yang, Y., Arias, F., Echegoyen, L., Chibante, L. P. F., Flanagan, S., Robertson, A., and Wilson, L. J. (1995) *J. Am. Chem. Soc.* **117**, 7801-7804.

PARALLEL PROCESSING IN AQUEOUS SOLUTION

S. FAULKNER, D. PARKER, J.A.G. WILLIAMS
Department of Chemistry
University of Durham
South Road
Durham DH1 3LE, U.K.

Luminescent lanthanide complexes have been prepared whose emission intensity and lifetime is a sensitive and selective function of pH, pO_2 and pCl^-. By appropriate choice of excitation wavelength, the output signal is predictably determined by the local environment and the composition of the mixture of lanthanide complexes chosen. Parallel processing in solution should be possible using multivariate methods of analysis. The benefits of addressing multicomponent systems that are amenable to selective recognition and quantification are discussed.

1. Introduction

The sophisticated neural systems used in Nature operate by monitoring perturbations to a set of balanced input parameters. The information content of the change is 'read-in', assessed and a response is stimulated. The nature of this response may involve multiple outputs, including signalling and triggering the release of certain messengers whose 'news' is acted upon or deciphered subsequently. The complexity of such parallel processing in solution may be contrasted with the simplicity of the binary logic systems, devised for digital information storage and transfer by solid-state scientists. In seeking to look ahead to developments in the next century, there may be much more to be gained by addressing multi-component systems. They would not exist in simple 'on/off' states, but would be characterised by the quantification of a set of independent input and output parameters, for each of which a 'threshold' value would limit or define their relative importance (or weighting), and its time dependence.

In the last 30 years, the vast majority of solution-based supramolecular chemistry has focused on single input/output systems.[1] Often a binding event has been signalled by an optical, electrochemical or spectroscopic change, usually in non-aqueous media and occasionally with good selectivity. A change in the concentration of a given ion or molecule may be signalled by changes in the absorption or emission spectrum parameters (I, ε, τ) of the receptor, by NMR typically involving different resonant frequencies for the 'free' and 'bound' states and by electroanalysis often relating to

R. Ungaro and E. Dalcanale (eds.), Supramolecular Science: Where It Is and Where It Is Going, 53–66.

perturbations to a redox potential or charge transport. Occasionally, reports have appeared where there are two 'input' parameters (*e.g.* two different ions) and only one measurable 'output' (*e.g.* an emission at a given wavelength). An analogy has been drawn with the 'logic gates' that characterise the operation of solid state electronic devices, and examples of 'OR',[2] NOR,[3] AND[4] and NAND[5] gates have been presented. Unlike the solid-state devices which they purport to emulate, more often than not the time-dependent output characteristics of such systems are much more 'sinusoidal' in form than the 'square-wave' output which is sought. Indeed a lot of information is lost if you can only address on/off states, even if appropriate threshold values are defined. Why be constrained by the binary logic of the solid state? ...... there is much more to be gained if parallel processing is attempted. Think of the benefits inherent in a diode array detector compared to fixed wavelength spectral analysis, to the efficiency gains that have arisen from the introduction of area detectors in X-ray crystallography and to the vast improvements that arose when pulsed FTNMR methods replaced continuous wave techniques. Some of these analogies may stretch the point a little, but the long-term benefits that may accrue from such an approach must outweigh the short-term logistical problems!

1.1 MULTICOMPONENT ANALYSIS AND ARTIFICIAL NEURAL NETWORKS

The simultaneous determination of several components in moderately selective analytical systems requires the application of multivariate modelling. As a simple example, consider spectroscopic multicomponent analysis. By assuming a constant thickness of the sampling cuvette, the absorbance (or emission intensity) at a particular wavelength for a single component is given by:

$$A_\lambda = k_\lambda c \qquad (1)$$

where A_λ is the absorbance at wavelength λ, c is the molar concentration of the absorbing species and k_λ is the normalised molar absorbtivity. For multicomponent systems, the absorbance at a specific wavelength is the sum of the absorbances of M individual components:

$$A_\lambda = k_{\lambda_1}c_1 + k_{\lambda_2}c_2 + \ldots k_{\lambda_M}c_M = \sum_{j=1}^{M} k_{\lambda_j}c_j \qquad (2)$$

With multiwavelength spectroscopy, the spectrum is acquired at K wavelengths giving rise to exactly determined linear equation systems if K = M, or to overdetermined systems when K > M:

$$A_1 = k_{11}c_1 + k_{12}c_2 + \ldots + k_{1M}c_M$$

$$A_2 = k_{21}c_1 + k_{22}c_2 + \ldots + k_{M}c_M$$

$$A_K = k_{K_1}c_1 + k_{K_2}c_2 + \ldots + k_{K_M}c_M$$

where A_l is the absorbance at wavelength i, k_{ij} is the absorptivity of the jth component at wavelength λ_i and c_j is the concentration of the jth component. Expressing this in matrix form:

$$\mathbf{a} = \mathbf{Kc} \qquad (3)$$

where a is a K·1 dimensional vector that represents the spectrum. The term **K** is a K x M matrix of normalised molar absorptivities and **c** is a M·1 dimensional vector of concentrations. Using such matrices, the analysis of unknown samples can be undertaken either directly (if all the absorptivities are known) or indirectly by estimating the calibration parameters from calibration mixtures. The different methods for multivariate calibration differ according to the mathematical model that is used. These either involve regressing the spectra on concentrations or use inverse calibration methods in which the regression of concentrations on spectra is carried out (*e.g.* with principal component regression (PCR) or partial least-squares regression (PLS)).

The most powerful of the chemometrics techniques used in multicomponent analysis now use a pattern recognition approach. Analyses based on artificial neural networks (ANNS) are particularly important.[6] Not only do they function effectively for data classification but it is being suggested that they may also be used generally when the recognition and quantification of components is necessary.[7,8] Artificial neural network models are made up of an interconnected mesh of non-linear computing elements and they strive to attain 'human-like' performance in cognitive science tasks. A neuron functions as the fundamental processing element of such networks, analogous to those found in biological neural networks. The individual neurons are linked together with a coefficient (or weight) associated with each link. The neurons receive an input signal externally or from other neurons, perform a weighted scan of the incoming inputs and process the resulting signal with non-linear or linear transfer function. An output is then given to the external world or to other neurons.

A particular feature of artificial neural networks is that they constitute a means of pattern classification: an input pattern (*i.e.* a vector whose values represent a data object) is transformed into an output pattern (a vector whose values define the desired classification of the data object). Moreover the correct values of the coefficients required to transform a given pattern to the required output pattern do not need to be pre-determined but can be developed by a 'training' or learning process. Many examples exist of the application of such 'back-propagation neural networks' including the quantitative modelling of non-linear kinetics[9] and the handling of fibre-optic fluorescence data.[10] However the method has not found widespread application because of unpredictable models that arise from overfitting the training set and the difficulties in training such networks. Such problems may be avoided in future using cascade correlation networks.[11]

2. Objectives and boundary conditions

Given that we should seek to emulate biological neural systems, our own crude first steps in this direction should at least operate in aqueous media and preferably in the high salt and protein background that characterises biological fluids. Of the several possibilities that exist for addressing the system, spectroscopic and in particular optical means perhaps have the most scope for multicomponent analysis. There is considerable scope for varying the nature of the input and output signals: the frequency, intensity, phase and duration of the input signal can be varied and the application of laser excitation or optical fibre techniques may allow further **selective** (time, space) excitation to occur. Many output signals can similarly be monitored in parallel or as a function of time, including emission intensity or its lifetime. Lifetime measurements are particularly attractive as they avoid problems of degradation or non-linearity that render both absolute and relative intensity measurements more problematic. For chiral systems, circular dichroism and circularly polarised luminescence methods may provide unique information on the ground and excited states respectively.

More complex parallel-processing makes it necessary for us to address a mixture of components (compounds/complexes) selectively or simultaneously rather than work with a system where a single 'intelligent' molecule or assembly of molecules is harnessed. The set of components may well exhibit rich structural homology but the individual complexes should each be capable of carrying out a unique function or transformation. In addition we must define the target species that are to be recognised and quantified. They must be examined individually and real interferences must be evaluated in competitive test procedures. The number of such inputs must be limited (so as to be manageable!), but must be sufficient to allow rigorous multicomponent analysis. In this respect, selective read-outs (channels) may be defined specifying the overall effect of a given input signal on each component (neuron) or combination of components. It is tempting to draw an analogy with a group theoretical character table: having specified the particular components and defined the various input signals, then the output is determined by the outcome of a set of allowed transformations on the system being examined.

2.1. REQUIREMENTS

Although it is very tempting to consider a system that operates in a static or gradient magnetic field because of the ease of frequency and spatial resolution that characterises the modern pulsed NMR experiment, problems of sensitivity (except in the case of course of ^{31}P, ^{1}H NMR) and cost favour consideration of an array of luminescent probes. As a starting point therefore we need to devise a set of structurally similar water-soluble, luminescent complexes which respond in a well-defined manner to changes in the local concentration of a limited number of species. Suitable variables could include pH, pO_2, pX^- ($X = Cl^-$ or HCO_3^-) and pM^{n+} ($M^{n+} = Mg^{2+}$, Zn^{2+}, Ca^{2+}). The luminescent complexes must be kinetically and thermodynamically stable with respect to degradation and should emit a signal which may be read without the need for expensive instrumentation.

3. Responsive lanthanide complexes for parallel processing

Following some detailed studies of kinetically inert complexes of yttrium - for targeted radiotherapy [12] - and of gadolinium, for use as contrast agents in magnetic resonance imaging,[13] we were attracted by the chemical versatility of single component, luminescent lanthanide complexes. The monoanionic tetracarboxylate and

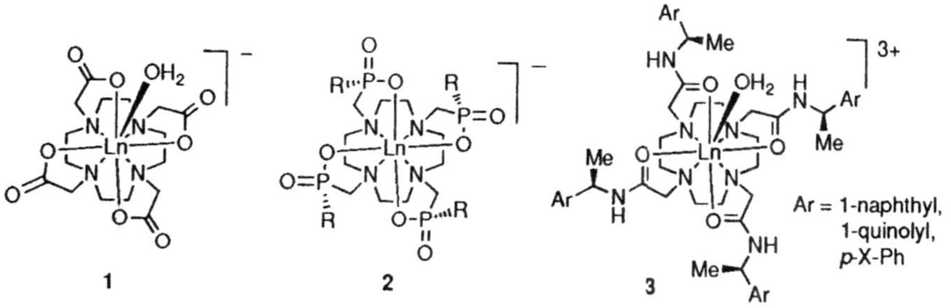

tetraphosphinate[14] complexes, **2**, and the cationic chiral tetra-amide complexes[15] **3**, are stable in the pH range 3 to 10 for prolonged periods, and varying the lanthanide ion has relatively little effect on this high kinetic stability. The tetraphosphinate and tetra-amide complexes exist as one predominant stereoisomer in aqueous solution. With the stereogenic centres at P and C of the same configuration in **2** and **3**, then concerted arm rotation - interconverting Λ and Δ stereoisomers - is frozen out. These three sets of complexes constitute a useful series from which a diverse array of related complexes can be designed by variation of the substituents at carbon or phosphorus. Furthermore, robust synthetic procedures have been devised that allow selective mono, or 1,7-disubstitution of the 12-N_4 ring,[16] affording an entry into mixed amide/acid ligand systems,[17] for example. Given this control in ligand structural diversity, coupled with the ease of lanthanide permutation, a wide range of structurally similar complexes can be prepared.

Excitation of lanthanide ions in solution may give rise to luminescence at very narrow frequencies which are characteristic of a given lanthanide. Thus Eu, Tb and Dy emit mostly in the visible, and Yb and Nd in the near-IR. The luminescence lifetime is also a parameter which characterises the behaviour of a given complex in solution. The long-lived (millisecond) luminescent lifetimes of Eu and Tb complexes have been exploited in certain bio-assays, and time-gating permits easy distinction from the shorter-lived (nanosecond) background present in most biological systems. Problems associated with autofluorescence and Rayleigh scattering are also obviated by time-gating. A consequence of the low extinction coefficients (typically $\varepsilon \sim 1$ $dm^3mol^{-1}cm^{-1}$) associated with Laporte-forbidden f-f transitions is that direct excitation is only practicable with lasers. It is often more convenient to build in a sensitising chromophore into the complex structure. This 'antenna' should preferably absorb light at wavelengths of ≥ 340 nm to avoid the need for quartz optics and eliminate co-excitation of biological molecules in the sample. Provided that the aromatic chromophore possesses a triplet energy that is at least 1,700 cm^{-1} above that of the emissive lanthanide state, then efficient intramolecular energy transfer should occur from the excited chromophore to the proximate lanthanide. Of course if the energy gap is either too large or very small, then quenching of the aryl triplet state by molecular oxygen may become competitive (Scheme 1). In the former case, a large energy gap may lead to a slow rate of energy transfer (poor wavefunction overlap),

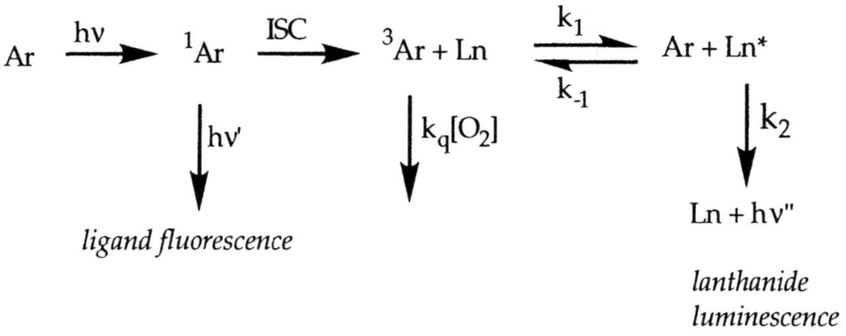

Scheme 1

while in the latter case thermally activated back-energy transfer may occur - proportional to $e^{-\Delta E/kT}$. Of course, in such a case a means of assessing pO_2 is provided as the metal (or ligand) luminescence is then a function of the dissolved oxygen concentration.[5]

The 'antenna' group may serve a dual function if it possesses functionality that is sensitive to local ionic or molecular environment and modulates the photophysical properties of the overall complex (**Scheme 2**).[18] A simple example is provided by a

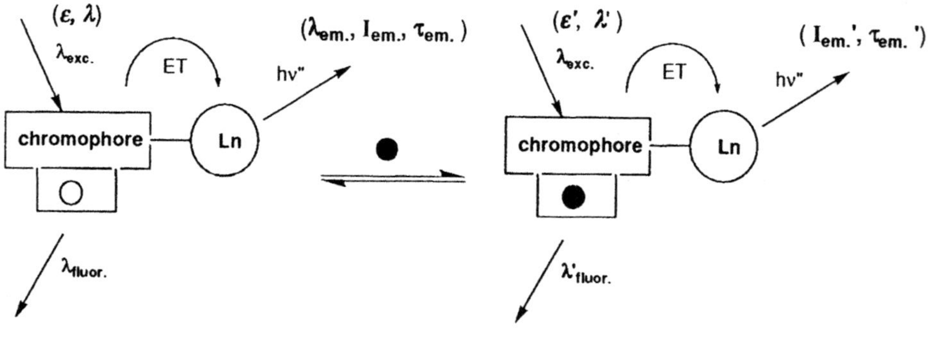

Scheme 2

site of protonation. The protonated chromophore would typically absorb light at a longer wavelength than the neutral species, perturbing therefore both the efficiency of energy transfer to the lanthanide which could lead - *inter alia* - to variations in I_{em} or τ_{em}. A suitable aromatic chromophore is the phenanthridine group ($E_T \approx$ 22,000 cm^{-1}), which is easily functionalised and the basicity is subject to control by varying the substituent at C-2 or C-5. The ease of N-alkylation allows the formation of phenanthridinium salts whose fluorescence is known to be quenched selectively by the presence of halide ions.[19] In addition, N-alkylphenanthridinium salts are attacked α to nitrogen by OH^- in a **reversible** reaction[20] which modifies the absorption characteristics of the aryl chromophore (**Scheme 3**).

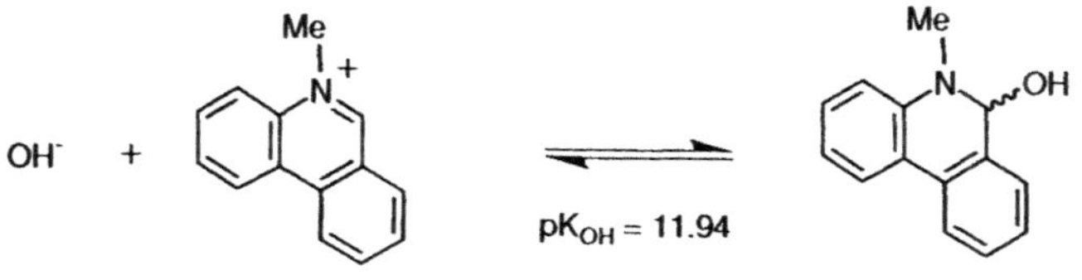

Scheme 3

Accordingly a series of neutral and cationic lanthanide complexes, **4** to **7**, has been prepared and the effect of variation in pH, pO_2 and $p(X^-)$ on the ligand and metal-based luminescence examined.

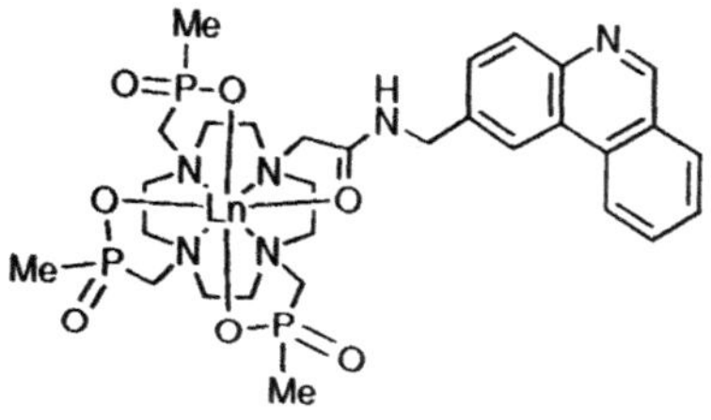

4 **a)** Ln = Eu
b) Ln = Tb
c) Ln = Yb

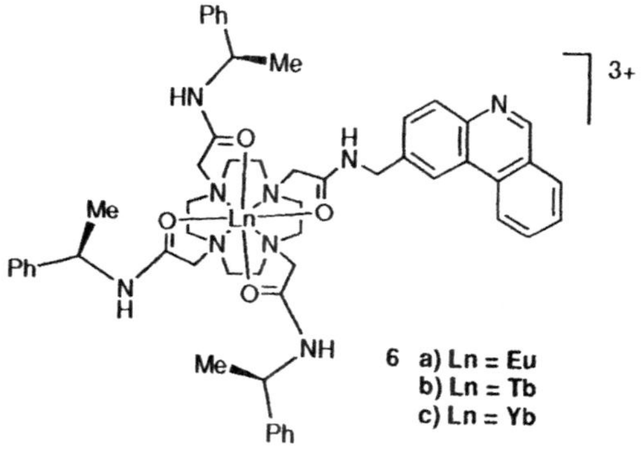

6 **a)** Ln = Eu
b) Ln = Tb
c) Ln = Yb

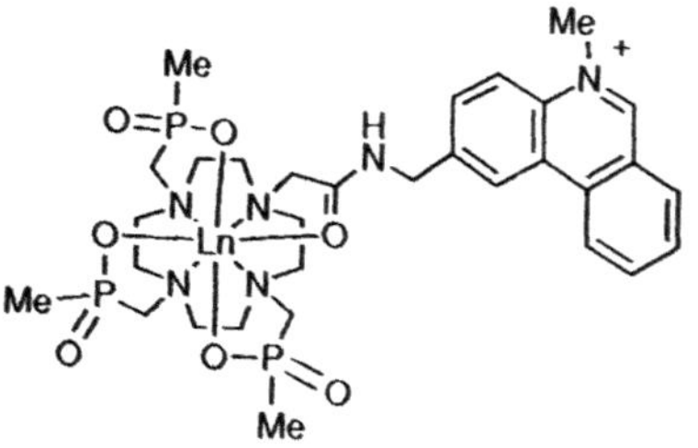

5 **a)** Ln = Eu
b) Ln = Tb
c) Ln = Yb

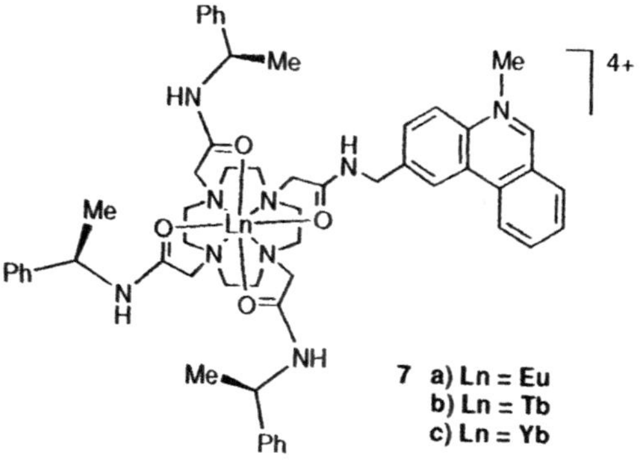

7 **a)** Ln = Eu
b) Ln = Tb
c) Ln = Yb

3.1. VARIATION OF COMPLEX ABSORPTION AND EMISSION WITH pH AND pO_2

The absorption spectrum of **4** and **6** at pH7 reveal distinct bands at 346, 330, 315, 304 and 295 nm. Protonation or N-methylation simplifies the absorption spectrum: only 2 distinctive and intense bands are observed at 370 and 330 nm. The N-substituted complex may therefore be excited **selectively** at wavelengths of greater than 375 nm. Ground-state pK_a's are 4.3(1) for **4** and 3.3(1) for **6**. Excitation of the europium complex at 370 nm gave rise to a fluorescence from the phenanthridine group and sensitised emission from the Eu centre that increased significantly as the pH was lowered. The excited state pK_a was 4.4 (I = 0.1 NMe_4ClO_4, 293 K) in each case, with a metal-based luminescence enhancement of at least 500. Indeed using an excitation wavelength of ≥ 378 nm, metal-based emission from the Eu complex could not be observed, so that a simple pH switch has been devised in which acidification switches on the red europium luminescence.[20] The Eu lifetime did not vary with pH, nor was it sensitive to changes in dissolved oxygen concentration.

The situation was quite different with the terbium complexes **4b** and **6b**. The ligand based fluorescence (λ_{em} 403 nm, pK_a 4.2 and 3.4 respectively (±0.1)) mirrored the behaviour of the europium complexes, but the metal based luminescence behaved quite differently, (**Figure 1**). The metal complex **4b** was quite strongly emissive (τ_{H_2O} =

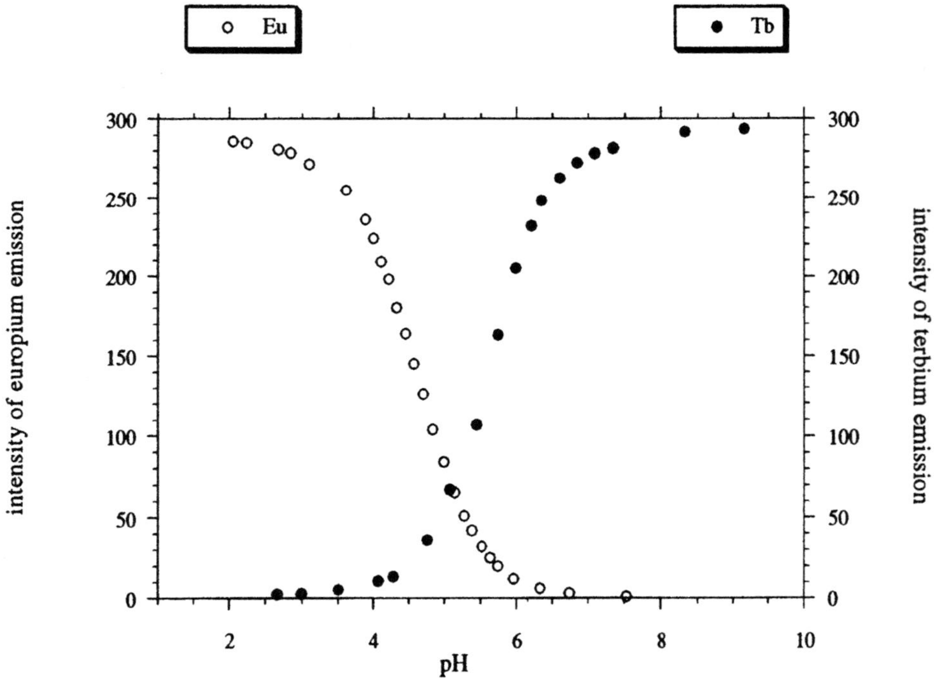

Figure 1 Effect of pH on the luminescence of the lanthanide in the europium (λ_{ex} 370, λ_{em} 594 nm) and terbium (λ_{ex} 304, λ_{em} 545 nm) complexes of ligand **4** (intensity measured 0.1 ms after the excitation flash with gate time 10 ms, I = 0.1 NMe_4ClO_4)

0.98 ms, φ_{H_2O} = 2.5%) but the luminescence was switched **off** following protonation (τ_{H_2O} = 0.11 ms, φ_{H_2O} = 0.091%). A luminescence enhancement of 125 was observed (293 K, excitation at isosbestic wavelength) with pK_a' of 5.7 (±0.1). Protonation lowers the triplet energy of the phenanthridinium chromophore by 700 cm^{-1}, so that there is only a small energy gap (800 cm^{-1}) between the triplet and emissive $^5D_4$ state of terbium. In the protonated complex only, significant energy transfer from the $^5D_4$ Tb state **back** to the phenanthridinium triplet occurs, and a photo-equilibrium is established (**Scheme 1**) such that the metal based emission reports the pK_a of the phenanthridyl triplet,[5,18] through variation of the emission intensity **and** the lifetime of the terbium emission.

The intensity of the terbium emission was also sensitive to the presence of dissolved oxygen: deaeration of the sample led to an increase in the overall quantum yield and the emission lifetime. For the neutral complex **4b** in the absence of O_2, φ_{H_2O} = 12%, τ_{H_2O} = 1.82 ms and, for the protonated complex under the same conditions, φ_{H_2O} = 5%, τ_{H_2O} = 0.83 ms. Thus both pH and pO_2 determine the lifetime of the terbium emission in a predictable manner. On the other hand, the N-methylated terbium complexes **5b** and **7b** display sensitivity to pO_2 but their emission is independent of pH in the range 2-9. Classical Stern-Volmer quenching behaviour was observed (**Figure 2**).

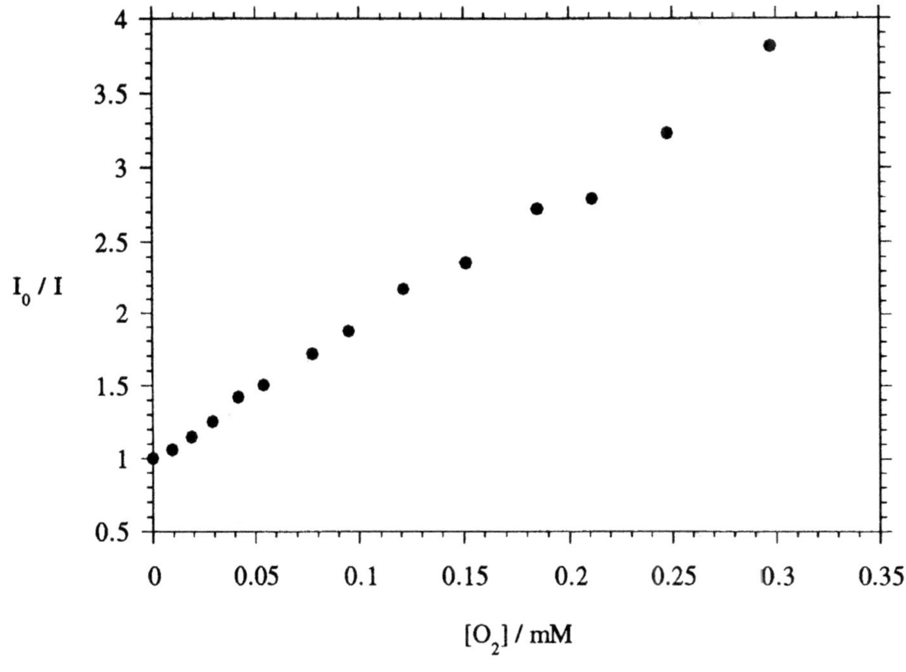

Figure 2 Stern-Volmer plot for the dependence of the terbium emission intensity in **4b** upon the concentration of dissolved molecular oxygen (λ_{ex} 370, λ_{em} 545 nm, emission measured 0.1 ms after the excitation flash with gate time 10 ms)

3.2 SENSITIVITY OF N-METHYLATED COMPLEXES TO OH^- AND Cl^-

N-Methylphenanthridinium ions are attacked by the hydroxide ion in a reversible reaction involving formation of a 'pseudo-base', leading to the disappearance of the 370 nm absorption bond.[21] For the europium complexes **5a** and **7a**, the emission intensity decreased with increasing $[OH^-]$ and was zero above pH 13.5 for **5a**. An effective dissociation constant of 16 mM (±3), equivalent to pK'_{OH} = 12.2 was measured and the pK_{OH} value was 10.6 for **7a**, consistent with Coulombic assistance to hydroxide attack. More complex behaviour was exhibited by **5b** and **7b**: addition of hydroxide **switched on** the green terbium emission (λ_{exc} 340 nm), reaching a maximum at a pH of 11.2 before reducing to a limiting value beyond pH 13 (**Figure 3**). The initial increase is readily rationalised from consideration of the triplet state energies: in the pseudo-base adduct, the triplet state of the chromophore is significantly higher than that of the N-methylphenanthridinium chromophore (21,300 cm^{-1}) and back energy transfer no longer competes.

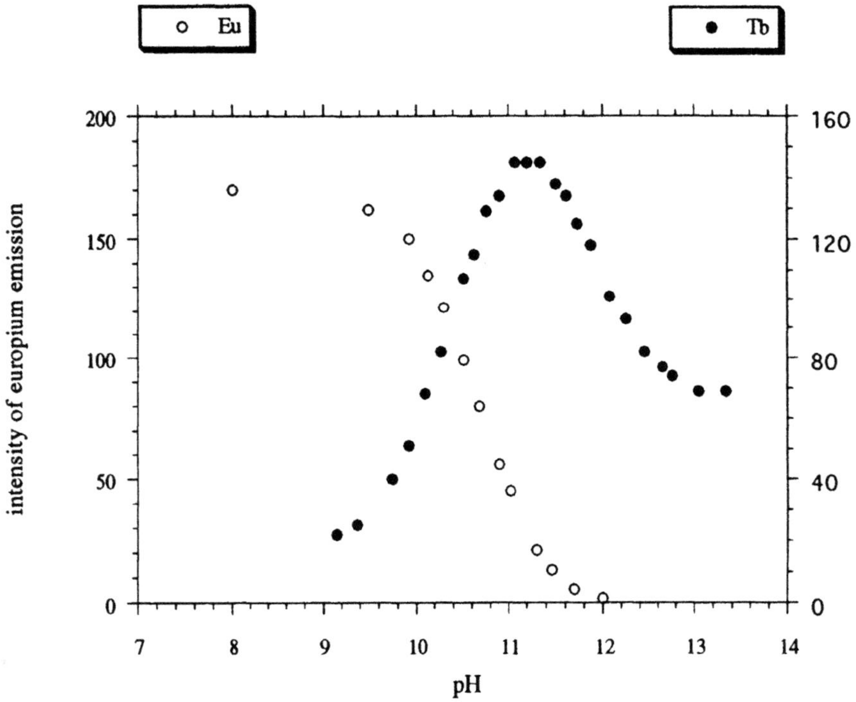

Figure 3 Variation of metal emission from **7a** and **7b** as a function of pH (λ_{em} 594 and 545 nm respectively, λ_{ex} 380 nm, delay time 0.1 ms, I = 0.1 NMe_4ClO_4)

The fluorescence of N-alkylquinolinium and phenanthridinium ions is quenched effectively by halide ions and this effect has been used in assays to determine chloride ion concentrations in biological fluids.[19] Quenching probably occurs by charge transfer from the halide ion to the singlet excited state of the electron-deficient aryl moiety and the quenching effect follows the sequence $I^- > Br^- > Cl^-$ consistent with the ease of anion oxidation. The quenching of the ligand fluorescence and the metal-based luminescence with the europium complexes **5a** and **7a** was studied, (**Table 1**). Chloride quenching was characterised by a Stern-Volmer quenching constant of 40 to 50 mmol dm^{-3} and was unaffected by the presence of added lactate, citrate, phosphate, and hydrogencarbonate. With the corresponding terbium complexes, the ligand based fluorescence was quenched in the same manner as for the Eu complexes.

TABLE 1. Stern-Volmer Quenching Constants[a,b] (K_{SV}^{-1}, mmol dm^{-3})[f] for the effect of halide ions on the Eu complexes, **5a** and **7a**.

Anion[c]	**5a**		**7a**	
	phenanthridine fluorescence	metal[d] emission	phenanthridine[e] fluorescence	metal[e] emission
Cl^-	40	50	6.0	26
Br^-	8.0	16	2.5	12
I^-	6.4	4.9	1.9	1.2

[a] Values given have a 15% error. [b] Concentration of the Eu complex used (as its triflate salt) was 5 x 10^{-5} mol dm^{-3} ($I = 0.1$ NMe_4ClO_4, 293 K). [c] Addition of added sodium dihydrogenphosphate (5 mmol dm^{-3}), citrate (1 mmol dm^{-3}), or hydrogencarbonate (30 mmol dm^{-3}) led to no change in the measured K_{SV} values - within the stated error. [d] No significant change in the europium emission lifetime was observed during the course of these measurements ($\tau_{H_2O} = 0.73$ ms, pH 6.8; $\varphi = 0.014$ for iodide salt, 0.035 for Cl^- salt). [e] With the terbium complex **7b**, at a concentration of 6.6 x 10^{-5} mol dm^{-3}, K_{SV} values for halide quenching were 4.5 mmol dm^{-3} for Cl^-, 1.9 mmol dm^{-3} for Br and 1.5 mmol dm^{-3} for I^- These values were derived from fluorescence intensity measurements: metal-based emission did reduce as a function of $[X^-]$, but was much less sensitive. [f] $I_o/I = \tau_o/\tau = 1 + K_{SV}[X^-]$, where I_0 and τ_0 are the intensity and lifetime in the absence of halide.

Halide addition caused little change to the metal-based emission. For example addition of 30 mM NaCl to the terbium complex **7b** caused only a 14% reduction in the intensity of the Tb emission at 545 nm: with the monocationic complex **5b**, an even smaller reduction was observed.

4. Matrices Of Response Characteristics

The array of structurally related lanthanide complexes has been shown to exhibit well-defined luminescence responses to variations in pH, pO_2, pX and pOH. Judicious choice of the excitation wavelength allows the N-alkylated and N-protonated complexes to be selectively excited, while the imposition of a time-delay in luminescence observation can gate out the emission from the shorter-lived components. This degree of control over the nature of the input and output signals, coupled with the ability to quantitatively measure intensities or lifetimes at a given wavelength (or set of wavelengths) leads to the formulation of a multitude of matrices that characterise the responses of a **mixture** of the separate lanthanide complexes in solution. Two representative matrices are shown below, in which 0 indicates insensitivity and + and - denote metal-based luminescence enhancements or diminutions in intensity or lifetime with increasing concentration of the species under investigation.

Matrix reflecting emission intensity variations

complex		pH^c	pO_2	pCl^-	pOH^-
Tb	**5b**[b]	0	- - -	0	++
	4b	- - -	-	0	0
Eu	**5a**	0	0	- -	- - -
	4a	+++	0	0	0

[a] For λ_{exc} = 370 nm. [b] If a delay time, t_d = 1.0 ms, is used then this row becomes all zeros.
[c] The pH range 2 to 9 is assumed.

Matrix reflecting emission lifetime dependence[a]

complex		pH	pO_2	pCl^-	pOH^-
Tb	**7b**	0	- -	0	++
	6b	- - -	-	0	0
Eu	**7a**	0	0	0	0
	6a	0	0	0	0
Yb	**7c**	0	-	0	+
	6c	0	-	0	0

[a] For λ_{exc} = 355 nm.

By building up response characteristics such as these - using suitably defined input and output signals - then it should be possible not only to model the output under a given set of conditions, but to quantitatively fit unknown data sets to these well-defined calibration matrices. Such methods would then allow truly responsive parallel processing in aqueous solution. To some extent, you could consider each individual lanthanide complex to behave as a neuron - stimulated and exchanging information by optical means - and the collection or assembly of different lanthanide complexes could then even constitute a neural network, processing the input signal into a defined spectral output pattern. Looking ahead, the recent definition of the controlled modulation of the frequency and polarisation of emitted light in solution with *enantiopure* complexes such as **7b** and **7c**, adds an extra dimension to the complexity of such systems allowing the output signal of these chiral complexes to be read selectively using circularly polarised luminescence[22] methods.

5. Conclusions and Prospects

A matrix of closely related lanthanide complexes has been prepared. The complexes are chemically robust, may be excited in the wavelength range 340 to 380 nm, and luminesce with an emission energy and lifetime that is readily varied by choice of the lanthanide ion. Certain complexes function as simple luminescent sensors for pH, pO_2 or pCl^-, whereas the emission intensity or lifetime of others is dependent on two or more variables. Selective excitation may be achieved by appropriate wavelength selection and selective observation of the emission is readily controlled by the use of time-gating. Further control in choosing an appropriate output signal is afforded by the defined enantiopure complexes: excitation of the inbuilt chromophore by ultra-violet light (e.g. 355 nm) followed by intramolecular energy transfer gives rise to a circularly polarised emission which may be detected selectively using circularly polarised luminescence (CPL) instrumentation.[22] Such complexes have also been examined for the time-resolved imaging of certain cellular systems and recently a 'lifetime' map of a labelled system has been obtained using the cationic europium complex **7a**.[23]

Multicomponent analysis is clearly possible in aqueous solution using mixtures of selected lanthanide complexes. Diversity in the response characteristics of the system is allowed by the ease of variation of the emissive lanthanide and by simple N or C-alkylation of the inbuilt chromophore. Control in selecting the desired output characteristics (I_{em}, τ_{em}) is aided by the ease of emission wavelength selection and by the application of time-gating procedures, allowing observation of a particular emission lifetime. The ease of calibration of each individual complex to a defined set of variables (e.g. pO_2, pH, pX^-) indicates that, different response matrices may be readily set up. Taken together with the relative ease of luminescence measurement it is likely that multivariate calibration and prediction will be fairly straightforward, even using well-defined 'direct' chemometric methods.

References

1. Lehn, J.M. (1995) *Supramolecular Chemistry*, VCH, Weinheim.
2. Ghosh, P., Bharadwaj, P.K., Mandal S., and Ghosh, S. (1996) *J. Am. Chem. Soc.*, **118**, 1553.
3. Gedi, A., Balzani, V., Langford, S.J. and Stoddart, J.F. (1997) *J. Am. Chem. Soc.*, **119**, 2679.
4. de Silva, A.P., Gunaratne, N.H.Q. and McCoy, C.P. (1997) *J. Am. Chem. Soc.*, **119**, 7891.
5. Parker, D. and Williams, J.A.G. (1998) *Chem. Commun.*, 245.
6. Brown, S.D., Sum, S.T., Despagne, F. and Lavine, B.K. (1996) *Anal. Chem.*, **68**, 21R-61R.
7. Amendolia, S.R., Doppiu, A., Ganadu, M.L. and Lubinu, G. (1998) *Anal. Chem.*, **70**, 1249.
8. Bos, M., Bos, A. and van der Linden, W.E. (1993) *Analyst*, **118**, 323.

9. Ventura, S., Siha, M., Pérez-Bendito, D. and Hervás, C. (1997) *J. Chem. Inf. Comput. Sci*, **37**, 287.
10. Sutter, J.M. and Jurs, P.C. (1997) *Anal. Chem*., **69**, 856.
11. Harrington, P. de B. (1998) *Anal. Chem*., **70**, 1297.
12. Parker, D. (1996) in J.M. Lehn, D.N. Reinhoudt, F. Vogtle, J.E.D. Davies, D.D. MacNicol and J.E. Atwood (eds.), *Comprehensive Supramolecular Chemistry*, Pergamon, volume 10, chapter 17, 487-536.
13. Parker, D. and Williams, J.A.G. (1996) *J. Chem. Soc., Dalton Trans*., 3613.
14. Aime, S., Batsanov, A.S., Botta, M., Dickins, R.S., Faulkner, S., Foster, C.E., Harrison, A., Howard, J.A.K., Moloney, J.M., Norman, T.J., Parker, D., Royle, L. and Williams, J.A.G. (1997) *J. Chem. Soc., Dalton Trans*., 3623.
15. Dickins, R.S., Howard, J.A.K., Lehmann C.A.W., Moloney, J.M., Parker, D. and Peacock, R.D. (1997) *Angew. Chem. Int. Ed. Engl*., **36**, 521.
16. Parker, D. (1996) *Macrocycle Synthesis*, OUP, Oxford.
17. Aime, S., Botta, M., Dickins, R.S. Maupin, C.L., Parker, D., Riehl, J.P. and Williams, J.A.G. (1998) *J. Chem. Soc., Dalton Trans*., 881.
18. Parker, D., Senanayake, K.P. and Williams, J.A.G. (1998) *J. Chem. Soc., Perkin Trans. 2*, in press.
19. Vasseur, M., Frangne, R. and Alvardo, F. (1993) *Am. J. Physiol*., (1993) **264**, C27.
20. Parker, D., Senanayake, P.K. and Williams, J.A.G. (1997) *Chem. Commun*, 1777.
21. Bunting, J.W. and Meathrel, W.G. (1974) *Can. J. Chem*., **52**, 981.
22. Dickins, R.S., Howard, J.A.K., Maupin, C.L., Moloney, J.M., Parker, D., Peacock, R.D., Riehl, J.P. and Siligardi, G. (1998) *New. J. Chem.*, in press.
23. Beeby, A., Clarkson, I.M., Faulkner, S., Heubest, K., Parker, A.W., Parker, D., Powell, T., Twist, V., Williams, J.A.G. *J. Photochem. Photobiol. B*, submitted.

CALIXARENES AND RESORCINARENES IN MOLECULAR RECOGNITION AND SUPRAMOLECULAR DEVICES

A. ARDUINI,[a] A. CASNATI,[a] E. DALCANALE,[a] A. POCHINI,[a] F. UGOZZOLI,[b] R. UNGARO.[a]
[a]Dipartimento di Chimica Organica e Industriale, e [b]Dipartimento di Chimica Generale, Inorganica, Analitica e Chimica Fisica dell'Università degli Studi, Parco Area delle Scienze 17/A, I-43100, Parma, Italy.

1. Introduction

Calixarenes (**1**) and resorcinarenes (**2**) (Figure 1), have been used as building blocks for the synthesis of a large variety of host molecules, having different supramolecular functions.

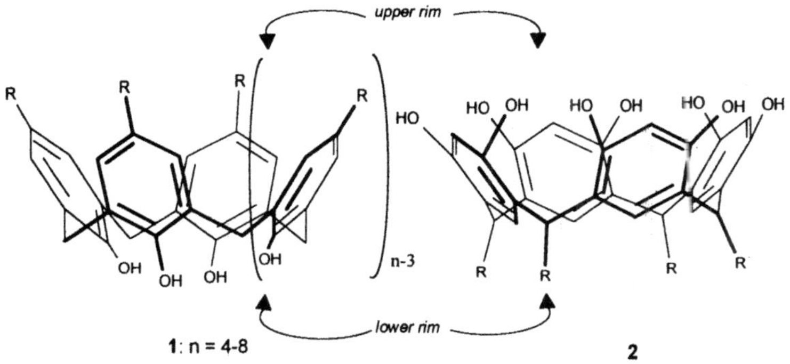

1: n = 4-8
2

Figure 1: Calix[n]arenes (**1**) and resorcinarenes (**2**).

From the early 80s, the emergence of Supramolecular Chemistry as an important area of research has caused an explosion of interest in the chemistry of these basket-shaped molecules, which still continues. Not only the number of scientific publications dealing with the chemistry and host-guest properties of calixarenes and resorcinarenes is growing with geometrical progression, but also the number of patents follows the same trend, thus indicating that the interest in these compounds is more than academic.[1]
Several recent and comprehensive review articles have been published on calixarenes[1-8] and resorcinarenes,[7-9] which cover both the chemistry and the host-guest properties of these macrocycles. The conformational properties of calix[4]arenes and resorcinarenes are particularly well studied and exploited. They can exist in four different conformations (Figure 2) named *cone*, *partial cone*, *1,3-alternate*, *1,2-alternate*[10] which strongly influence their complexation properties.

R. Ungaro and E. Dalcanale (eds.), Supramolecular Science: Where It Is and Where It Is Going, 67–94.

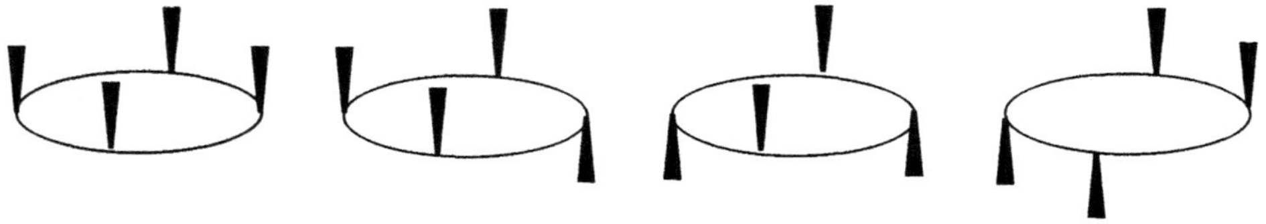

Figure 2: The four possible conformations of calix[4]arenes and resorcinarenes.

In calix[4]arenes, when the substituent at the lower rim is larger than ethyl, the conformational interconversion among these structures does not occur and it is possible to isolate the four stereoisomers.[11,12] Synthetic protocols for the selective functionalization of the OH groups of calix[4]arenes (*regioselective* alkylation) and for the selective synthesis of certain stereoisomers of tetraalkoxy calix[4]arenes (*stereoselective* alkylation) have been developed.[4,12]
We would like to summarise in this account the most recent achievements of our groups in the use of calixarenes and resorcinarenes for the design and synthesis of new hosts for molecular recognition and sensing, and indicate some new goals we are pursuing in our laboratories.

2. Calixarenes in cation complexation

The regio- and stereoselective control obtained in the alkylation of calix[4]arenes allowed us to synthesize ligands for metal ion complexation by introducing chelating chains such as amides, esters, hydroxamic acids or polyether bridges at their phenolic OH groups.[13-16] The selectivity observed with these ligands depends on the type and number of the chelating groups and on the size of the hydrophilic cavity thus created. However, a clear stereochemical control of the selectivity is observed in cation binding. The *cone*, being the most polar conformation of a calix[4]arene, prefers to bind hard metal ions such as Na^+, Ca^{2+} or trivalent cations such as iron (III) and lanthanides (III). On the contrary, derivatives fixed in the *1,3-alternate* structure are selective toward metal ions with lower charge density (K^+, Rb^+, Cs^+).
Calix[4]arenes **3-6** in the *cone* conformation, bearing four acetate or acetamide binding groups at the lower rim, possess a hydrophilic pseudo-cavity created by four carbonyls and four ether oxygen atoms and are able to complex alkali and alkaline-earth metal ions with an efficiency comparable with that of cryptands.[13] Using these ligands, we also developed ion selective electrodes (ISE's) which show a high selectivity for Na^+, being K^+ the main interfering ion.[17] A systematic study on these derivatives, changing the nature of the chelating chains and the substituent at the upper rim of the macrocycle, indicates that the best selectivity ($pK^{pot}_{Na,K} = 2.65$) on ISE's is obtained when compound **5** is incorporated into the membrane.

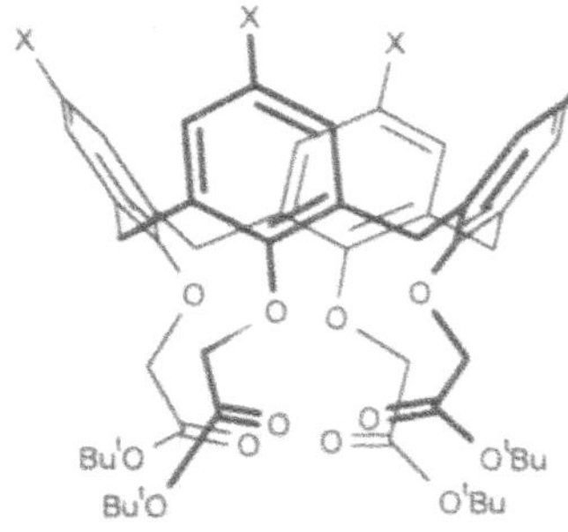

3: X = H
4: X = t-Bu

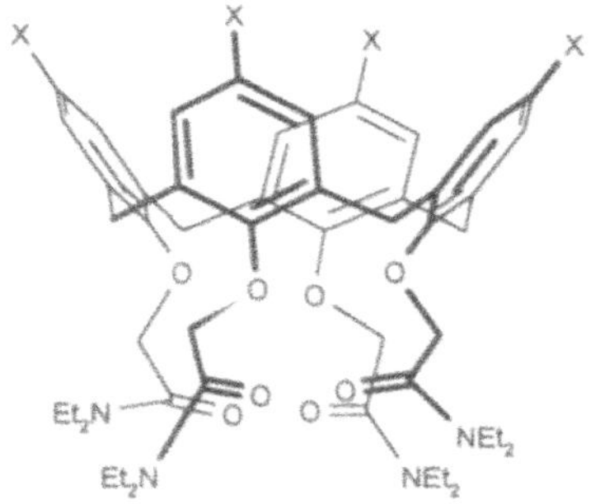

5: X = H
6: X = t-Bu

Molecular Dynamics studies on complexes of tetramide **6** with alkali and alkaline earth metal ions, together with the X-ray crystal structure of $[Sr\subset\mathbf{6}]^{2+}$, evidence that the cation is encapsulated by the ligand and no interactions with the counterion or the solvent molecules are present.[18] This property has been exploited in the preparation of luminescent complexes with lanthanide cations and in particular with Eu^{3+} and Tb^{3+}. These cations, in fact, give water-soluble and luminescent complexes with **6**.[19,20] On excitation of the ligand at 270 or 280 nm it is possible to observe an emission spectrum of the metal ion, which indicates that energy transfer takes place between the two species in the supermolecule (*antenna effect,* Figure 3b).

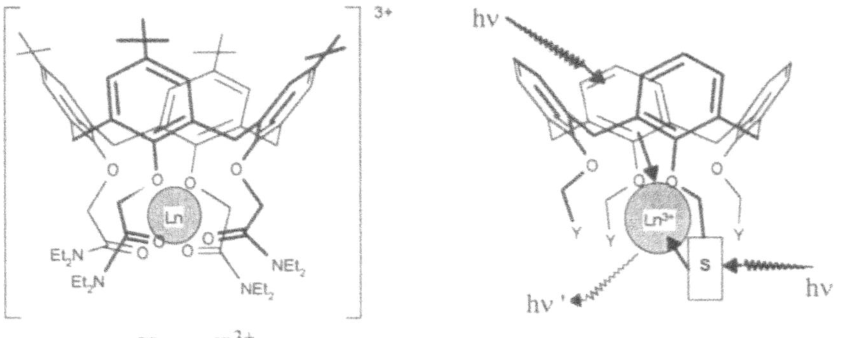

$[Ln\subset\mathbf{6}]^{3+}$
(a)

(b)

Figure 3: a) Complex of ligand **6** with a lanthanide (III) metal ion and b) antenna effect between a calixarene and a lanthanide ion: S = sensitizer; Y = binding group.

The application of the Horrock and Sudnick equation[21] to the terbium complex of ligand **6**, reveals that only one water molecule is still coordinated to the complexed cation.[19] Consequently, the lifetime (τ = 1.5 sec) and the luminescence quantum yield (ϕ = 0.2) of this complex are very high and make it attractive for applications as luminescent probes in time-resolved fluoro-immunoassays (TR-FIA).[20] In order to further improve the luminescence properties of these ligands by increasing their molar absorbitivities (ε) we have synthesized calix[4]arenes (e.g. **7**) with four bipyridines or phenantrolines at the lower rim.[22,23] A significant increase in the value of ε of these derivatives has been obtained, but the low binding ability of pyridines decreases the stability of the complexes.

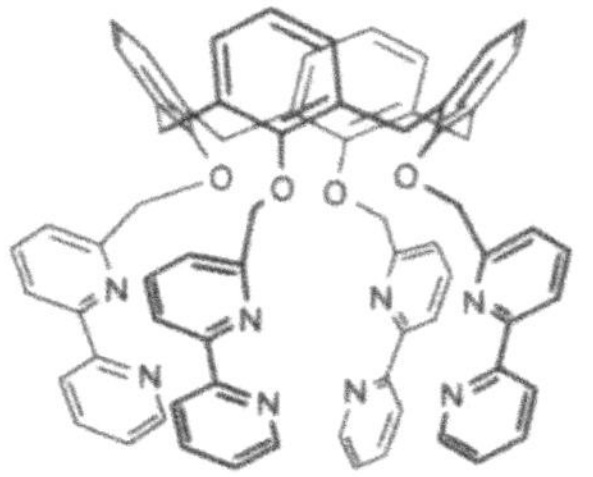

7

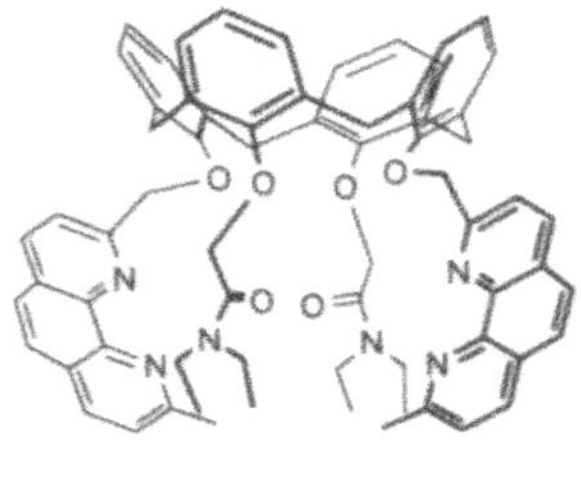

8

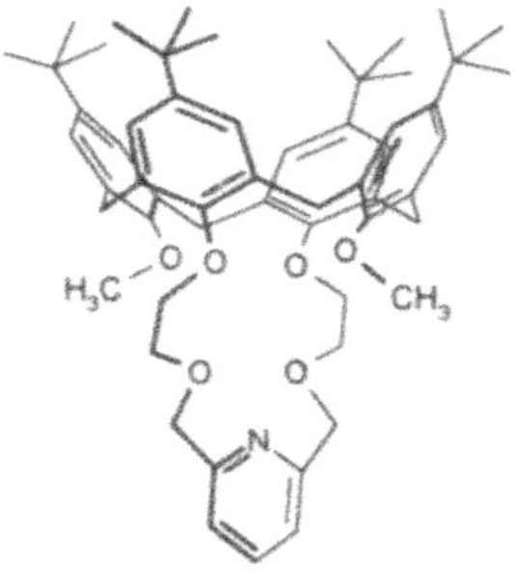

9

The synthesis of ligands bearing mixed type of chelating groups such as amide-bipyridine or amide-phenantroline (e.g. **8**)[24] or having a macrobicyclic structures such as **9**,[25] allows a good compromise between stability and luminescence properties of the corresponding complexes with lanthanide ions.

An interesting class of ionophores, which have been developed using a calix[4]arene skeleton, is that of the calixcrown-n (n = 5, 6).[26] These macrobicyclic ligands can be obtained by bridging the distal (1,3) positions of the lower rim of a calix[4]arene with a tetra- or pentaethylene glycol chain. The dimethoxy derivatives **10** and **11**, are mainly present in the *cone* conformation in solution at room temperature, but they are conformationally mobile and isomerize to the *partial cone* or *1,3-alternate* at high temperature. Compounds **10** and **11** change their structure to the *1,3-alternate* (Figure 4) upon K^+ and Cs^+ complexation.[27,28]

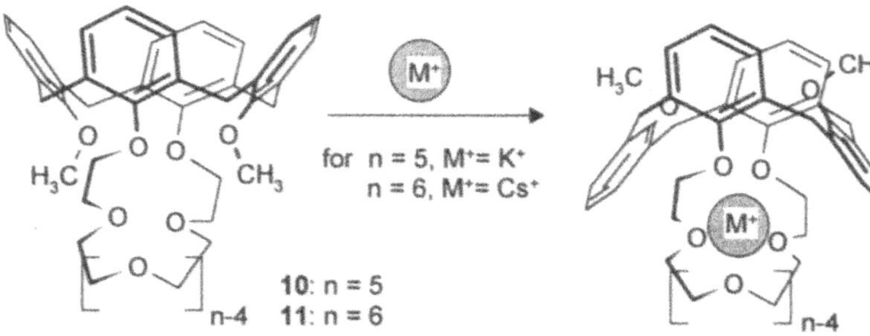

Figure 4: Conformational change of the dimethoxycalix[4]arene-crown-5 (**10**) and -crown-6 (**11**) upon complexation of potassium or cesium ions.

As clearly evidenced by NMR spectroscopy and also confirmed by the X-ray crystal structure of the [Cs⊂**11**]$^+$ complex (Figure 5), the cation is coordinated not only by the ether oxygen atoms of the crown, but also by the two opposite aromatic nuclei through cation -π interactions (η^2 and η^3).[28-30] This evidence prompted us to study the effect on selectivity of the ligand preorganization in alkali cation complexation and therefore to synthesize calixcrowns fixed in the *1,3-alternate (a), partial cone (p)* or *cone (c)* structures.

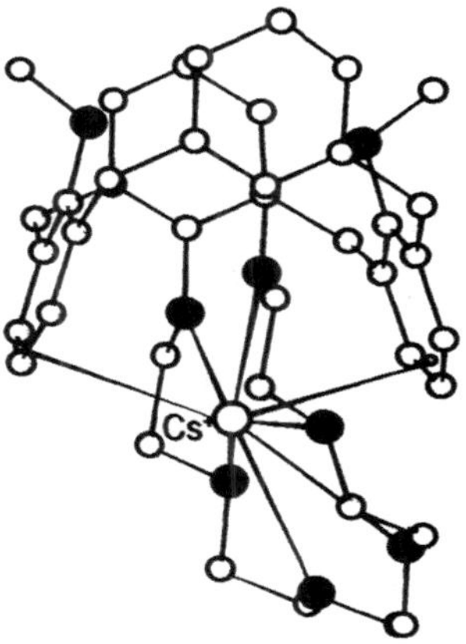

Figure 5: X-Ray crystal structure of the [Cs⊂**11**]$^+$ complex.

Using very efficient stereoselective methodologies we synthesized the conformationally fixed isomers **12** or **13** (**a, p, c**).[27,29]

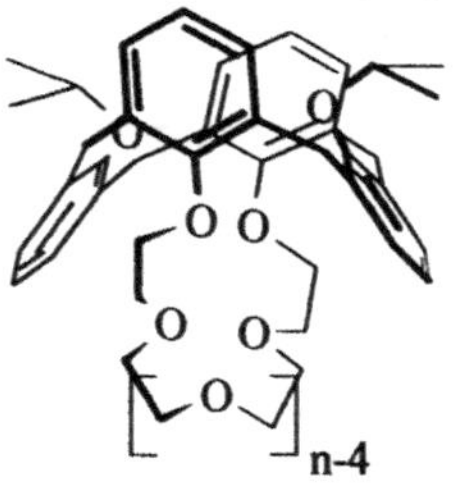

12a: n = 5
13a: n = 6

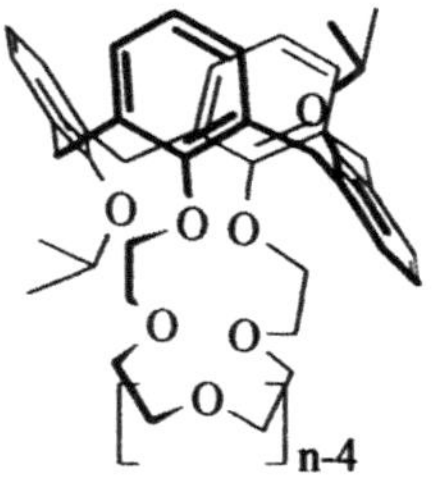

12p: n = 5
13p: n = 6

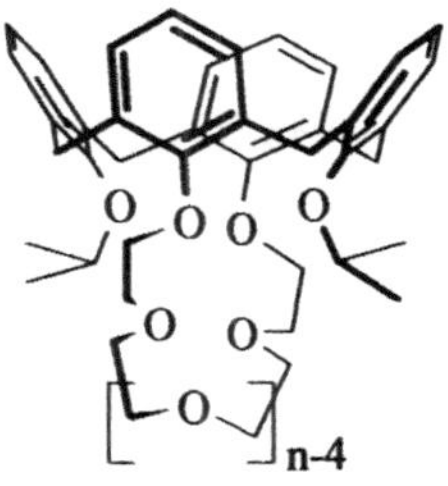

12c: n = 5
13c: n = 6

Although all calixcrown-5 (**10** and **12a,p,c**) are selective for potassium, and all calixcrown-6 (**11** and **13a,p,c**) for cesium, the efficiency and selectivity of complexation are strongly dependent on the calixarene stereochemistry.

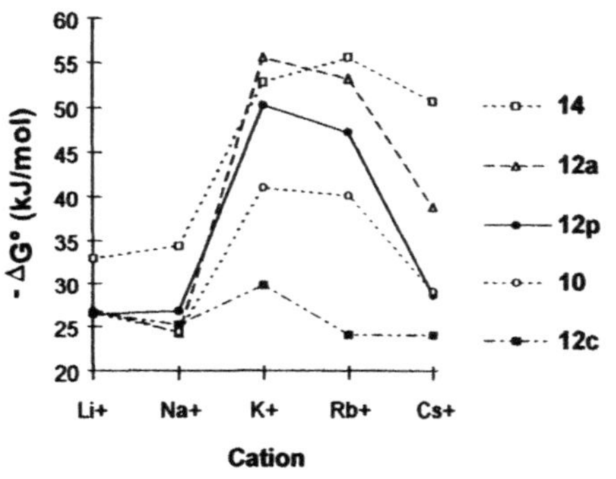

Figure 6: Free energies of association (-ΔG°) of calixcrown-5 and alkali picrates in $CHCl_3$ at 25° C.

The K^+/Na^+ selectivity for crowns-5 (Figure 6) and Cs^+/Na^+ for crowns-6 decrease in the order *1,3-alternate* > *partial cone* > *mobile* > *cone*, emphasizing how the preorganization of the ligand in the conformation preferred for complexation is a crucial point for obtaining ion selectivity. In particular, the calixcrown-5 **12a** in the *1,3-alternate* structure shows the highest K^+/Na^+ selectivity ($\Delta\Delta G°$ = -31.2 kJ/mol in $CHCl_3$) ever observed for a synthetic ionophore and also higher than valinomycin **14**, the best potassium selective natural ionophore known so far.

In supported liquid membranes (SLMs), the transport of potassium salts is always dependent on the diffusion coefficient of the complex through the membrane with the only exception for the derivative (**12p**), in the *partial cone* structure, where the rate limiting step is the release of the salt in the receiving phase. Both in transport (SLM) and in detection with ion selective field-effect transistors (ISFETs) the derivative in the *1,3-alternate* structure (**12a**) shows higher selectivity than valinomycin.[27]

Also for calixcrown-6 the stereoisomer (**13a**), fixed in the *1,3-alternate* structure (Figure 7), shows the highest Cs^+/Na^+ selectivity ever observed ($\Delta\Delta G° = -20.2$ kJ/mol in $CHCl_3$). This has allowed us to study the removal of the long-lived radionuclide ^{137}Cs from nuclear waste, where sodium nitrate and nitric acid are present at a much higher concentration than cesium nitrate ($[NaNO_3] = 4M$, $[CsNO_3] = 0.001M$, $[HNO_3] = 1M$).[31,32]

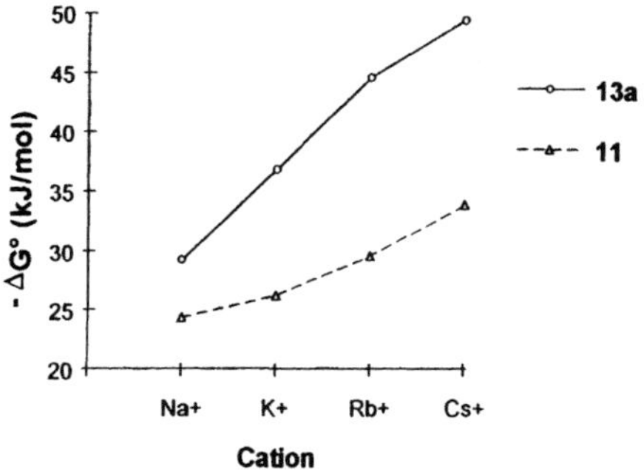

Figure 7: Free energies of association (-ΔG°) of calixcrown-6 and alkali picrates in $CHCl_3$ at 25° C.

The use of **13a** in the extraction process[28] or in the transport through supported liquid membranes (SLMs)[29,32] allows to recover more than 98% of the cesium cation present in solution, making this derivative extremely attractive for declassification of nuclear wastes. Ligand **13a** was also used for the selective detection of cesium in ISEs and ISFETs with very high selectivity and low detection limit.[33,34] Very recently, we anchored calix[4]arene-crown-5 and -crown-6 derivatives on silica-gel *via* hydrosilanization and we were able to separate by chromatography potassium or cesium from other alkali metal ions with high efficiency.[35]

Compared with calix[4]arenes, the calix[6]arenes (**1**: n = 6) are larger and have a much higher conformational mobility in solution. The cone structure is maintained in the solid state[36] and in monolayers at the water-air interphase,[37] as proven by the determination of the limiting area and of the dipole moment in Langmuir-Blodgett films, which also show the complementarity of this macrocycle for cesium cation. A significant control of the conformational properties of calix[6]arenes was obtained in few derivatives such as the 1,3,5-trimethoxy-2,4,6-trialkoxy-p-*tert*-butyl calix[6]arenes (**15**), which bear three methoxy and three bulky alkoxy groups at the lower rim and *tert*-butyl groups in para positions. This substitution pattern reduces the conformational flexibility of the calix[6]arene derivatives, which mainly adopt a *flattened cone* conformation.[38,39]

For compounds **16**, this conformation is in slow exchange ($\Delta G^{\neq} = 70$-88 kJ/mol at 328K in $CDCl_3$) with the *1,2,3-alternate* conformation, the latter being less stable ($\Delta G° = 2$-7 kJ/mol).[40] Therefore it is possible to conclude that three important factors stabilize the *flattened cone* structure, *viz.* (i) the self-inclusion of the methoxy groups in the apolar cavity driven by CH-π interactions, (ii) the presence of three bulky groups in 2,4,6 positions at the lower rim, and (iii) the presence of the *tert*-butyl groups at the upper rim.[40]

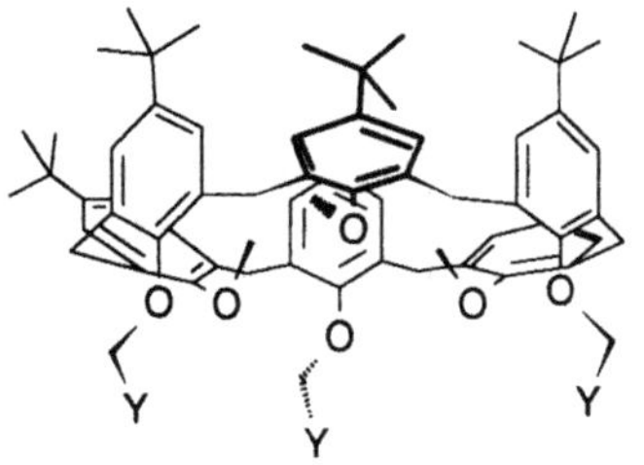

15 : Y = alkyl, aryl
16a: Y = COONEt$_2$
b: Y = COO^{t}Bu
Flattened Cone

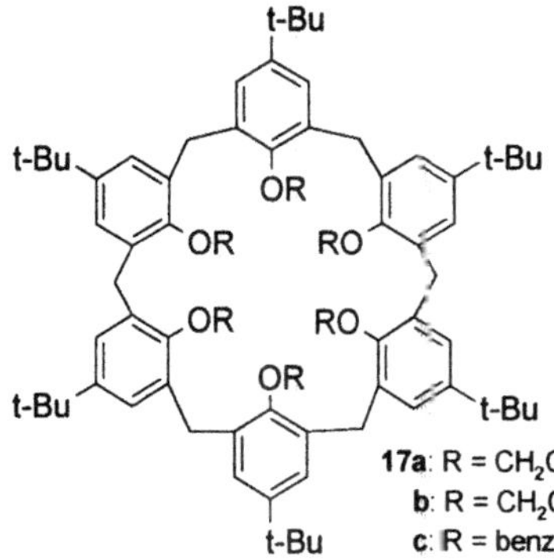

17a: R = CH$_2$CONEt$_2$
b: R = CH$_2$COOBu^t
c: R = benzyl
Mobile

The complexing properties of the triamide **16a** toward alkali and guanidinium cations in $CDCl_3$ were determined and compared with those of hexamide **17a**.[39] The latter is nearly 350 times more efficient than the former toward guanidinium ion (Gua^+) complexation, but is less selective since also sodium and potassium are well complexed. On the contrary, the triamide **16a**, with the amide chains organized in a C_{3v} geometry, is not able to complex alkali cations and therefore is much more selective toward the highly complementary guanidinium cation. In the transport through supported liquid membranes (SLMs), the triamide **16a** is also less efficient but more selective than hexamide **17a**, both in single ion and competitive transport experiments. NMR experiments with hexamide **17a** and sodium picrate in $CDCl_3$ evidenced the presence of the 1:2 (ligand:metal) complex. The first complexed sodium cation exerts a positive allosteric effect and preorganizes the ligand in the *cone* conformation and therefore favors the coordination of a second cation.[39] We also studied the behaviour of the hexamide **17a** in monolayers at the water-air interphase, changing the composition of the aqueous phase from pure water to NaCl, KCl, CsCl or GuaCl solutions. The properties of the resulting monolayers strongly depend on the affinity of the ligand for the cation and on the structure of the ligand in the different complexes.[41]

The rigidification of calix[6]arenes can be also achieved *via* the introduction of a bridge between the aromatic nuclei in 1,4 positions.[42] The reaction of calix[6]arenes with *t*BuOK and pentaethyleneglycol ditosylate in benzene gives the calix[6]arene-crown-5 (**18** and **19**) in 36 and 42% yield respectively. These derivatives are still conformationally flexible in solution, and only by the introduction of four acetamide groups on the remaining phenolic OH of compound **18** bearing *tert*-butyl goups at the upper rim, it is possible to obtain a derivative (**20**) which is conformationally frozen. In $CDCl_3$ solution, NOESY experiments clearly evidence that this derivative is present in the *cone* conformation. The ligand adopts this conformation also in the solid state. Compound **20**, having a hydrophilic region for cation complexation at the lower rim and a hydrophobic cavity for the complexation of apolar molecules, is potentially a ditopic receptor.

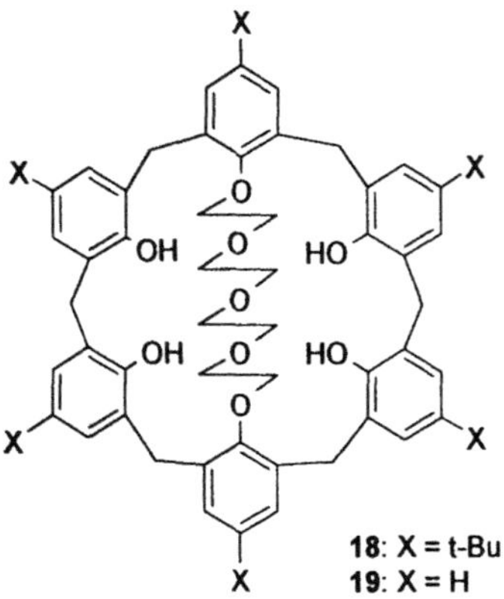

18: X = t-Bu
19: X = H

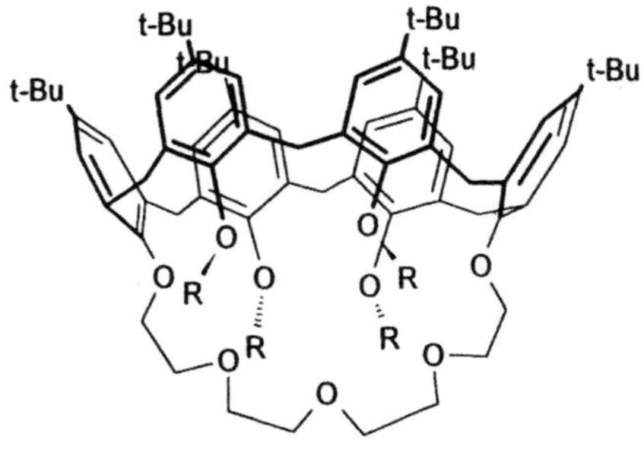

20: R = CH_2CONEt_2

Therefore, we asked ourselves in which of these two regions would a quaternary ammonium salt be complexed. In $CDCl_3$, the tetramide **20** is able to complex, with a 1:1 stoichiometry, $Me_4N^+Cl^-$ but not $Et_4N^+Cl^-$ or n-$Bu_4N^+Cl^-$. The exchange of the cation is slow on NMR time scale at 400 MHz and therefore, from the integration of the signals of the free ligand and of the complex, a $K_{ass} = 750\ M^{-1}$ for the complex of **20** and $Me_4N^+CH_3COO^-$ was determined. Large shifts are observed in the signals of the methylene groups of the crown moiety and of the methylene α to the amide carbonyls, thus indicating that the ammonium cation is preferentially bound in the hydrophilic cavity at the lower rim.[42]

3. Calixarenes for anion complexation

The selective recognition of anions by synthetic receptors is a less developed area of research compared with metal ion complexation. Only recently the Supramolecular Chemistry of anions has been investigated more systematically with a variety of host molecules[43] including calixarene-based receptors.[44] Particularly intense has been the study of carboxylate anion recognition.[45] Urea and thiourea moieties are well known as complementary binding groups for carboxylate anion recognition.[46] Therefore, by means of selective functionalization at the upper rim we synthesized calix[4]arenes (**21-23**) bearing one or two urea or thiourea moieties.[47] Compounds **21** and **22** are able to selectively complex carboxylate anions in DMSO-d_6, while spherical anions are not recognized. Carboxylates are complexed through hydrogen bonding with the NH protons, as indicated by the ^1H NMR spectra of the complexes which show large down-field shifts for the urea protons. Binding constants for the 1:1 complexes in DMSO-d_6 at 300K range from 50 to 2200 M^{-1}. Thiourea derivatives (**21b** and **22b**) are more efficient than ureas (**21a** and **22a**), as expected from the higher acidities of their NH protons.

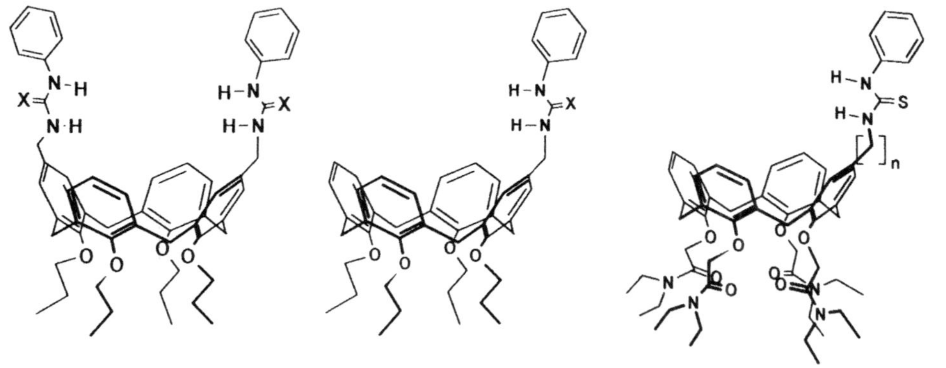

21 **a**: X = O; **b**: X = S **22** **a**: X = O; **b**: X = S **23** **a**: n = 0; **b**: n = 1

The diphenylthiourea **21b** is the receptor with the highest values of the association constants, probably because the two chelating chains can cooperate in the binding of carboxylates. The complexes with acetate and benzoate anion with monothiourea **22b** are also stabilized by attractive CH_3/π or π/π interactions.[47] A further development of the urea-derivatized calix[4]arenes was the synthesis of the tetraamide-monothiourea compounds **23**. These macrocycles present a ditopic structure able to complex both components of a salt in two distinct regions of the receptor. Besides the complexation of the anion at the upper rim due to thiourea groups, also the cation is strongly coordinated at the lower rim by four acetamide chains. In the case of compound **23a**, the cation bound at the lower rim induces an increase in the binding of carboxylates at the upper rim, as a consequence of inductive effect of the metal ion which increases the acidity of the urea NH group. The overall binding of sodium acetate by the ditopic receptors **23a** and **23b** is very high.[48]

Three urea or thiourea groups were also introduced at the lower rim of 1,3,5-trialkoxy calix[6]arenes, thus obtaining ligands for complexation of anionic guests. These derivatives form 1:1 complexes with halide, mono-, di- and tricarboxylate anions with a high selectivity for tricarboxylates and particularly for the trianion of trimesic (1,3,5-benzentricarboxylic) acid, which is strongly complexed in $CDCl_3$ (K_{ass}= 290x10^3 M^{-1}). Quite important is the complementarity between receptor and substrate, since the trianions of trimellitic (1,2,4-benzentricarboxylic) and of hemimellitic (1,2,3-benzentricarboxylic) acid show much lower binding (K_{ass} ~ 2.5x10^3 – 47x10^3 M^{-1}).[49]

4. Calixarenes and resorcinarenes for neutral molecules complexation

The binding ability of the apolar cavity defined by the four aromatic nuclei of p-*tert*-butylcalix[4]arene toward neutral molecules was evidenced during the early elucidation of its solid state structure, since a toluene molecule was found into the cavity.[50]

The systematic studies that followed have shown that calix[4]arenes having alkyl groups at the upper rim are able to recognize aromatic guest molecules in the solid state and are selective toward benzene and p-xylene.[51]

X-ray crystal structure determination together with theoretical calculations and molecular modeling carried out on the complex between a p-*tert*-butylcalix[4]arene derivative and pyridine[52] have evidenced the important role played by weak CH/π attractive interactions[53] in stabilizing the complexes.

It was therefore of great interest to exploit the π – donor apolar cavity of calix[4]arene for the complexation of neutral molecules both in solution and in the gas-phase. The binding ability of calix[4]arenes (**1**: n = 4) toward neutral molecules is however very poor in solution and the reason for this is that the intermolecular CH/π interactions involved in the complex formation are not strong enough to block the *cone-to-cone* ring inversion (see Figure 8) experienced by these macrocycles in solution.[3]

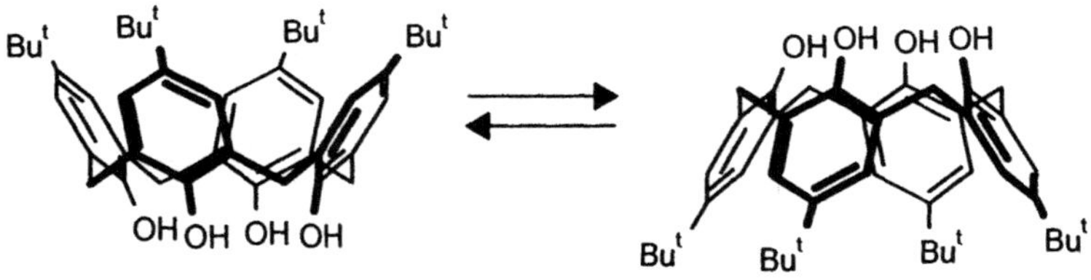

Figure 8. Cone-to-cone ring inversion of p-*tert*-butylcalix[4]arene.

As previously mentioned, the introduction of bulky substituents at the lower rim of the calix prevents the interconversion among the four possible stereoisomers: *cone, partial cone, 1,2-alternate* and *1,3-alternate* (see Figure 2).

However, most of the tetra-O-alkylated calix[4]arene *cone* isomers adopt a *flattened cone* (pinched cone)[54] conformation in the solid state,[55] showing a C_{2v} symmetry. This conformation has also been shown to be the most stable by molecular modeling.[56] In solution, a residual conformational mobility of the *cone* isomer still exists, and the C_{4v} symmetry, usually observed in the ^{1}H NMR spectra of these compounds, is considered to be the result of a fast interchange between two C_{2v} structures[56] (see Figure 9).

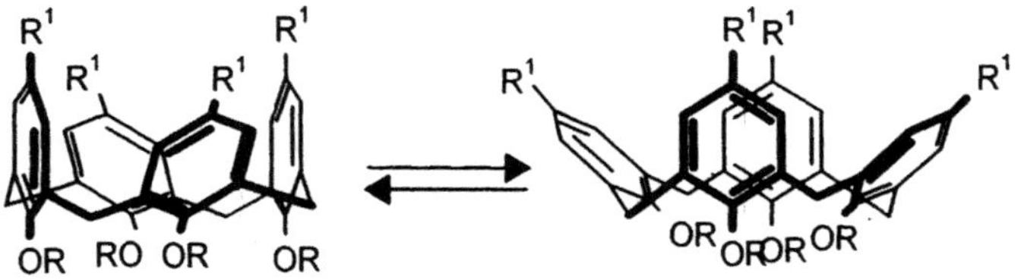

Figure 9. Dynamic stereochemistry of tetraalkoxy calix[4]arenes in the *cone* conformation.

We have obtained[57] direct experimental evidence of this residual conformational flexibility by studying the temperature and solvent dependence of ^{1}H NMR spectra of *cone* tetrakis(n-octyloxy)calix[4]arene **24** which, in CD_2Cl_2, gives a spectrum compatible with a C_{4v} symmetry at room temperature. Decreasing the temperature to 213 K the spectrum of **24** changes, indicating a C_{2v} structure. These changes clearly show the freezing of the molecular motion into a preferred *flattened cone* conformation.

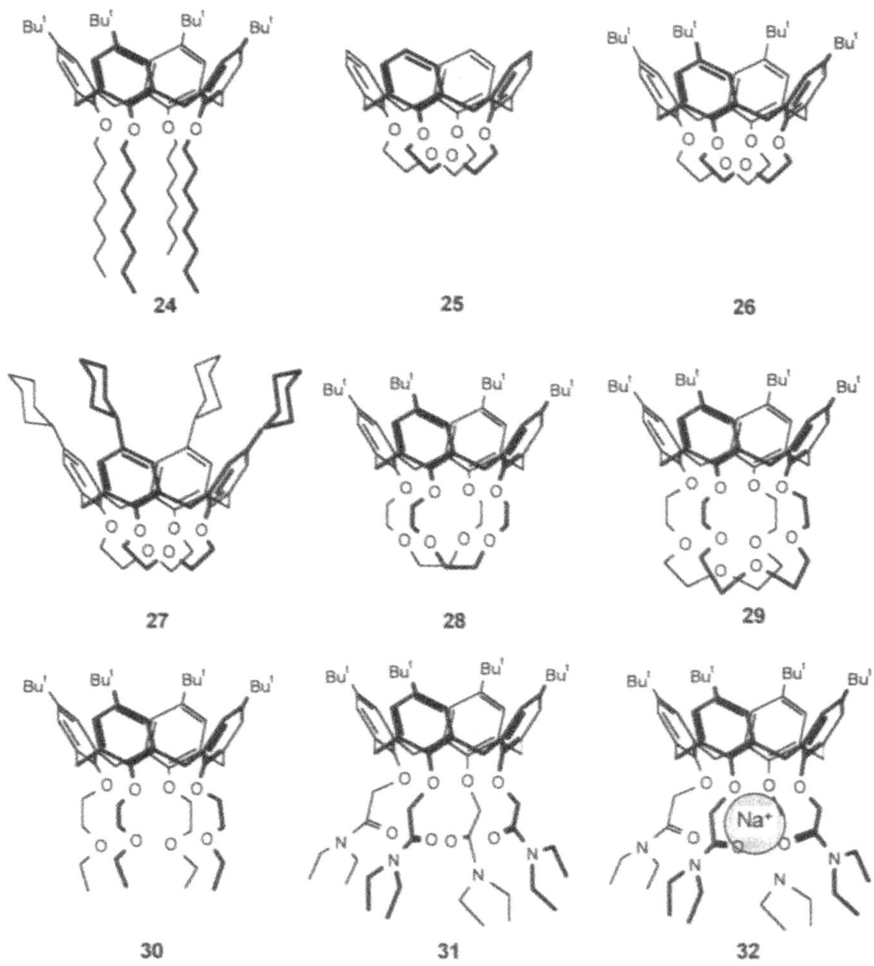

Figure 10. Calix[4]arene hosts having different rigidity

For most of the uses of calix[4]arene *cone* isomers, this more subtle conformational interconversion is usually neglected. However, simple tetraalkoxycalix[4]arenes are not able to bind neutral organic molecules in solution and we argued that the residual flexibility experienced by these receptors in solution could be responsible for the lack of their binding ability. CPK molecular models show that an attractive target for a very rigid and not distorted calix[4]arene *cone* derivative is the linkage of two proximal phenol rings with short diethylene glycol bridges at the lower rim. We previously developed an indirect multi-step procedure for the selective 1,2 (proximal) functionalization of calix[4]arenes at the lower rim,[58] but more recently we have found a direct one-step synthetic procedure to obtain a series of calix[4]arene biscrowns **(25-29)** where the conformational flexibility of the calix[4]arene cavitands (see Figure 10) can be tuned varying the length of the bridging unit.[59]

In fact, while p-*tert*-butylcalix[4]arene-biscrown-5 (**29**) is flexible and adopts a *flattened cone* conformation in the solid state,[60] calix[4]arene-biscrown-3 derivatives (*e.g.* **27**) possess a rigid *cone* structure. The recognition properties of rigidified p-*tert*-butylcalix[4]arene-biscrown-3 (**26**) in solution were evaluated in apolar organic solvents (chloroform, carbon tetrachloride) using nitromethane as guest.[61] By adding variable amounts of guest to a solution of **26** a significant upfield shift of the CH protons of the guest and fast exchange conditions are observed. These shifts clearly show an interaction of the acidic protons of the guest with the π electrons of the calixarene cavity, excluding a possible interaction of these protons with the crown-ether region which should result in a downfield shift of these signals.

TABLE 1. Associations constants (K_{ass}, M^{-1}) for 1:1 complexation of nitromethane and malononitrile by rigidified calix[4]arene at 300 K.

	CH_3NO_2		$CH_2(CN)_2$
Calix[4]arene	$CDCl_3$	CCl_4	$CDCl_3$
25	5±2	28±7	17±2
26	27±4	230±60	6±2
27	36±8	123±25	23±5
28	b	50±10	b
29	b	a	b
30	a	a	b
31	a	b	b
32	34±7	b	b

[a] no significant variation of the chemical shift observed; [b] not determined.

The analysis of the complexation data shows the formation of 1:1 complex and provides quantitative information on the Host-Guest interactions (see Table 1).

In order to gain further insight into the role of rigidity on the molecular recognition properties of calix[4]arene receptors we studied the behaviour of the more mobile tetrakis-(2-ethoxyethoxy)-p-*tert*-butylcalix[4]arene (**30**) and p-*tert*-butylcalix[4]arene biscrown-5 (**29**). Interestingly, with both hosts no variation of the chemical shift of the guest is observed, while with **28** complexation does occur only in CCl_4, but with a lower binding constant than with the more rigid **26** (see Table 1). These results demonstrate the importance of rigidity in determining the complexation properties of the π donor cavity of calix[4]arenes. Another interesting observation, which confirms the importance of rigidity and, consequently, of the preorganization of the *cone* conformer of calix[4]arenes in determining their complexation ability, is obtained by the comparison between p-*tert*-butylcalix[4]arene tetraamide (**31**) and its sodium picrate complex (**32**)[62] in the complexation of nitromethane in $CDCl_3$. In fact, while the conformationally mobile **31** does not show any significant complexation, its sodium complex **32**, which is rigid, strongly binds CH_3NO_2 (see Table 1).

As expected, the groups present at the upper rim of calix[4]arene-biscrown-3 strongly affect the K_{ass} values. In fact, increasing the extension of the cavity the complexation efficiency increases, as deduced by comparing the K_{ass} values of **25**, **26**, and **27**.

Interestingly, malononitrile which is more acidic than nitromethane, is bound at a lesser extent, indicating that not only the acidity but also the shape of the guest included plays an important role and that steric hindrance strongly reduces the binding.

The X-ray crystal structure of the **CH_3NO_2⊂ 27** complex (Figure 11)[61] further supports the data obtained in solution and proves the symmetrical cone structure of the host. In the solid state the guest molecule lies inside the cavity and orients its N-C bond along the axis of the cone whereas the CH_3 group face the aromatic nuclei of the host.

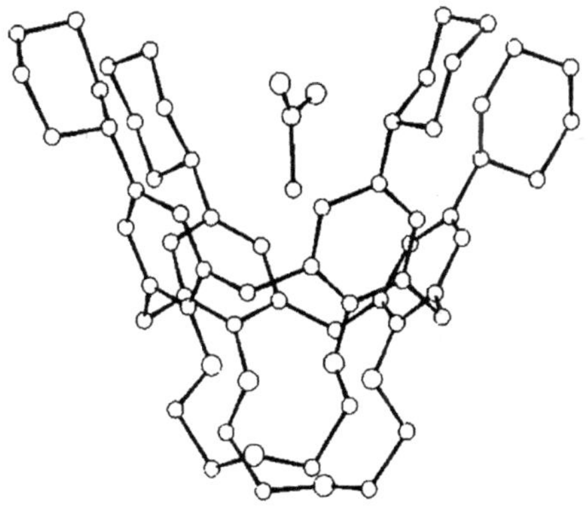

Figure 11. X-ray crystal structure of the **CH_3NO_2 ⊂ 27** complex.

Following a complementary approach for reducing the conformational flexibility of calix[4]arenes, but preserving their *cone* structure, we introduced rigid bridges containing aromatic or other π donor groups at the upper rim of the calix.[63]

We used mass spectrometry techniques to study the ability of a series of diametrically bridged calix[4]arenes with a 2,4-hexadiynyl (**33**) or a p-xylyl (**34**) moiety to complex neutral molecules in the gas phase.[64]

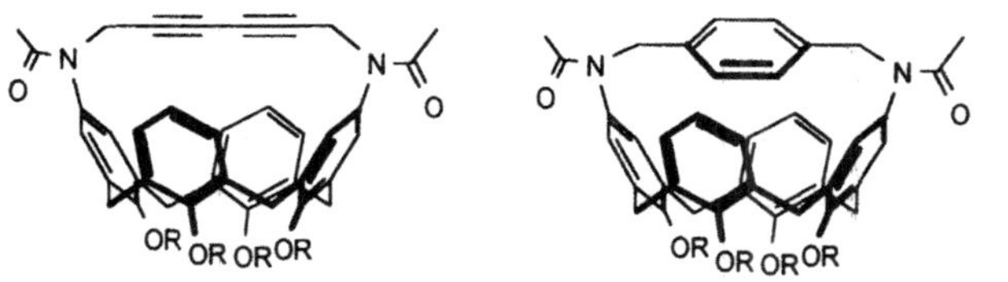

33 R = $CH_2CH_2OCH_2CH_3$ **34**

Figure 12. Calix[4]arene cavitands able to recognize guests in the gas phase.

These cavitands show in the gas phase high complexation efficiency and selectivity toward guests having acidic CH groups. However, no evidence of complexation was obtained in solution, probably because the interactions involved are too weak to overcome solvation effects.

Therefore we designed and synthesized new cavitands (*e.g.* **35**) to improve the complexation ability by introducing more efficient additional binding sites, using pyridine containing bridges. By comparing the data obtained with **35** with those of the isophthaloyl analogue **36**, it was possible to investigate the role of the pyridine basic group located in close proximity to the calixarene cavity on complexation.

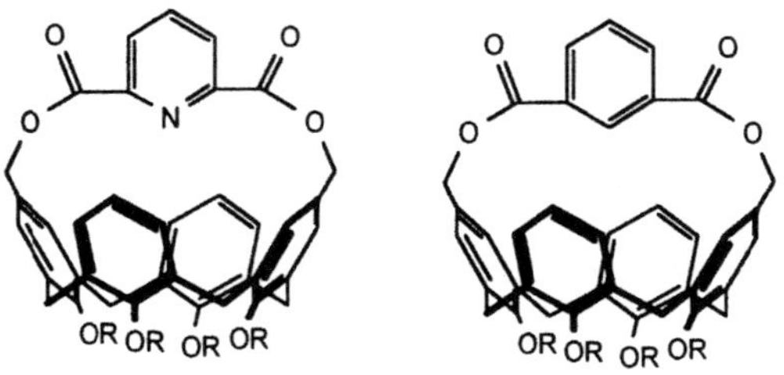

35 R = $CH_2CH_2OCH_2CH_3$ **36**

Figure 13. Upper rim bridged calix[4]arene derivatives.

Nitromethane, malononitrile and acetonitrile were chosen as guests and the complexes studied by ^1H NMR spectroscopy in $CDCl_3$ and CCl_4.[65]
Significant upfield shifts of the methyl or methylene protons of the guest were observed by adding the guest to a solution of host **35**. This indicates that the acidic protons of the guest interact with the π electrons of the host cavity. In fact, any interaction of these protons involving only the nitrogen of the pyridine moiety and the oxygen atoms present in the bridge should result in a downfield shift of these signals. By applying the continuous variation method a 1:1 stoichiometry for the complexes was established. As expected the *iso*phthalic derivative **36** shows no complexation for any of the guests studied, clearly demonstrating that the presence of the pyridine nitrogen donor atom in the bridge is essential for the efficiency of the receptor. From the values of the stability constants listed in Table 2 it emerges that the acidity of the guest, its shape and the polarity of the solvent affect the degree of complexation.

TABLE 2. Associations constants (K_{ass}, M^{-1}) for 1:1 complexation of bridged calix[4]arene with neutral guests at 300 K.

Calix[4]arene	CH_3CN CCl_4	CH_3NO_2 CCl_4	$CH_2(CN)_2$ $CDCl_3$
35	36±10	57±5	79±20
36	a	a	a

[a] no significant variation of the chemical shift observed.

As expected, the stability of the complexes is higher in the less polar CCl_4 and increases with the acidity of the guest (malononitrile > nitromethane > acetonitrile), while it is less sensitive (compared with the biscrown series) to the steric demand of the guest.
The structures of the complexes **$CH_2(CN)_2 \subset$ 35** and **$CH_3NO_2 \subset$ 35** obtained from the crystallographic study are reported in Figure 14.[65] The most interesting and common feature of the two complexes is that both guests are held by the host *via* the cooperation of two different hydrogen bonds: one "multifurcated" with the pyridine bridge and the other with the aromatic ring of the calixarene *via* a CH-π interaction. In particular, in the complex **$CH_2(CN)_2 \subset$ 35** the orientation of the pyridine bridging group with respect to the aromatic nuclei of the calixarene allows the host to act as "tongs" toward the malononitrile guest, which is complexed *via* a trifurcate hydrogen bond involving the hydrogen atom H(1G) of the malononitrile and the O(1*), N(1*), O(3*) acceptor atoms of the biscarbonyl pyridine bridge. The guest forms a weaker bifurcate hydrogen bond between H(2G) and the π orbitals on C(4A) and C(5A) atoms on the phenolic ring A with a bond distance of 2.85 and 2.87 Å respectively.

The structural properties of $CH_3NO_2 \subset$ **35** complex are significantly different. The calixarene host shows a more symmetric and less distorted *flattened cone* conformation, whereas the biscarbonyl pyridine group is more bent towards the reference plane of the calixarene. The complexation of the guest occurs mainly *via* a bifurcated hydrogen bond involving the H(3G) of the nitromethane and the two acceptor atoms O(3*) and N(1*) of the biscarbonyl pyridine bridge. The guest is thus blocked in the observed position by the additional hydrogen bond between H(2G) and the π cloud of aromatic nucleus A (distance H(2G) - C(3A) = 2.71 Å).
The X-ray crystal structure then confirms the participation in the binding process of both "hard" donor groups on the bridge and the "soft" aromatic cavity.

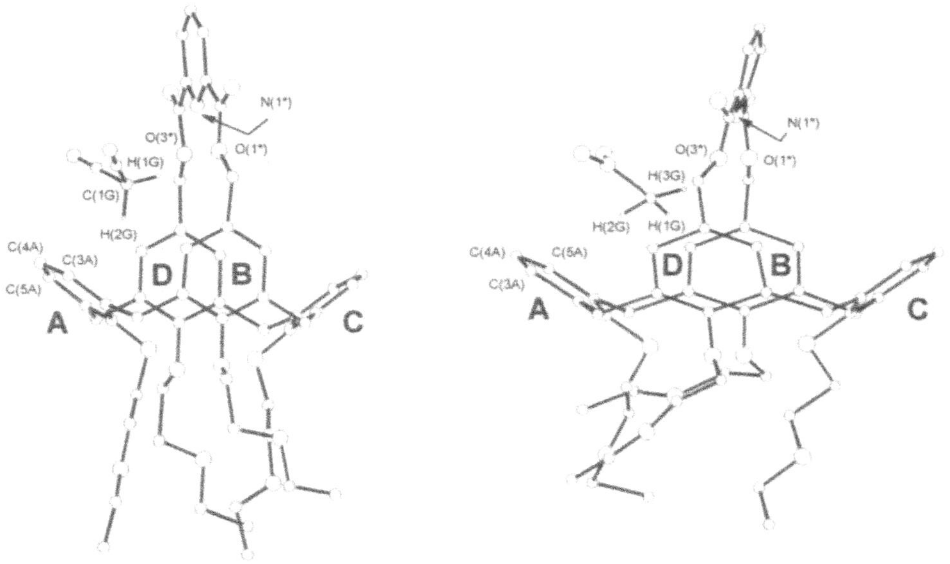

Figure 14. X ray crystal structure of the $CH_2(CN)_2 \subset$ **35** and $CH_3NO_2 \subset$ **35** inclusion complexes

Important information on weak intermolecular interactions have been also derived from the study of several host-guest complexes involving cavitands, obtained through the functionalization of resorcinarenes.[7-9] They have been investigated in the solid state,[66] in solution[67] and in the gas phase.[68] The main specific interactions responsible for recognition are CH-π, dipole - dipole and, in water or polar solvents, solvophobic interactions. By changing the nature of the bridging groups at the upper rim it is possible to control shape and dimensions of the formed cavity and therefore the selectivity in complexation. Instead, the choice of the R groups at the lower rim imparts different solubility properties to the molecules.

5. Supramolecular sensors

The possible analytical applications of cavitands developed so far are purge and trap chromatography[63,69] and sensors.[70] The direct translation of the molecular recognition properties of a given receptor from solution studies to the solid devices is not trivial, since non specific interactions such as dispersion forces, and material properties, like surface morphology, come into play. Sensors are devices which allow the continuous measurement of physical or chemical parameters. The need of accurate real-time analyses is the driving force for the development of sensor technology. There is an increasing demand for chemical sensors in the field of environmental analysis, industrial process control and work-place monitoring. Chemical sensing is part of an acquisition process in which some information on the chemical composition of a system is obtained in real time. In this process an amplified electric signal results from the presence of some chemical species. The acquisition process consists of two distinct steps: recognition and amplification. The recognition is given by a material capable of interacting with the desired analyte, while the amplification is provided by a transducer on which the material is coated. The most commonly used transducers are electrochemical, optical and piezoelectric.[71]

The quartz-crystal microbalance (QCM) piezoelectric sensor operating system is based on interactions between thin organic layers, coated on the surface of a quartz crystal, and analytes. The ability of a QCM sensor to selectively recognize some molecules in a complex mixture depends on how selective and sensitive is the coated receptor. In order to obtain selective responses the coating of the quartz must be stable and capable of specific interactions with the desired analyte. Reversibility of the responses is another essential feature which requires to resort to weak interactions, since the formation of covalent or ionic bonds would lead to irreversible saturation of the sensitive layer. On the other hand pure dispersion forces are unsuitable due to their aspecificity. Sensitivity in mass sensors depends mainly on the transduction mechanism employed. Surface acoustic wave devices (SAW) are usually at least two order of magnitude more sensitive than QCM ones with the same coating.

The QCM sensor is based on interactions between the surface of an oscillating quartz disc covered with the testing receptor layers and gaseous analytes. The resulting mass changes on the sensor surface (Δ m) produce a frequency drop (-Δf) of the fundamental resonance frequency of the oscillating quartz disc according to Sauerbrey equation:

$$-\Delta f = k \cdot \Delta m \qquad (1)$$

where k is a constant depending on the sensor configuration.

All the reported frequency shifts are reproducible ± 5% and have been normalised with respect to acetonitrile, the lowest molecular weight compound of the series.

All analytes are injected at the same concentration of 3 mM in N_2. For all the details regarding frequency shifts, coating technique and measuring system configuration we report to the literature.[72]

For measurements in water solution, having a very high dielectric constant and medium viscosity, only one side of the quartz crystals is exposed to the liquid solution kept at constant temperature (24°C). All test mixtures are prepared in advance by mixing the analyte with drinking water at a concentration of 100 ppm. The flow rate is kept constant at 3.2 l/h. Every test run starts with a purge, using the respective "pure" water. After a time of stabilization the liquid source is switched to the test mixture of known analyte concentration. The typical measuring time is 10 min. Every measuring step is followed by a purge.

Here we summarise the results of a comparative study on the gas-solid and gas-liquid interactions between two classes of cavitand receptors derived from resorcinarenes and a series of organic vapors using mass sensitive QCM (Quartz Crystal Microbalance) sensors.

The hosts employed have rigid preorganized cavities of different size and shape obtained by bridging the phenolic OHs of the building blocks with methylene and diazanaphthalene groups. In the case of methylene-bridged cavitands **37-39** a bowl-shaped cavity is present, approximately 9.0 Å wide and 3.3 Å deep.[73] In the case of 1,4-diazanaphthalene-bridged cavitands **40-42** the vase-like cavity formed is 7.8 Å wide and 8.3 Å deep.[66,67]

Different substituents at the lower ring are introduced to each class of cavitands in order to modify the morphological structure of the resulting layers. The criteria adopted in the choice of the analytes are to represent a wide range of chemical classes, whose characteristics in terms of molecular dimension and structure could be compatible with the inclusion within the cavities. The molecular recognition properties exhibited by these two classes of cavitands both in the solid state and in the gas phase suggested their use as selective layers in QCM gas sensors for the detection of neutral analytes, where solid-gas interactions are present.[74] In particular studies in the gas phase have shown that, in the absence of interfering solvation effects, CH-π interactions are the dominant interactions among cavitands and neutral guests, such as aromatic compounds, 2-ketones, acetates and others. In this respect the **40-42** series has proven to be more effective than the **37-39** one due to its deeper, vase-like cavity, where the guests can be completely embedded. In the gas phase the different substituents at the lower rim do not exert any influence on the complexation properties of the corresponding cavitands. In order to evaluate the selectivity of our cavitands, we compared their behaviour with that of two organic polymers, poly-isobutylene (PIB) and poly-epichlorhydrine (PECH), toward the same analytes. With such measurements we obtained a detection pattern with two coatings having different polarity (PECH is more polar than PIB), in which no molecular recognition phenomena are present.

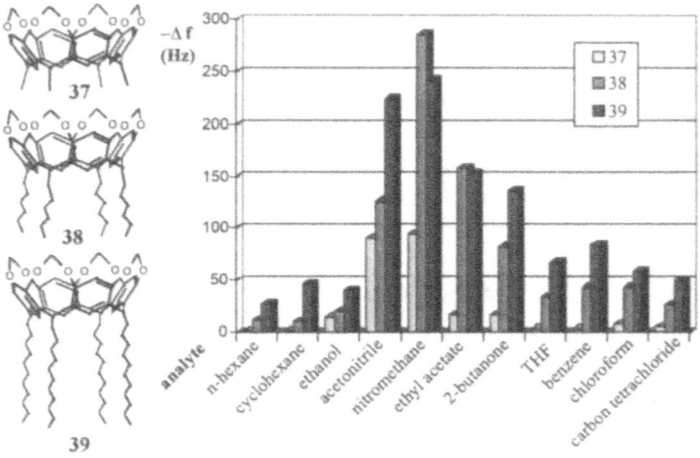

Figure 15: Histogram representing the response patterns of cavitands **37-39** to eleven organic vapors

In Figures 15 and 16 are reported the responses in Hz of QCM sensors coated with cavitands **37-42** to different analytes at the same concentration. In both series the responses increase with increasing the chain length of the substituents at the lower rim, in particular between C_1 and C_6.

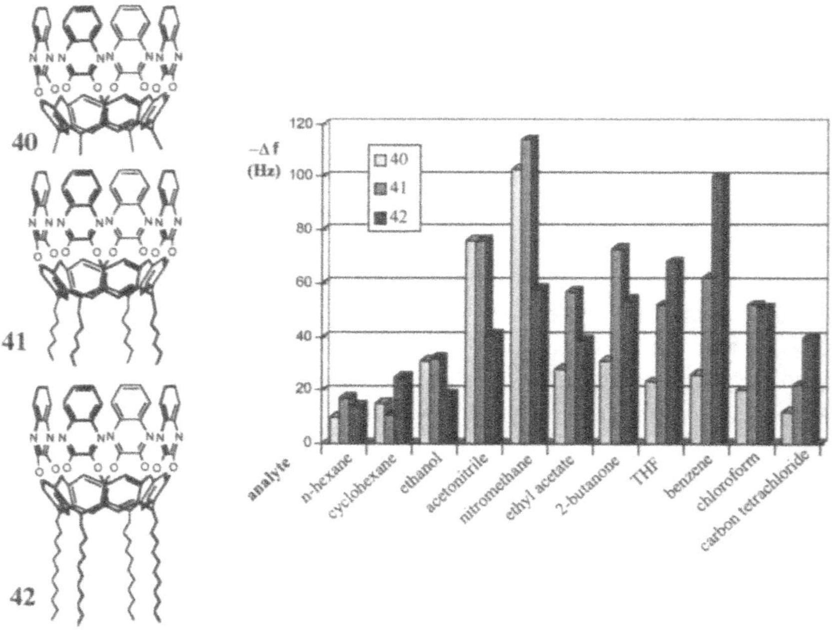

Figure 16: Histogram representing the response patterns of cavitands **40-42** to eleven organic vapors

This behaviour can be explained by recalling that more dispersed and porous layers are formed in the presence of long alkyl chains, with better diffusion properties and easier access to the bulk of the layer.[75]

This increase in sensitivity comes at selectivity expense, since it is due to aspecific interactions such as van der Waals forces. In the case of cavitands **37-39** for nitromethane, acetonitrile, ethyl acetate and 2-butanone, all molecules having an "acidic" CH_3, selective responses are observed also in the presence of long alkyl chains, while this trend is not observed for **40-42**. The trend is more evident in Figure 17, where the responses of sensors coated with PIB, PECH, **41** and **38** are compared.

Only **38** exhibits a selective response toward three analytes capable of CH-π interactions, while for the other ones the responses are comparable or even lower to those of the polymer coatings, where no host-guest interactions are present. Therefore the dimensions of the preorganized cavity do not constitute a determinant factor for sensor selectivity, which depends on the presence of a π-basic cavity, easily attainable from the gaseous analyte.

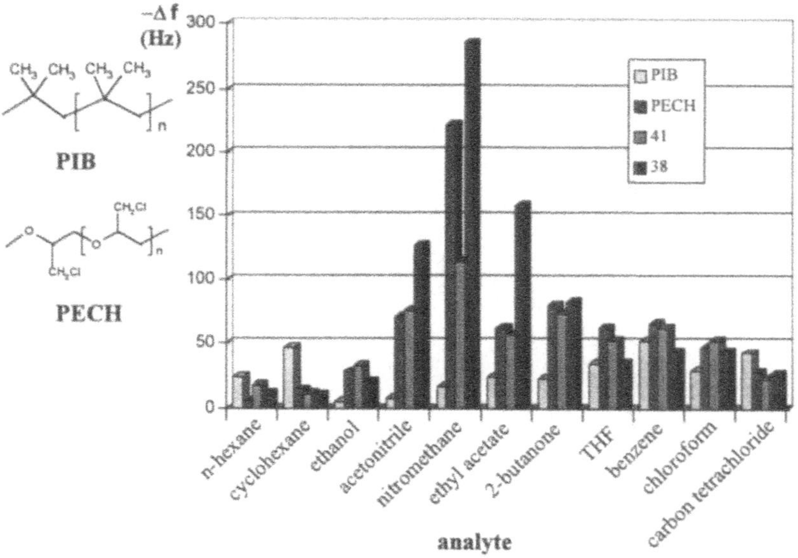

Figure 17: Histogram representing the response patterns of **PIB**, **PECH** and cavitands **38, 41** to eleven organic vapors

From these measurements and many others reported in the cited literature the following conclusions can be drawn regarding the behaviour of cavitands as selective layers for QCM sensors:

- the overall response of cavitands to the various analytes is the results of three factors: the *chemical constitution* of the cavitand which determines the *polarity* of the layer; the *permeability* of the layer which influences the sensibility; the presence of a *preorganized cavity* which can change the selectivity providing specific interaction with analytes.
- the porosity of the layer and therefore its permeability can be increased by introducing long alkyl chains at the lower rim of the cavitands. This results in an increase of sensibility at expenses of selectivity.
- the presence of a preorganized cavity in itself does not give raise to a selective sensor unless specific interactions, like CH-π interactions, are operating. This requires not only the presence of a π-basic cavity for recognition, but also its availability both

macroscopically (porous layer) and microscopically (depth of the cavity).
As a corollary of these results we anticipate that cavitands capable of two or more synergistic interactions with a chosen analyte will achieve an high level of selectivity when used as selective layers in QCM sensors.[76]
The key to the successful use of QCM in liquids is the development of sensitive and stable sensor coatings on the oscillating surface. The material of choice must be insoluble in water with no swelling behaviour. We studied the responses of the QCM sensors coated with cavitands **38-39** and **41-42** toward several different analytes (Figure 18), and compared the resulting sensor signals with their partition coefficients log P_{oct} (octanol/water) as function of their hydrophobicity (Table 3).[77]

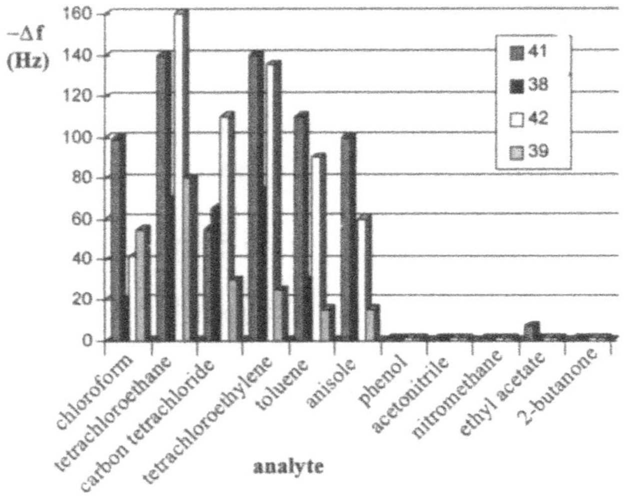

Figure 18: Histogram representing the response patterns of cavitands **38, 39, 41, 42** to eleven organic analytes in water

TABLE 3: Used analytes and their logP_{oct} values[78]

Analyte	log P_{oct}	Analyte	log P_{oct}
Chloroform	1.97	Phenol	1.46
Tetrachlorethane	2.39	Acetonitrile	-0.34
Carbon tetrachloride	2.64	Nitromethane	-0.33
Tetrachloroethylene	2.88	Ethyl acetate	0.73
Toluene	2.69	2-Butanone	0.29
Anisole	2.11		

Also in this case in order to evaluate the contribution to selectivity given by molecular recognition phenomena we compared the behaviour of our cavitands with that of polybutadiene (PBD) toward the same analytes. Figure 19 shows the responses to anisole of QCMs uncoated, coated with PBD, cavitands **39** and **42**. Diazanaphthalene-bridged cavitand **42** is the most sensitive, while methylene-bridged cavitand **39** responds only slightly more than PBD. This trend is followed by all detected analytes, where either **41** or **42** are the most sensitive coatings. By looking to Figure 18 a clear cut-off in

response is observed. Analytes like acetonitrile, nitromethane, ethyl acetate which are selectively detected by methylene-bridged cavitands in the gas-phase, are totally ignored by all cavitand coatings in water.
In this case the driving force for complexation is the hydrophobicity of the analytes which prefer the liphophilic cavity to water solvation. For a given molecule the log P_{oct} is a quantitative measure of its hydrophobicity.[78] Only analytes with log P_{oct} greater than 1.5-1.6 are detectable by cavitand-coated QCM in water (Table 3). This effect is clearly illustrated by the different behaviour of anisole and phenol. Both are aromatic derivatives bearing an electrodonating group but the more polar phenol (log P_{oct} = 1.46) prefers water solvation to inclusion in the cavity of either **41** or **42**. The opposite is observed for anisole (log P_{oct} = 2.11).

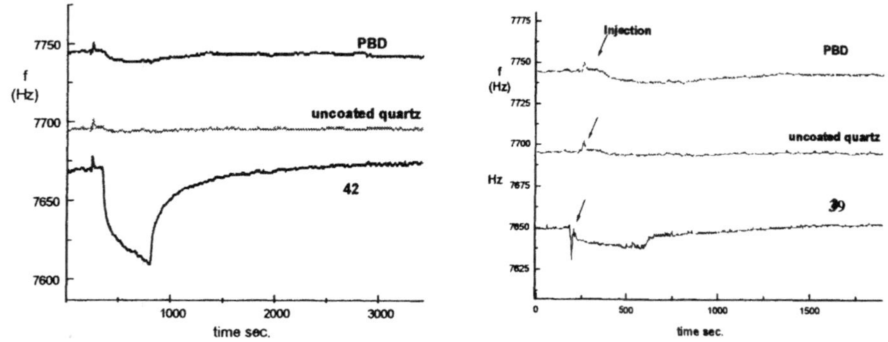

Figure 19: Detection of anisole (100 ppm) in water by QCM sensors coated with **PBD**, cavitands **39** and **42**. Uncoated QCM sensors are also monitored as control

In conclusion the selective response of cavitand-coated QCM sensors in water to neutral analytes depends on the size and shape of the liphophilic cavity and on the hydrophobicity of the analyte. The presence of the strong hydrophobic effect gives rise to a great enhancement in sensitivity (at least one order of magnitude with respect to QCM in air), balanced by a reduction in selectivity, since specific interactions like CH-π ones, are much weaker.

6. Where we are going

We report in this section the most recent results obtained in our groups in fields which are directly connected with those described in the previous section and which will probably keep us busy in the next few years.
The knowledge developed in the study of *carboxylate anion recognition* has been partly exploited for the synthesis of calixarene receptors which mimics the action of the vancomycin family of antibiotics. These natural compound are able to inhibit the formation of the cell wall of Gram-positive bacteria, because they strongly bind the D-alanyl-D-alanine terminal part of peptidoglycan. The selective and strong binding is due to the formation of a series of hydrogen bonds between the antibiotic and the dipeptide, terminating with a carboxylate anion, inside a hydrophobic environment (*the carboxylate binding pocket*)[79] created by the vancomycin apolar skeleton. Inspired by the mode of action of these antibiotics we designed and synthesized a series of open

chain (e.g. **43**)[80] and macrobyciclic (e.g. **44**)[81] *peptido*calixarenes, with the aim of developing biologically active compounds based on molecular recognition. Whereas the cleft-like compounds **43** show no or very modest biological activity, the macrobyciclic *peptido*calixarene **44** shows a remarkable vancomycin-type activity *in vitro*, being able to inhibit the growth of Gram-positive bacteria.[81] Preliminary binding studies carried out in $CDCl_3$ clearly indicate that compound **44** is able to recognize and strongly bind N-lauroyl-D-alanine and N-lauroyl-D-alanyl-D-alanine, which have been used as model for the terminal part of peptidoglycan.
An important role in complexation is played by the amine nitrogen which, on protonation, interacts with the carboxylate group of the guest.

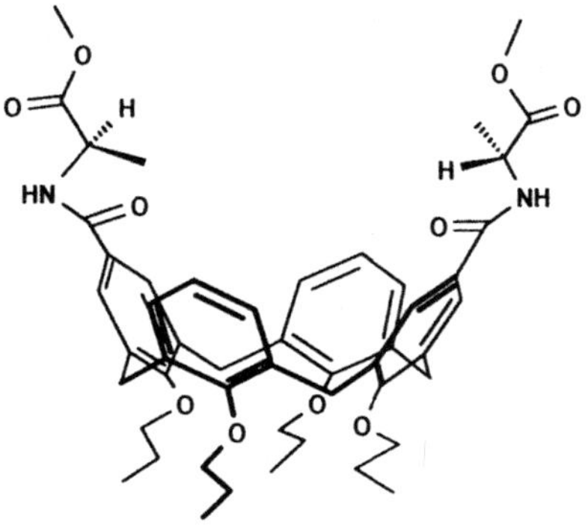

43

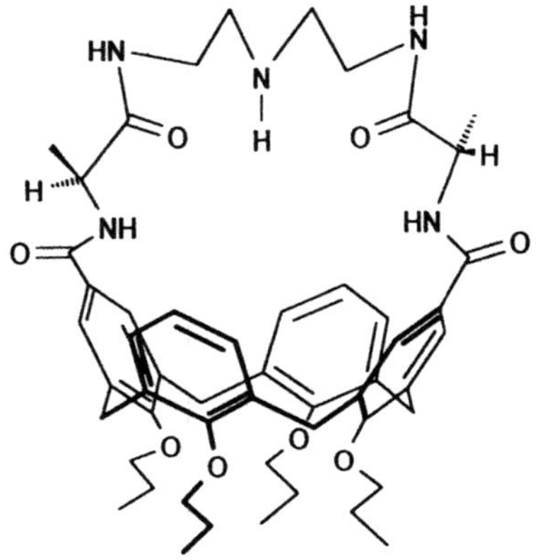

44

Evidence have been collected that beside the electrostatic interaction between ammonium and carboxylate ions, additional hydrogen bonds contribute to stabilize the complexes. This is also confirmed by NMR studies which show low-field shifts for the amide protons of the receptor and by Electrospray-Mass spectrometry which demonstrates for **44**·N-acetyl-D-Ala-D-Ala the presence of the 1:1 host/guest species.[82]
Within this context we have also synthesized *glyco*calixarenes (calixsugars),[83,84] whose binding properties toward carbohydrates are, however, modest. In the field of anion recognition we are exploring new binding groups, in order to enlarge the synthetic possibilities in the design of new host molecules for these important species. Recently, we have reported the synthesis of *cone* calix[4]arenes, functionalized at the upper rim with perfluoroalcoholic functions.[85] Preliminary binding studies indicate that these functions are able to interact with anions with efficiency comparable with that of amide groups but weaker than that with urea groups.
We are also interested in the synthesis of new water soluble calixarene receptors, especially those designed for the binding of biologically relevant species such as anions, carbohydrates and peptides. At the moment we have obtained simple water soluble calixarene hosts functionalized at both rims with charged groups such as phosphonates,[86] carboxylates,[80,87] sulphonates,[88] or a combination of them.[89,90] Interesting information on the selective binding of alcohols,[91] aminoacids,[92] and organic cations[90] in water solution have been obtained using these receptors.

Polymacrocyclic container molecules constitute a very interesting class of organic compounds with peculiar host-guest properties.[93] The confined space within carcerands, for example, can be regarded as a new phase of the matter for which reactivity[94] and physical properties[95] of the guest molecules are different from those in the bulk phases. Following a general trend in Supramolecular Chemistry, where more complex systems are obtained *via* self-assembly rather then by the more classical covalent synthesis (see the chapter by Reinhoudt and coworkers in this book) we have recently obtained new cage molecules based on self-assembly of calixarene and resorcinarene cavitands. The first approach exploits hydrogen bonding, the second one metal coordination. Calix[4]arene diacid derivatives in the *cone* structure were already shown to be able to form a dimeric molecular capsule *via* the intermolecular association of two carboxylic acid groups.[57] We have extended these results to conformationally more mobile calix[6]arene derivatives in order to test the possibility of rigidifying these hosts and increasing their binding properties through self-assembly. By exploiting recently developed general methodologies for the selective functionalization of calix[6]arenes at both rims[96] we have obtained the 1,3,5-calix[6]arene-tricarboxylic acid derivative (**45**), which is able to form a dimeric capsule (**45·45**) in apolar media (Figure 20).[97]

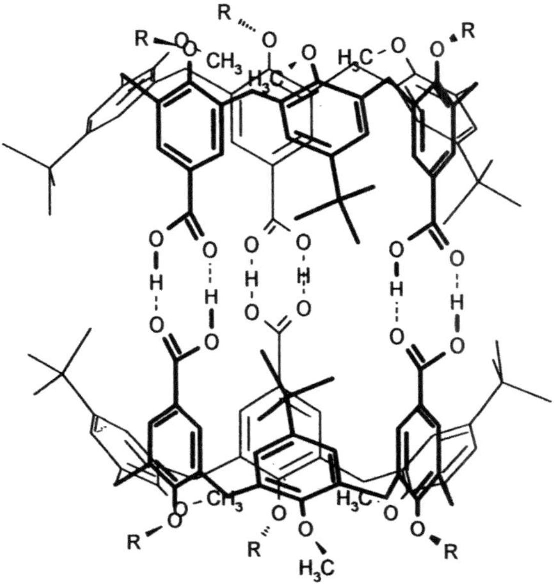

45 · 45
R= n-C_8H_{17}

Figure 20. Self-assembled hydrogen bonded molecular cage of calix[6]arene tricarboxylic acid derivatives

Metal-directed self-assembly is emerging as one of the most promising approaches to the generation of complex supramolecular architectures.[98] The quantitative self-assembly of stable organopalladium and organoplatinum cage molecules composed of two tetracyanocavitand derivatives connected *via* four Pd (II) or Pt (II) square planar complexes represents one example of this approach.[99] Cages of general structure **46** have been assembled by simply mixing the two components in a 1:2 molar ratio at room temperature in chlorinated solvents (Scheme 1). This self-assembly process is highly cooperative: the cage is the only product observed even in the presence of an excess of either one of the reactants. This is consistent with the rapid formation of the thermodynamically favoured cage.

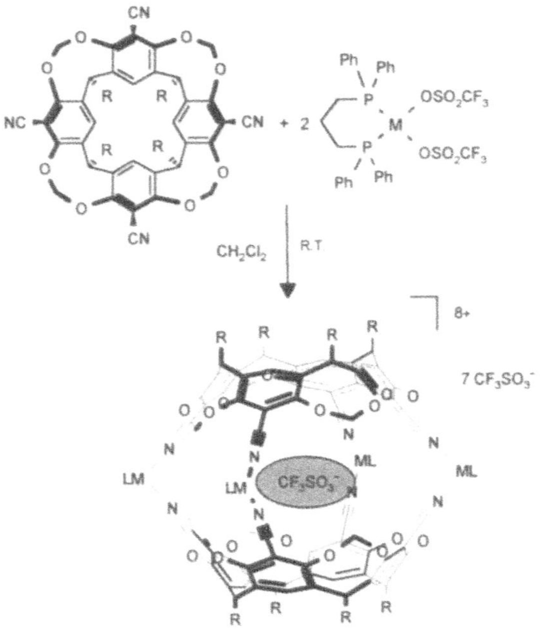

46

Scheme1: Self-assembly of Pd and Pt cage molecules (L = dppp)

The necessary requirements for the formation of the cages have been recently throughly investigated: preorganization and stiffness of the cavitand ligand, coordination and geometry of the metal complex precursor, type of counterion. In particular, the tetrahedral coordination around the metal center is not compatible with the formation of the cage. The counterions must be weakly coordinated to the metal center in order to exchange with the nitriles of the cavitand ligand: the best choices are $CF_3SO_3^-$, BF_4^-, PF_6^-. Encapsulation of anions during the self-assembly process was proven in the case of triflate, tetrafluoborate and hexafluorophosphate anions by ^{19}F-NMR. The X-ray crystal structure of a Pt-dppp cage confirms the presence of an included triflate anion in each cage and evidences the perfect square planar coordination of the ligands around the metal centers, delimiting a closed cavity of 4.5 x 4.5 x 4.5 Å, with a cross section in the equatorial region of 13.7 Å.[100] Control of the self-assembly process can be achieved *via* metal-ligand exchange (Scheme 2): this result opens the way for the generation of reversible on-off cage molecules from different sets of molecular components.

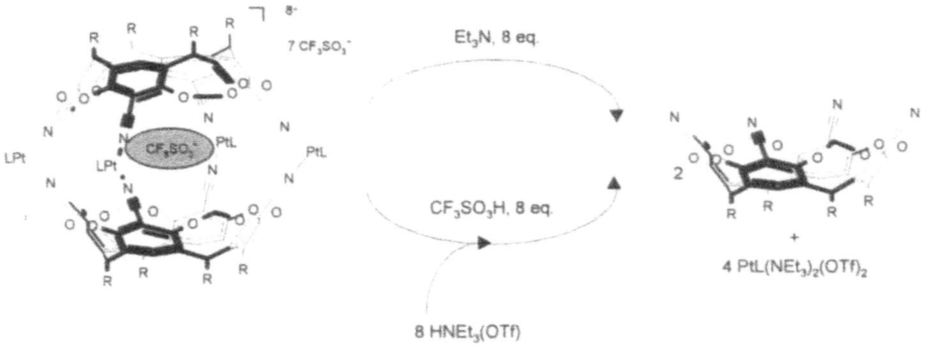

Scheme 2: Control of the self-assembly process *via* ligand exchange (L = dppp).

The possibility of anion recognition and encapsulation in the cage during the self-assembly process has been also investigated: it emerges that there is a high preference for the triflate and tetrafluoroborate anions versus the hexafluorophosphate one, with an interesting effect of the temperature on the selectivity of the encapsulation.

7. Acknowledgements

This work was supported by M.U.R.S.T (Ministero dell'Università e della Ricerca Scientifica e Tecnologica), Progetto "Dispositivi Supramolecolari", CNR and NATO Supramolecular Chemistry Project.
The authors are grateful to the Centro Interdipartimentale di Misure "G. Casnati" dell'Università di Parma for the use of NMR and mass spectrometry instruments, and to the Centro di Studio per la Strutturistica Diffrattometrica del CNR, Parma for X-ray diffraction facilities.

8. References

1. Casnati, A. (1997) *Gazz. Chim. Ital.* **127**, 637.
2. Ikeda, A.; Shinkai, S. (1997) *Chem. Rev.* **97**, 1713.
3. Gutsche, C.D. (1998) *Calixarenes Revisited*, The Royal Society of Chemistry, Stoddart J.F. Ed., Cambridge.
4. Arduini, A.; Casnati, A. (1996) "Calixarenes" in *Macrocyclic Synthesis: a Practical Approach*, D. Parker, Ed., Oxford University Press, Oxford, p. 145.
5. Roundhill, D.M. (1995) *Prog. Inorg. Chem.* **43**, 533.
6. Shinkai, S. (1996) *Supramol. Sci.* **3**, 189.
7. Böhmer, V. (1995) *Angew. Chem., Int. Ed. Engl.* **34**, 717.
8. Pochini, A.; Ungaro R. (1996) in *Comprehensive Supramolecular Chemistry*, Vol. 2, Vögtle F. Ed., Pergamon Press, Oxford, p. 103.
9. Timmermann, P.; Verboom, W.; Reinhoudt, D.N. (1996) *Tetrahedron* **52**, 2663.
10. Gutsche, C.D.; Dhawan, B.; Levine, J.A.; No, K.H.; Bauer, L.J. (1983) *Tetrahedron* **39**, 409.
11. Groenen, L.C.; van Loon, J.D.; Verboom, W.; Harkema, S.; Casnati, A.; Ungaro, R.; Pochini, A. Ugozzoli, F.; Reinhoudt, D.N. (1991) *J. Am. Chem. Soc.* **113**, 2385.
12. Groenen, L.C.; Ruël, B.H.M.; Casnati, A.; Timmerman, P.; Verboom, W.; Harkema, S.; Pochini, A.; Ungaro, R.; Reinhoudt, D.N. (1991) *Tetrahedron Lett.* **32**, 2675.
13. Arduini, A.; Casnati, A.; Fabbi, M.; Minari, P; Pochini, A.; Sicuri, A.R. Ungaro, R. (1992) "New Shape the Selective Molecular Recognition from Calixarenes" in "*Supramolecular Chemistry*" Balzani V. and De Cola L., Eds., Kluwer Academic Publishers, Dordrecht, p. 31.
14. Ungaro, R.; Arduini, A.; Casnati, A.; Ori, O.; Pochini, A.; Ugozzoli, F. (1994) "Complexation of Ions and Neutral Molecules by Functionalized Calixarenes" in *Computational Approaches in*

Supramolecular Chemistry, Nato ASI Series C 426, Wipff, G. Ed., Dordrecht, p. 277.

15. Ungaro, R.; Casnati, A.; Fochi, M.; Minari, P.; Fabbi, M.; Reggiani, M.; Pochini, A.; Ugozzoli, F. (1995) "Complexation of Cations and Anions by Functionalized Calix[4]arenes" in *Macrocyclic and Supramolecular Chemistry in Italy*, Savelli F. Ed., Perugia, p. 11.
16. Arduini, A.; Casnati, A.; Fabbi, M.; Minari, P.; Pochini, A.; Sicuri, A.R.; Ungaro, R. (1993) *Supramol. Chem.* **1**, 235.
17. Careri, M.; Casnati, A.; Guarinoni, A.; Mangia, A.; Mori, G.; Pochini, A.; Ungaro, R. (1993) *Anal. Chem.* **65**, 3156.
18. Muzet, N.; Wipff, G.; Casnati, A.; Domiano, L.; Ungaro, R.; Ugozzoli, R. (1996) *J. Chem. Soc., Perkin Trans. 2*, 1065.
19. Sabbatini, N.; Guardigli, M.; Mecati, A.; Balzani, V.; Ungaro, R.; Ghidini, E.; Casnati, A.; Pochini, A. (1990) *J. Chem. Soc., Chem. Commun.*, 878.
20. Casnati, A.; Ghidini, E.; Guardigli, M.; Sabbatini, N.; Ungaro, R. *Italian Patent* n° 42508 A/89, 1st August 1989.
21. Horrocks, W.D.; Sudnick, D.R. (1981) *Acc. Chem. Res.* **14**, 384.
22. Sabbatini, N.; Guardigli, M.; Manet, I.; Ungaro, R.; Casnati, A.; Ziessel, R.; Ulrich, G.; Asfari, Z.; Lehn, J.-M. (1995) *Pure & Appl. Chem.* **67**, 135.
23. Sabbatini, N.; Guardigli, M.; Manet, I.; Ungaro, R.; Casnati, A.; Fischer, C.; Ziessel, R.; Ulrich, G. (1995) *New J. Chem.* **19**, 137.
24. Casnati, A.; Fischer, C.; Guardigli, M.; Isernia, A.; Manet, I.; Sabbatini, N.; Ungaro, R. (1996) *J. Chem. Soc., Perkin Trans. 2*, 395.
25. Sabbatini, N; Casnati, A.; Fischer, C.; Girardini, R.; Guardigli, M.; Manet, I.; Sarti, G.; Ungaro, R. (1996) *Inorg. Chim. Acta.* **252**, 19.
26. Ungaro, R.; Arduini, A.; Casnati, A.; Pochini, A.; Ugozzoli, F. (1996) *Pure & Appl. Chem.* **68**, 1213.
27. Casnati, A.; Pochini, A.; Ungaro, R.; Bocchi, C.; Ugozzoli, F.; Egberink, R.J.M.; Reinhoudt, D.N. (1996) *Chem. Eur. J.* **2**, 436.
28. Ungaro, R.; Casnati, A.; Ugozzoli, F.; Pochini, A.; Dozol, J.-F.; Hill, C.; Rouquette, H. (1994) *Angew. Chem. Int. Ed. Eng.* **33**, 1506-1509. *Angew. Chem.* **106**, 1551.
29. Casnati, A.; Pochini, A.; Ungaro, R.; Ugozzoli, F.; Arnaud, F.; Fanni, S.; Schwing, M..-J.; Egberink, R.J.M.; de Jong, F.; Reinhoudt, D.N. (1995) *J. Am. Chem. Soc.* **117**, 2767.
30. Ugozzoli, F.; Ori, O.; Casnati, A.; Pochini, A.; Ungaro, R.; Reinhoudt, D.N. (1995) *Supramol. Chem.* **5**, 179.
31. Dozol, J.-F.; Rouquette, H.; Ungaro, R.; Casnati, A. *French Patent* n. 93 04566, 19th April 1993.
32. Hill, C.; Dozol, J.-F.; Lamare, V.; Rouquette, H.; Tournois, B.; Vicens, J.; Asfari, Z.; Ungaro,R.; Casnati, A. (1995) *J. Incl. Phenom. Mol. Recogn. Chem.* **19**, 399.
33. Lugtenberg, R.J.W.; Brzozka, Z.; Casnati, A.; Ungaro, R.; Engbersen, J.F.J.; Reinhoudt, D.N. (1995) *Anal. Chim. Acta* **310**, 263.
34. Bocchi, C.; Careri, M.; Casnati, A.; Mori, G. (1995) *Anal. Chem.* **67**, 4234.
35. Arena, G.; Casnati, A.; Contino, A.; Mirone, L.; Sciotto, D.; Ungaro, R. (1996) *J. Chem. Soc., Chem. Commun.* 2277.
36. Andreetti, G.D.; Ugozzoli, F.; Casnati, A.; Ghidini, E.; Pochini, A.; Ungaro, R. (1989) *Gazz. Chim. Ital.* **119**, 47.
37. Dei, L.; Casnati, A.; Lo Nostro, P.; Baglioni, P. (1995) *Langmuir* **11**, 1268.
38. Casnati, A.; Minari, P.; Pochini, A.; Ungaro, R. (1991) *J. Chem. Soc., Chem. Commun.*, 1413.
39. Casnati, A.; Minari, P.; Pochini, A.; Ungaro, R.; Nijenhuis, W.F.; de Jong, F.; Reinhoudt, D.N. (1992) *Isr. J. Chem.* **32**, 79.
40. van Duynhoven, J.P.M.; Janssen, R.G.; Verboom, W.; Franken, S.M.; Casnati, A.; Pochini, A.; Ungaro, R.; de Mendoza, J.; Nieto, P.M.; Prados, P.; Reinhoudt, D.N. (1994) *J. Am. Chem. Soc.* **116**, 5814.
41. Dei, L.; Casnati, A.; Lo Nostro, P.; Pochini, A.; Ungaro, R.; Baglioni, P. (1996) *Langmuir* **12**, 1589.
42. Casnati, A.; Jacopozzi, P.; Pochini, A.; Ugozzoli, F.; Cacciapaglia, R.; Mandolini, L.; Ungaro, R. (1995) *Tetrahedron* **51**, 591.
43. Bianchi, A.; Bowman James, K.; Garcia Espana, E. Eds. (1997), *Supramolecular Chemistry of Anions*, John Wiley and Sons Inc., New York. Schmidtchen, F.P.; Berger, M. (1997) *Chem. Rev.* **97**, 1609. Beer, P.D.; Schmitt, P. (1997) *Curr. Opin. Chem. Biol.* **1**, 475. Antonisse, M.M.G.; Reinhoudt, D.N. (1998) *Chem. Commun.* 443.
44. For recent examples of anion complexation by calixarenes see: (a) Beer, P.D.; Goulden, N.J.; Grieve, A.; Hezek, D.; Szemes, F.; Wear, T. (1994) *J. Chem. Soc., Chem. Commun.* 1269; (b) Scheerder, J.;

Fochi, M.C.; Engbersen, J.F.J.; Reinhoudt, D.N. (1994) *J. Org. Chem.* **59**, 7815; (c) Boerrigter, H.; Grave, L.; Nissink, J.W.M.; Christoffels, L.A.J.; van der Maas, J.H.; Verboom, W.; de Jong, F.; Reinhoudt, D.N. (1998) *J.Org. Chem.* **63**, 4174; (d) Beer, P.D.; Gale, P.A.; Hesek, D. (1995) *Tetrahedron Lett.* **36**, 767; (e) Redman, J.E.; Beer, P.D.; Dent, S.W.; Drew, M.G.B. (1998) *Chem. Commun.* 231; (f) Beer, P.D.; Cooper, J.B. (1998) *Chem. Commun.* 129; (g) Staffilani, M.; Hancock, K.S.B.; Steed, J.W.; Holman, K.T.; Atwood, J.L.; Juneja, R.K.; Burkhalter, R.S. (1997) *J. Am. Chem. Soc.* **119**, 6324.

45. For recent examples of carboxylate complexation see: Cameron, B.R.; Loeb, S. J. (1997) *J. Chem. Soc., Chem. Commun.* 573. Beer, P.D.; Shade, M. (1997) *Gazz.Chim. Ital.* **127**, 651. Hawley, J.C.; Bampos, N.; Abraham, R.J.; Sanders, J.K.M. (1998) *Chem.Commun.* 661. Prohens, R.; Tomas, S.; Morey, J.; Deya, P.M.; Ballester, P.; Costa, A. (1998) *Tetrahedron Lett.* **39**, 1063. Eblinger, F.; Schneider, H.J. (1998) *Angew.Chem. Int. Ed. Eng.* **37**, 826.
46. Erkang, F.; van Armen, S.A.; Kincaid, S.; Hamilton, A.D. (1993) *J. Am. Chem. Soc.* **115**, 369; Haman, B.C.; Brande, N.R.; Rebek J. Jr. (1993) *Tetrahedron Lett.* **34**, 6837.
47. Casnati, A.; Fochi, C.; Minari, P.; Pochini, A.; Reggiani, M.; Ungaro, R.; Reinhoudt, D.N. (1996) *Gazz. Chim. Ital.* **126**, 99.
48. Pelizzi, N.; Casnati, A.; Friggeri, A.; Ungaro, R. (1998) *J.Chem. Soc., Perkin Trans. 2*, 1307.
49. Scheerder, J.; Engbersen, J.F.J.; Casnati, A.; Ungaro, R.; Reinhoudt, D.N. (1995) *J. Org. Chem.* **60**, 6448.
50. Andreetti, G.D.; Ungaro, R.; Pochini, A. (1979) *J. Chem. Soc., Chem. Commun.* 1005.
51. Ungaro, R.; Pochini, A.; Arduini, A. (1990) in *Inclusion Phenomena and Molecular Recognition*, Atwood, J.L. Ed., Plenum Press, New York and London, 135-144.; Vicens, J.; Armah, A.E.; Fujii, S.; Tomita, K.-I. (1990) *J. Incl. Phenom. Mol. Recogn. Chem.* **10**, 159.
52. Andreetti, G.D.; Ori, O.; Ugozzoli, F.; Alfieri, C.; Pochini, A.; Ungaro, R. (1988) *J. Incl. Phenom. Mol. Recogn. Chem.* **6**, 523.
53. Umezawa, Y.; Tsuboyama, S.; Honda, K.; Uzawa, J.; Nishio, M. (1998) *Bull. Chem. Soc. Jpn.* **71**, 1207 and references therein.
54. For indirect evidence in solution see: Yamada, A.; Murase, T.;Kikukawa, K.; Arimura, T.; Shinkai, S. (1991) *J. Chem. Soc. Perkin Trans. 2*, 793.
55. Andreetti, G.D.; Ugozzoli, F.; Pochini, A.; Ungaro, R. (1991) in *Inclusion Compounds*; Atwood, J.L., Davies, J.E., Mac Nicol, D.D., Eds.; Oxford University Press: Oxford; Vol.4.
56. Grootenhuis, P.D.J.; Kollman, P.A.; Groenen, L.C.; Reinhoudt, D.N.; van Hummel, G.J.; Ugozzoli, F.; Andreetti, G.D. (1990) *J. Am. Chem. Soc.* **112**, 4165.
57. Arduini, A.; Fabbi, M.; Mantovani, M.; Mirone, L.; Pochini, A.; Secchi, A.; Ungaro, R. (1995) *J. Org. Chem.* **60**, 1454.
58. Arduini, A.; Casnati, A.; Dodi, L.; Pochini, A.; Ungaro, R. (1990) *J. Chem Soc. Chem. Commun.* 1597.
59. Arduini, A.; Domiano, L.; Pochini, A.; Secchi, A.; Ungaro, R.; Ugozzoli, F.; Struck, O.; Verboom, W.; Reinhoudt, D.N. (1997) *Tetrahedron* **53**, 3767.
60. Pèpe, G.; Astier, J.-P.; Estienne, J.; Bressot, C.; Asfari, Z.; Vicens, J. (1995) *Acta Crystallogr., Sect. C* **51**, 726.
61. Arduini, A.; McGregor, W.M.; Paganuzzi, D.; Pochini, A.; Secchi, A.; Ugozzoli, F.; Ungaro, R. (1996) *J. Chem. Soc., Perkin Trans. 2*, 839.
62. Arduini, A.; Ghidini, E.; Pochini, A.; Ungaro, R.; Andreetti, G.D.; Calestani, L.; Ugozzoli, F. (1988) *J. Incl. Phenom. Mol. Recogn. Chem.* **6**, *119.*
63. Vincenti, M.; Minero, C.; Pellizzetti, E.; Secchi, A.; Dalcanale, E. (1995) *Pure & Appl. Chem.* **67**, 1075.
64. Arduini, A.; Cantoni, M.; Graviani, E.; Pochini, A.; Secchi, A.; Sicuri, A.R.; Ungaro, R.; Vincenti, M. (1995) *Tetrahedron* **51**, 599.
65. Arduni, A.; McGregor, W.M.; Pochini, A.; Secchi, A.; Ugozzoli, F.; Ungaro, R. (1996) *J. Org. Chem.* **61**, 6881.
66. Dalcanale, E.; Soncini, P.; Bacchilega, G.; Ugozzoli, F. (1989) *J. Chem. Soc., Chem. Commun.* 500.
67. Soncini, P.; Bonsignore, S.; Dalcanale, E.; Ugozzoli, F. (1992) *J. Org. Chem.* **57**, 4608.
68. (a) Vincenti, M.; Dalcanale, E.; Soncini, P.; Guglielmetti, G. (1990) *J. Am. Chem. Soc.* **112**, 445; (b) Vincenti, M.; Pelizzetti, E.; Dalcanale, E.; Soncini, P. (1993) *Pure & Appl. Chem.* **65**, 1507; (c) Vincenti, M.; Dalcanale, E. (1995) *J. Chem. Soc Perkin Trans. 2*, 1069; Vincenti, M. (1995) *J. Mass Spectrom.* **30**, 925.
69. Careri, M.; Dalcanale, E.; Mangia, A.; Ruffini, M. (1997) *Anal. Commun.* **34**, 13.

70. (a) Nelli, P.; Dalcanale, E.; Faglia, G.; Sberveglieri, G.; Soncini, P. (1993) *Sens. Actuators B* **14**, 302; (b) Dickert, F.L.; Haunschild, A. (1993) *Adv. Mater.* **5**, 887; (c) Schierbaum, K.D.; Weiss, E.U.; Thoden van Velzen, Engbersen, J.F.J., Reinhoudt, D.N., Göpel, W. (1994) *Science* **265**, 1413.
71. Janata, J. (1989) *Principles of Chemical Sensors*, Plenum Press, New York.
72. Hartmann, J.; Auge, J.; Lucklum, R.; Rösler, S.; Hauptmann, P.; Adler, B.; Dalcanale, E. (1996) *Sens. Actuators B* **34**, 305.
73. Cram, D.J.; Karbach, S.; Kim, H.-E.; Knobler, C.B.; Maverick, E.F.; Ericson, J.L.; Helgeson, R.C. (1988) *J. Am. Chem. Soc.* **110**, 2229.
74. (a) Dalcanale, E.; Hartmann, J. (1995) *Sens. Actuators B* **24**, 39; (b) Hartmann, J.; Hauptmann, P.; Levi, S.; Dalcanale, E. (1996) *Sens. Actuators B* **35**, 154.
75. Dickert, F.L.; Bäumler, U.P.A.; Zwissler, G.K. (1993) *Synth. Met.* **61**, 47.
76. Pinalli, R.; Dalcanale, E.; Ugozzoli, F. unpublished results.
77. Dalcanale, E.; Costantini, G.; Soncini, P. (1992) *J. Incl. Phenom. Mol Recogn. Chem.* **13**, 87.
78. Leo, A.; Hansch, C.; Elkins, D. (1971) *Chem. Rev.* **71**, 525.
79. Waltho, J.P.; Williams, D.H. (1991) "The natural design of vancomycin family antibiotics to bind their target peptides" in "*Host-guest molecular interactions: from chemistry to biology*", Ciba Foundation Symposium *158*, J. Wiley & Sons, Chichester, p. 73.
80. Sansone, F.; Barboso, S.; Casnati, A.; Fabbi, M.; Pochini, A.; Ugozzoli, F.; Ungaro, R. (1998) *Eur. J. Org. Chem.*, 897.
81. Casnati, A.; Fabbi, M.; Pelizzi, N.; Pochini, A.; Sansone, F.; Ungaro, R. *Bioorg. & Med. Chem. Lett.* **1996,** *6*, 2699.
82. Casnati, A.; Favretto, D.; Sansone, F.; Traldi, P.; Ungaro R. unpublished results.
83. Marra, A.; Scherrmann, M.C.; Dondoni, A.; Casnati, A.; Minari, P.; Ungaro, R. (1994) *Angew. Chem. Int. Ed. Eng.* **33**, 2479. *Angew. Chem.* **106**, 2533.
84. Dondoni, A.; Marra, A.; Scherrmann, M.C.; Casnati, A.; Sansone, F.; Ungaro, R. (1997) *Chem. Eur. J.* **3**, 1774.
85. Pelizzi, N.; Casnati, A.; Ungaro, R. (1998) *Chem. Commun.* 2607.
86. Almi, M.; Arduini, A.; Casnati, A.; Pochini, A.; Ungaro, R. (1989) *Tetrahedron* **45**, 2177.
87. Arduini, A.; Pochini, A.; Reverberi, S.; Ungaro, R. (1984) *J. Chem. Soc., Chem. Commun.* 981.
88. Casnati, A.; Ting, Y.; Berti, D.; Fabbi, M.; Pochini, A.; Ungaro, R.; Sciotto, D.; Lombardo, G.G. (1993) *Tetrahedron* **49**, 9815.
89. Arena, G.; Calì, R.; Lombardo, G.G.; Casnati, A.; Rizzarelli, E.; Sciotto, D.; Ungaro, R. (1992) *Supramol. Chem.* **1**, 19. Arena, G.; Bonomo, R.P.; Calì, R.; Gulino, F.G.; Lombardo, G.G.; Sciotto, D.; Ungaro, R.; Casnati, A. (1995) *Supramol. Chem.* **4**, 287.
90. Arena, G.; Casnati, A.; Contino, A.; Lombardo, G.G.; Sciotto, D.; Ungaro, R. (1999) *Chem. Eur. J.* **5**, 738.
91. Arena, G.; Casnati, A.; Contina, A.; Sciotto, D.; Ungaro, R. (1997) *Tetrahedron Lett.* **38**, 4685.
92. Arena, G.; Contino, A.; Gulino, F.B.; Magrì, A.; Sansone, F.; Sciotto, D.; Ungaro, R. submitted for publication.
93. Cram, D.J.; Cram, J.M. (1994) *Container Molecules and their Guests, Monographs in Supramolecular Chemistry,* Stoddart J.F. Ed.; Royal Society of Chemistry, Cambridge, Vol. 4, chapters 7-10.
94. Cram, D.J.; Tanner, M.E.; Thomas, R. (1991) *Angew. Chem. Int. Ed. Engl.* **30**, 1024.
95. Pina, F.; Parola, A.J.; Ferreira, E.; Maestri, M.; Armaroli, N.; Ballardini, R.; Balzani, V. (1995) *J. Phys. Chem.* **99**, 12701.
96. Casnati, A.; Domiano, L.; Pochini, A.; Ungaro, R.; Carramolino, M.; Magrans, J.O.; Nieto, P.N.; Lopez-Prados, J.; Prados, P.; de Mendoza, J.; Janssen, R. G.; Verboom, W.; Reinhoudth, D.N. (1995) *Tetrahedron* **51**, 12699.
97. Arduini, A.; Domiano, L.; Ogliosi, L.; Pochini, A.; Secchi, A.; Ungaro, R. (1997) *J. Org. Chem.* **62**, 7866.
98. Stang, P.J. (1998) *Chem. Eur. J.* **4**, 19.
99. Jacopozzi P.; Dalcanale, E. (1997) *Angew. Chem. Int. Ed. Engl.* **36**, 613.
100. Dalcanale, E.; Rissanen, K. unpublished results.

SIMULATIONS OF LIQUID-LIQUID INTERFACES: A KEY BORDER IN SUPRAMOLECULAR CHEMISTRY

F. BERNY, N. MUZET, L. TROXLER AND G. WIPFF *

UMR 7551, Université Louis Pasteur, Institut de Chimie,
4, rue B. Pascal,
67 000 Strasbourg (France)

Abstract. The question of ion reçognition by macrocyclic hosts, a founding theme of supramolecular chemistry, is revisited, with a particular focus on interfacial phenomena in liquid-liquid ion extraction or transport. Based on molecular dynamics simulations we investigate the behaviour of ions and ionophores at a liquid-liquid interface, represented explicitly. The surface activity of free and complexed ligands of macrocyclic (crown-ethers, cryptands, calixarenes, valinomycin ...) or acyclic type is pointed out. Ions like Pic^-, $B\phi 4^-$, $ClO4^-$, $Guanidinium^+$, $As\phi 4^+$ or $NMe4^+$ are also surface active, although they do not possess, like classical surfactants, a polar head linked to a hydrophobic tail. Beyond the behaviour of monomeric solutes, collective properties are observed when the concentration is increased. Some ionophores like 18-crown-6 are found at the unsaturated interface, in equilibrium with molecules diluted in the organic phase. Others like alkylphosphates or calixarenes remain completely adsorbed at the interface. Some like-ions which are surface active, display a marked tendency to "attract each other" and to stack in the water phase. The formation of the interface upon demixing of "homogeneous" binary water / "oil" mixtures containing species involved in the assisted ion extraction process has also been simulated. The results have implications concerning the mechanism of ion recognition at the interface, of assisted ion transfer, interfacial electrochemistry, and supramolecular organization induced by the asymmetrical nature of the liquid-liquid interface.

Keywords: liquid-liquid extraction; interface; ionophore; calixarene; counterion; solvation; molecular dynamics; immiscible liquids; phase separation; recognition; organization.

R. Ungaro and E. Dalcanale (eds.), Supramolecular Science: Where It Is and Where It Is Going, 95–124.

I- INTRODUCTION

The question of ion recognition by macrocyclic molecules is a founding theme of supramolecular chemistry [1-3]. In the early seventies, as the first crown ethers, or cryptands were designed, computational chemistry was still at the "prehistoric age" of punched cards and listing outputs with myriads of numbers to be deciphered. Quantum chemists developed the "supermolecule" approach to investigate small systems involving non-covalent interactions (hydrogen bonds with water, interactions with small ions, like Li^+, Mg^{2+} or H^+)[4-7] to mimick key interactions in host-guest systems. In parallel appeared the first simulations on simple solvents (Lennard Jones particles, water), and molecular mechanics studies on organic molecules [8], proteins, nucleic acids, etc ....

Since, enormous progress has been achieved, as illustrated by the presentations given at the NATO ARW "Computational Approaches in Supramolecular Chemistry", held in 1994 [9]. Today, thanks to the computer technology and sofware developments, the energy and stereochemistry of non-covalent interactions can be determined "routinely" with high quality quantum mechanical (QM) methods. Full geometry optimization can be performed *ab initio* on "real systems" (*e.g.* cation complexes of 18-crown-6 [10-12], or actinide complexes[13, 14], etc ....). The results can be compared with gas phase data [15], and serve as a reference for force field methods [16] used for fast calculations on large systems moving in complex environments.

The solvation properties of host-guest systems have been simulated by molecular dynamics or Monte Carlo methods [17-21]. The main issues concern the structure and dynamics of the solvent (mostly water) around the solute, the question of shielding of the complexed guest by the host, the status of the uncomplexed ions, the solvent effect on preorganization and conformational state of the host, and the dynamic features of host-guest complementarity. The many simulations on 18-crown-6, a popular test molecule for computational methodologies and concepts in supramolecular chemistry, illustrate the continuous progress achieved in the last two decades: see *e.g.* molecular mechanics calculations in the gas phase on selected structures [22], conformational sampling in the gas phase [23] and in solution [24], solvation properties [25-28], ion recognition in water [29] and in non-aqueous solvents [30-32], energy profiles for ion capture in solution [33, 34], simulation on the pure liquid [35], etc ...

In this paper, we focus on a problem of inherent complexity: the interfacial behaviour of species involved in the assisted ion transfer from an aqueous to an organic phase. In liquid-liquid extraction experiments, the source phase containing the hydrophilic ions is mixed with an organic solution of extractant molecules (hereafter noted **L**). The nature of the mixture is ill-defined, as it may range from a perfectly mixed homogeneous system to droplets or microemulsions of the two liquids. Then, phase separation takes place, due to gravitational forces or centrifugation, and differences in surface tension between the two liquids. In transport experiments, the ions are transported by carrier molecules **L** through an organic liquid phase or an organic membrane, from the aqueous source phase to the aqueous receiving one [36-38]. The precise mechanism of ion capture, extraction and release is still unclear but it is stressed that the interfacial region between water and the organic solvent plays a crucial role [39-41].

The precise nature of the interface between immiscible liquids is expected to depend on the way the experiment is carried out. In the case of static interfaces, indirect information comes from interfacial tension measurements in the presence of extractants [42-44]. For instance, the interfacial tension of bi-phasic systems decreases when the concentration of organophosphorus molecules [45, 46], carbonic acids and ethers or macrocyclic compounds [41, 47] is increased. This suggests that these ionophores adsorb and accumulate like surfactants at the interface. At the microscopic level, adsorption of extractants supposedly facilitates the cation capture, because their hydrophilic binding sites likely point to the water phase, *i.e.* to the approaching cation. From a phenomenological point of view, the thermodynamics of adsorption is described by the Gibbs's adsorption theorem and related Langmuir isotherms [48], widely used in solvent extraction, phase transfer catalysis [49, 50], or electrochemistry [51-53], though without providing microscopic pictures of the interface. Physical surface sensitive optical methods such as "second harmonic generation" [54], resonance laser Raman spectroscopy and reflectance spectroscopy [55] provide insights into structural aspects, but data are still rather limited, especially in relation with the question of ion extraction. Indirect information also comes from extraction kinetics models, where the adsorption / desorption of extractants and their complexes at the interface is explicitely included [40, 56]. Here, the systems are not static but agitated. Generally speaking, the extent of adsorption depends on the conformation dependent hydrophilic / hydrophobic balance of the ionophore, related to its charge, size, topology, nature and orientation of binding sites, flexibility, which are difficult to assess.

Computational chemistry may contribute to understand what happens at the liquid-liquid interface in the assisted ion extraction. There have been simulations of neat liquid-liquid interfaces [57] and on liquid-liquid or liquid-air interfaces with ionic [58-61], or neutral solutes [62-68]. Recent accounts on microscopic views of chemical reactions and solvation at liquid interfaces can be found in ref. [60, 69]. Simulations involving extractant molecules have been reported in ref. [70-82]. Some results are summarized here, together with new data, obtained from molecular dynamics (MD) simulations.

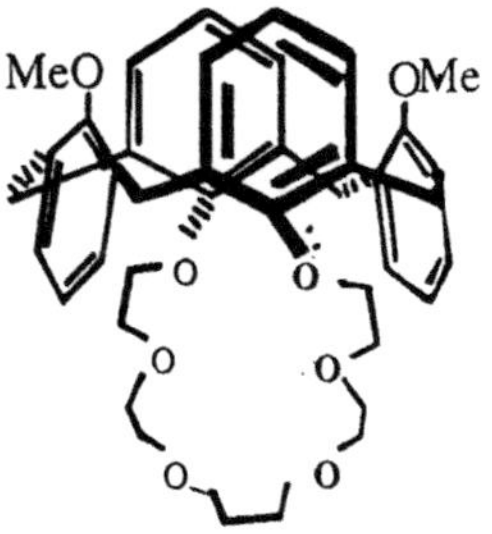

CalixC6

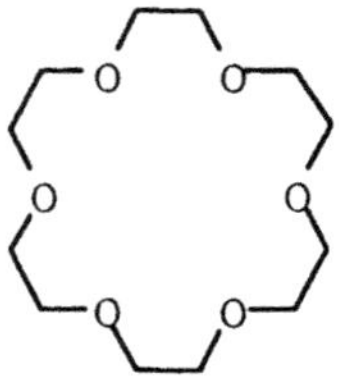

18C6

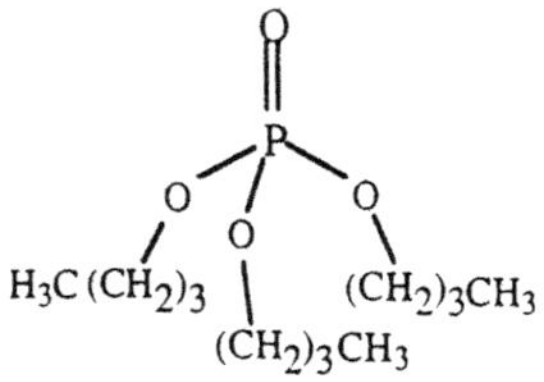

TBP

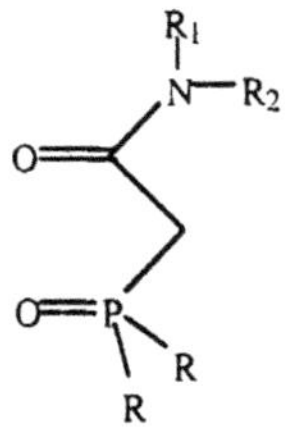

CMPO

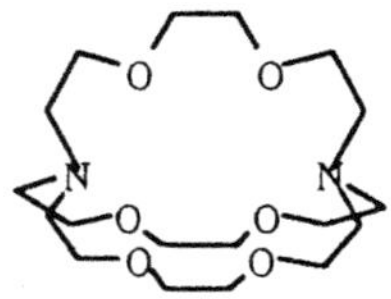

222

K^+ Cl^-
K^+ ClO_4^-
NMe_4^+ Cl^-
$NTMA^+$ Cl^-
Gu^+ Cl^-
K^+ Pic^-
Cs^+ Pic^-

Chart 1: Examples of simulated extractant molecules and salts.

We first compare different uncomplexed salts at the interface, with ions of different shapes and hydrophilicities, in concentration of about 0.4 to 0.04 mol/l (Table 1): K^+Cl^-, $K^+ClO_4^-$, the $NMe_4^+Cl^-$ and $NTMA^+Cl^-$ salts of quaternary ammonium cations, and the Guanidinium$^+Cl^-$ and K^+Picrate$^-$ salts. Some are simulated at the interface, as well as in pure water and chloroform phases for comparison. Then, we consider monomers of extractant molecules of macrocyclic type (calixarenes,18-crown-6, the *222* cryptand) or of acyclic type (TBP) (Chart 1), which are shown to display a high interfacial activity. Next, the effect of ligand concentration is illustrated, in relation with the possible formation of "monolayers" or aggregates at the interface. Finally, we report "computer demixing experiments" where the evolution of binary water - chloroform mixtures which contain species involved in the assisted ion tranfer, is followed as a function of time. As organic liquid, we chose chloroform, used experimentally in extraction processes. On the computational side, potentials are available for this liquid, which is simpler to simulate than large solvent molecules. The results suggest that the nature of the organic solvent is not critical for a number of properties, because the interfacial behaviour of ions and ionophores is mostly driven by the liquid with the highest surface tension (here, water).

II- METHODS

1- Energy representation of the system. The potential energy U of the system is calculated via the computerized "ball and stick representation" (force field model). We use the modified AMBER4.1 software [83] with the following representation of U:

$$U = \Sigma_{bonds} K_r (r-r_{eq})^2 + \Sigma_{angles} K_\theta (\theta-\theta_{eq})^2 + \Sigma_{dihedrals}\Sigma_n V_n(1+\cos n\phi)$$
$$+ \Sigma_{i<j} (q_iq_j/R_{ij} - 2\varepsilon_{ij}(R_{ij}^*/R_{ij})^6 + \varepsilon_{ij}(R_{ij}^*/R_{ij})^{12})$$

The bonds and bond angles are treated as harmonic springs and a torsional term is associated to the dihedral angles. The non-covalent interactions between atoms separated by at least three bonds, and those involving the ions are described by pairwise additive 1-6-12 contributions, without explicit term for polarization or many body effects. Most parameters are taken from the AMBER force field [84] and from previous studies in pure homogeneous solvents [81, 82]. The parameters of alkali [85], and of halide ions have been fitted to reproduce their hydration energies. The atomic charges on ClO_4^-, Pic^-, $Guanidinium^+$ have been fitted from ESP potentials. The solvents are represented explicitly, using the TIP3P model for water [86] and the OPLS model for chloroform [87]. In order to reduce the computational costs, the non-bonded interactions are calculated within a residue based cutoff of 12Å in most cases.

In the MD simulations, the temperature is controlled by coupling to a thermal bath with a relaxation time of 0.2 ps. All C-H, O-H, H···H, C-Cl and Cl···Cl "bonds" are constrained with SHAKE, using a time step of 1 fs.

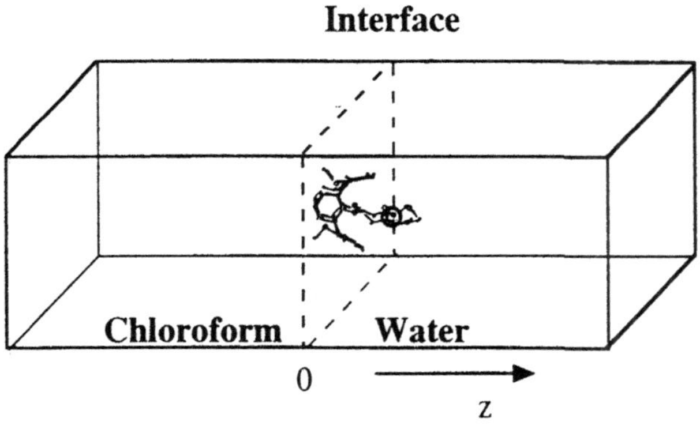

Chart 2: Representation of the simulation box.

2- The solvent system. The solvent system consists of two adjacent cubic boxes of pure liquids (Chart 2 and Table 1). After immersion of the solute near the interface, each system is first energy minimized and equilibrated at 300K under a constant pressure of 1 atm, to ensure that both liquid phase reach densities close to their experimental values (of 1.0 and 1.49, respectively) "in the bulk region". Most simulations use periodic boundary conditions in the three (3D) dimensions, where the system corresponds to alternating slabs of water and of chloroform. In other simulations (2D) periodicity is applied along the x and y directions only, while the Z-edges of the box correspond to a liquid - air interface, where solvent evaporation is prevented by using a restraining potential (see Table 1 and details in ref [76]).

Solute	Box Size (Å^3)	Solvent $N_{chl} + N_{wat}$	2D / 3D	(Mol/l)	Time (ns)
$(K^+ Cl^-)_{20}$	34*32*(36+36)	279 + 1237	2D	0.43	1
$(K^+ ClO_4^-)_{10}$	38*49*(28+27)	375 + 1575	2D	0.16	1
$(NMe_4^+ Cl^-)_{10}$	30*41*(42+42)	372 + 1237	2D	0.16	1
$(NTMA^+ Cl^-)_{10}$	52*42*(39+32)	581 + 2208	2D	0.11	1
$(Gu^+ Cl^-)_{10}$	49*43*(32+33)	497 + 2256	2D	0.12	1
$(K^+ Pic^-)_{10}$	45*44*(41+32)	581 + 2039	2D	0.11	1
$222K^+Pic^-$	23*33*(45+24)	252 + 808	3D	0.03	0.85
$(\mathit{18C6})_{10}$	33*33*(44+27)	251 + 808	3D	0.21	1
$(\mathit{18C6}.K^+Pic^-)_6$	33*33*(42+33)	325 + 1088	3D	0.12	1
$(\mathit{18C6})_6/(K^+Pic^-)_6$	34*34*(42+27)	262 + 923	3D	0.12	1
222	25*24*(34+27)	140 + 538	3D	0.04	0.6
$(\mathit{CalixC6}\cdot Cs^+Pic^-)_6$	48*44*(34+35)	494 + 2252	3D	0.10	1
"Mixture-1" [a]	39*43*(37+37)	446 + 1846	3D	0.04 / 0.11 [b]	1
$(TBP)_{30}$	40*40*(35+32)	363 + 1469	3D	0.46	1
$(TBP)_{30}\,[UO_2(NO_3)_2]_5$	40*40*(35+32)	357 + 1459	3D	0.46 / 0.08 [c]	5.1

Table 1: Simulation conditions at the interface: Box size $V_x*V_y*(V_{z\text{-chlor}}+V_{z\text{-wat}})$. See Chart 2. Number of chloroform and water molecules. Periodicity (2D / 3D) of the simulated system. Concentration of the solute (mol/l) and simulated time (ns). [a] *Mixture-1* contains two free *CalixC6*, one *CalixC6*.Cs^+Pic^- complex, $4(Cs^+ Pic^-)$, $4(Cs^+ NO_3^-)$ ions as solute. [b] *CalixC6* / Cs^+ concentrations. [c] TBP / UO_2^{2+} concentrations.

III - RESULTS

The interface between the aqueous and the organic liquids is more or less flat, of about 10 Å wide. Its precise position, defined as the intersection between the density curves of the two solvents, fluctuates during the dynamics [70]. In the following, we mostly focus on the location and solvent environment of the solutes. Because of the dynamics nature of the simulated systems, it is not easy to depict shortly their evolution during the simulations. Some remain close to their initial position. For others, major changes may take place. They are illustrated by selected snapshots or cumulated views extracted from the trajectories.

1- Simulation of M^+ X^- salts at the interface: some ions are "repelled", others are "attracted" by the interface. Like-ions may "attract each other".

A series of MD experiments starts with a grid of $(M^+ X^-)_n$ salts placed right at the interface, equally shared between the two solvents (Figure 1). Rapidly, the salts dissolve as expected in the water phase. However, the ions display different distributions and proximity with the interface, depending on their hydrophilic / hydrophobic balance.

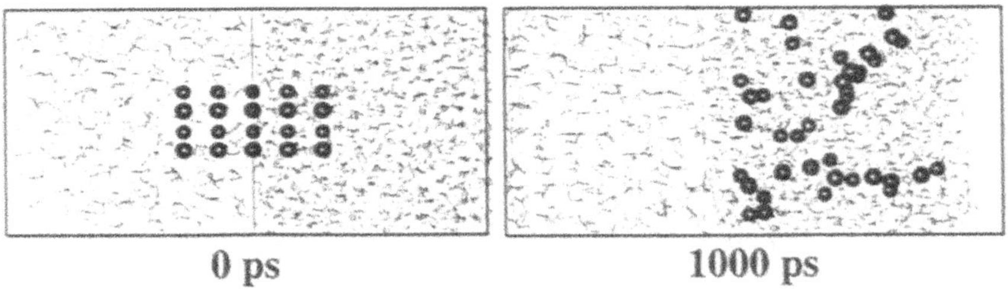

Figure 1: The $(K^+ Cl^-)_{20}$ salt at 0 ps, and after 1 ns of MD.

1.1 The $(K^+Cl^-)_{20}$, $(K^+ClO_4^-)_{10}$, $(NMe_4^+Cl^-)_{10}$ and $(K^+Pic^-)_{10}$ salts. It is generally considered that small cations are more repulsed than anions of similar size by the interface [69, 89]. Simulation of the $(K^+Cl^-)_{20}$ salt indicates that all ions are spread in the water slab, none of them being close to the interface (Figure 1). The density profiles (Figure 2) are almost identical for cations and anions: they peak at the center of the water slab and are almost zero close the interface. Contrary to expectations, there is thus no marked difference in interfacial behaviour of Cl^-, compared to K^+. On the other hand, a comparison of F^- to I^- halides showed that their lifetime at the interface increases with their size [80].

In the case of $(K^+ClO_4^-)_{10}$, $(K^+Pic^-)_{10}$ and $(NMe_4^+Cl^-)_{10}$ salts, the ions distribution differs as, compared to K^+Cl^-, these salts contain more lipophilic anions (ClO_4^-; Pic^-) or cations (NMe_4^+). Most ions sit in water but, near the interface, there is a significant fraction of ClO_4^- or Pic^- anions or of NMe_4^+ cations (see cumulated positions and density curves in Figure 2). We calculated the relative population P_{itf}^X of X ions found within 7 Å from the interface during the dynamics (i.e. up to a distance where the water density becomes close to 1.0). For the anions, $P_{itf}^{ClO_4^-}$ is 31 ± 9 % and $P_{itf}^{Pic^-}$ is 42 ± 9 % , while for the Cl^- ions in the K^+Cl^- salt, $P_{itf}^{Cl^-}$ is 7 ± 6 %). Similarly, comparison of the K^+Cl^- and $NMe_4^+Cl^-$ solutions shows that the NMe_4^+ cations are more concentrated than K^+ near the interface ($P_{itf}^{NMe_4^+}$ = 17 ± 6 % and $P_{itf}^{K^+}$ = 7 ± 6 %).

1.2 The $(Guanidinium^+Cl^-)_{10}$ salt at the interface and in the pure liquids: Guanidinium+ cations are surface active and tend to stack in the aqueous phase.

Two independent simulations with $(Gu^+ Cl^-)_{10}$ as solute ("Gu^+" = Guanidinium$^+$, $C(NH_2)_3^+$) reveal that these cations display a noticeable concentration near the interface (P_{itf}^{Gu+}= 40 ± 1 %), in addition to the center of the water slab (Figure 2). They exchange between these two regions.

The graphical animation of the two trajectories reveals interesting relationships between Gu^+ cations, which exchange between π-stacking arrangements (Figure 3), predicted to correspond to a local energy minimum in water [90, 91]. Near the interface or in bulk water, we notice in addition the stabilizing role of Cl^- anions around transient stacks of two to four Gu^+ ions. Such a proximity is never observed in the simulations with K^+ or NMe_4^+ ions.

The Gu^+ cation displays analogies with the picrate "Pic^-" anion: both are flat, possess two hydrophobic faces, and a hydrophilic periphery. According to simulations, the Pic^- ions are also surface active [77], and display a marked tendency to self-stack in water [92]. The $(Pic^-)_2$ stacked dimer is unstable in acetonitrile solution, be counterions present, or not [92]. Similarly, the $(Gu^+)_2$ dimer which stacks in water rapidly dissociates in acetonitrile [93].

The $(Gu^+ Cl^-)_{10}$ salt also displays contrasted behaviours in pure water versus pure chloroform solutions. In the organic solvent, the Gu^+ and Cl^- ions rapidly collapse to an aggregate of "molten salt" type (Figure 3), analogous to the one observed for the $(Pic^-K^+)_{27}$ salt in acetonitrile [92]. In pure water, the salt slowly dissolves "randomly", where some Cl^- ions are completely solvated, while others are loosely paired with Gu^+ cations, which form transient stacks of two to four ions (Figure 3). The average radius of gyration R_{gyr} of the solute is a simple index to measure its dilution. In chloroform solution, $R_{gyr(Cl-)}$ = $R_{gyr(Gu+)}$ = 5±1Å) whereas in water $R_{gyr(Cl-)}$ is 14±2Å and $R_{gyr(Gu+)}$ is 13±1Å. Thus

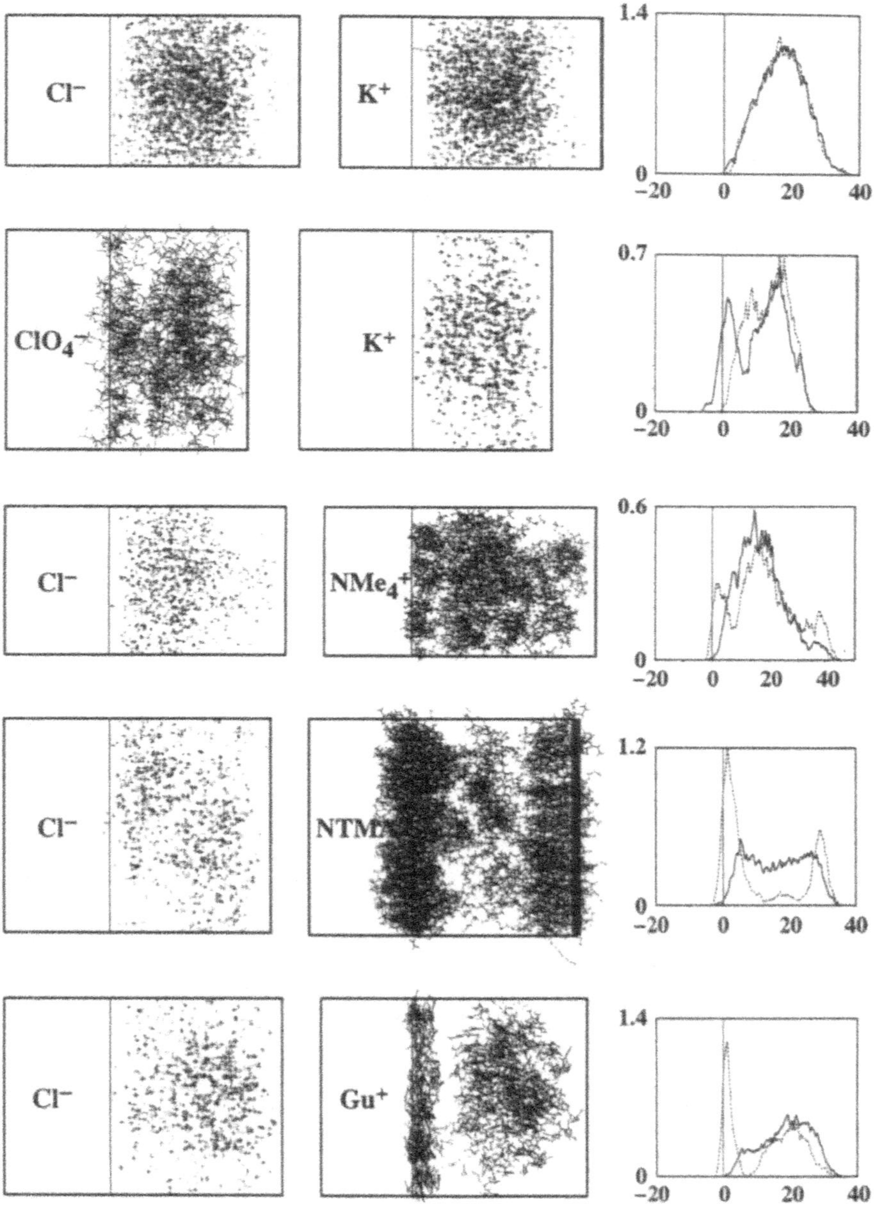

Figure 2: The $(K^+Cl^-)_{20}$, $(K^+ ClO_4^-)_{10}$, $(NMe_4^+ Cl^-)_{10}$, $(NTMA^+ Cl^-)_{10}$ and $(Gu^+ Cl^-)_{10}$ salts at the interface. Cumulated views of the anions (left) and of the cations (middle), between 0.2 ns and 1 ns. Solvents are not shown. Right: density curves of the cations (dotted line) and anions (full line) as a function of z (see Chart 2).

the comparison of the pure solvent with the interfacial behaviour of $Gu^{+}Cl^{-}$ clearly demonstrates the *crucial role of solvent environment on the effectiveness of ionic interactions, on the status of ion pairing, and the remarkable property of water to induce self aggregation and ordering (stacking) of like-ions, which may be viewed as the primarily process of crystal nucleation.*

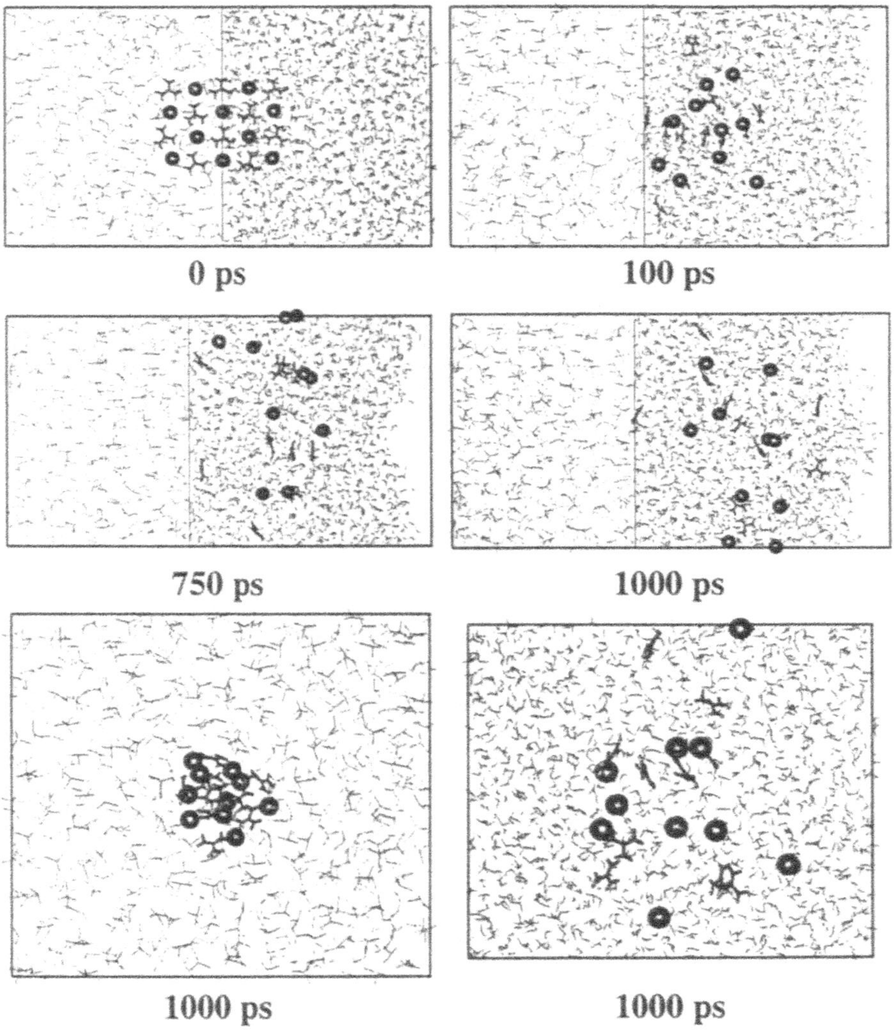

Figure 3: The $(Gu^{+} Cl^{-})_{10}$ salt. ***Top***: at the interface (at 0, 100, 750 and 1000 ps). ***Bottom***: in chloroform solution (left) and pure water (right) after 1 ns.

1.3 The $(NTMA^+Cl^-)_{10}$ salt at the interface and in pure solvents. Salts of quaternary ammonium cations have been used in extraction processes, in relation with the high interfacial activity of these cations [94]. They have been simulated in assemblies to model monolayers, micelles, or as single ions [95-101] at the water interface.

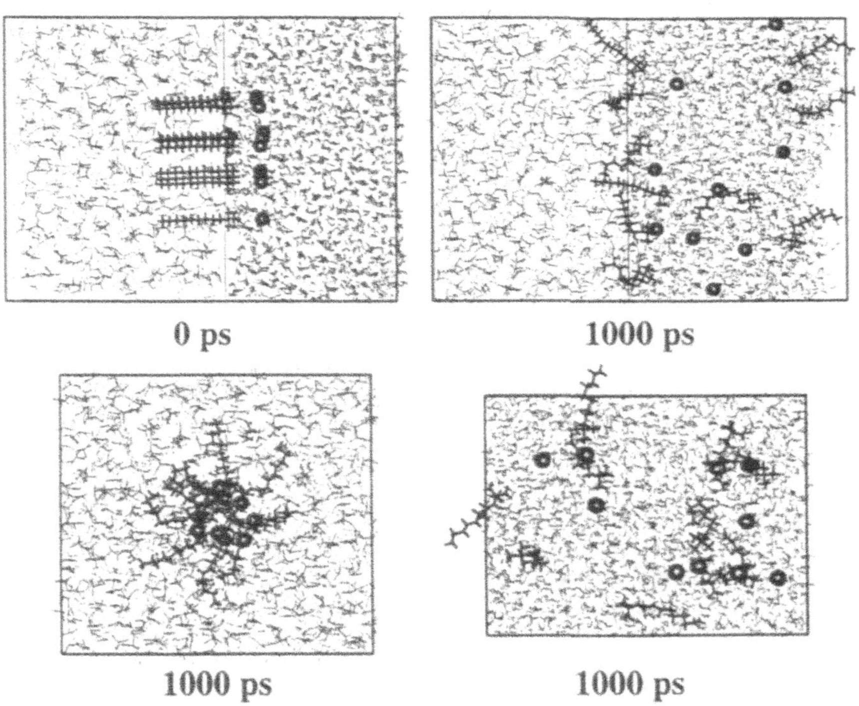

Figure 4: The $(NTMA^+ Cl^-)_{10}$ salt..***Top***: at the interface at 0 and 1 ns.
Bottom : in chloroform (left) and in water (right) after 1 ns.

For a purpose of comparison with the above salts, we have simulated $(NTMA^+Cl^-)_{10}$ at the water / chloroform interface, and in the pure water and chloroform phases, respectively. The $NTMA^+$ (Nonyl-trimethylammonium, $(H_3C)_3NC_9H_{19}^+$) amphiphilic cations have been modeled with atomic charges which amplify their amphiphilic character, i.e. with an "ammonium head" of +1 charge, flanked by a neutral alkyl C_8H_{17} tail. As for the $(Gu^+Cl^-)_{10}$ system, the outcome of the three simulations is completely different. In pure chloroform, the system rapidly forms an "inverted micelle", with an ionic core, and alkyl chains pointing outwards to chloroform (Figure 4). In water, the ion pairs completely dissociate and the ions are diluted in the simulation box (Figure 4). As a result, the average radii of gyration are larger in water ($R_{gyr}(Cl^-)$ = 17±2 Å and $R_{gyr}(N^+)$ = 18±1 Å) than in pure chloroform ($R_{gyr}(Cl^-) = R_{gyr}(N^+)$ = 7±1 Å). One also notices transient pairing of $NMTA^+$ cations, whose ammonium head is close to Cl^- anions.

At the interface, the system organizes differently. All anions are immersed in the aqueous phase, while the cations are at, or close to the two interfaces, which become positively charged. Thus, during the simulation, some cations diffuse through the organic phase (from the liquid-liquid interface to the water / air interface), presumably to reduce their mutual electrostatic repulsions. Although Cl^- and $NTMA^+$ never form intimate ion pairs, they display some attractions, as evidenced by the cumulated positions or density curves of Cl^- anions of $NTMA^+Cl^-$, compared to those of K^+Cl^- (Figure 2).

2- Single ionophores, free or complexed, adsorb at the interface.

Our first simulations involving interfaces concerned ionophores **L**, free or complexed by cations [70]. As these molecules are not soluble in water, but are soluble in organic liquids or membranes, it was anticipated that, when placed initially on the border between the two liquids, they would diffuse to chloroform. This turned out not to be the case. Whatever the nature or conformation of **L**, it adsorbed on the chloroform side of the interface, anchored via the polar moities solvated by a few water molecules ("water fingers"[59]). Some of the ligands studied (calixarene derivatives [72, 81, 82], podants [77], CMPO's [71], alkyl phosphates [78], ..) are amphiphilic and possess polar+hydrophobic moities. Others (valinomycin [102], *18C6* [77], cryptand *222* [74], calix-*bis*-crown6 [73]) of nearly zero dipole moment, are also surface active, in relation with specific hydration features at the interface. Their surface activity has been confirmed by several computational tests (initial state, solvent representation, thermodynamic conditions, etc ..)[82].

Complexed ionophores also display a strong affinity for the interface. For instance the *222*K^+Pic^- cryptate [74] or the *calixC6*·Cs^+Pic^- complex [82], initially immersed in chloroform, diffuse to the interface and adsorb on the chloroform side (see Figure 5). Migration to the interface seems to be initiated by the anionic charge, or the cation - anion dipole, interacting with the water dipoles. Adsorption of other $\mathbf{LM^+}$ alkali complexes (**L** = calix[4]arene$^-$ anion [72], calix[4]arene-tetraamide [70], *18C6* [77], valinomycin [102], *calixC6* [77], calix[4]arene-*bis*-crown6 [73], podants [74]) has also been reported, in the absence as well as in the presence of counterions. Some counterions like Pic^- or BPh_4^- adsorb at the interface [80], while more hydrophilic ones like Cl^- or NO_3^- do not.

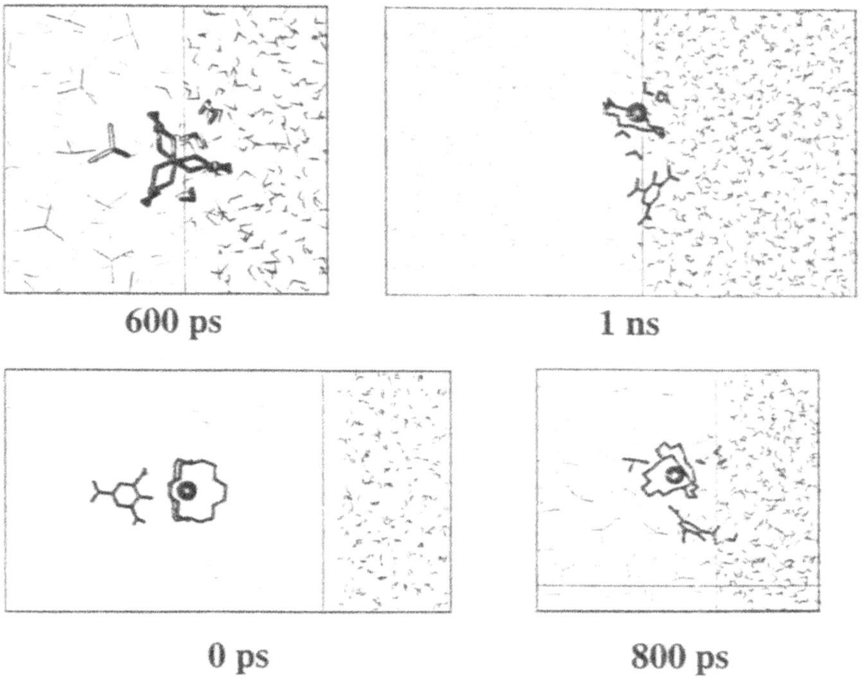

Figure 5: *Top:* The *222* cryptand after 600 ps and the *18C6*K^+Pic^- complex after 1 ns. *Bottom*: The 222K^+Pic^- cryptate at 0 ps (in chloroform) and after 800 ps.

More quantitative insights into the energetic features of interface crossing have been obtained by "PMF" (Potential of Mean Force) calculations corresponding to the departure of ions and of ionophores from the interface to the water and organic phases, respectively [76]. These free energy profiles fully support trends observed in the free MD simulations.

3- Assemblies of ionophores at the water / chloroform interface: from monolayers to diluted aggregates. Implications for the assisted ion extraction process.

The high affinity of free or complexed ionophores for the interface, and the fact that they are more soluble in organic solvents than in water, led us to simulate more concentrated solutions, to test whether spontaneous migration to the organic phase would take place. The concentration was increased more or less arbitrarely, without performing systematic studies. Interestingly, different situations were obtained, depending on the nature of the solute.

3.1 Assemblies of uncomplexed ionophores at the interface.

TBP (see Chart 1), a typical phosphoryl containing extractant molecule for lanthanides and actinides, adsorbs at the water interface [45]. Simulations on the decamer $(TBP)_{10}$ starting with different arrangements converged to a similar situation where the TBP molecules form an irregular unsaturated monolayer at the interface, pointing their Op atoms towards water, "ready to capture" an approaching cation. Details are given in ref.[78]. Thus, although TBP is more soluble in chloroform than in water, it does not migrate from the interface to chloroform (Figure 6). A similar accumulation of pyridine-containing TPTZ molecules has been reported earlier [80].

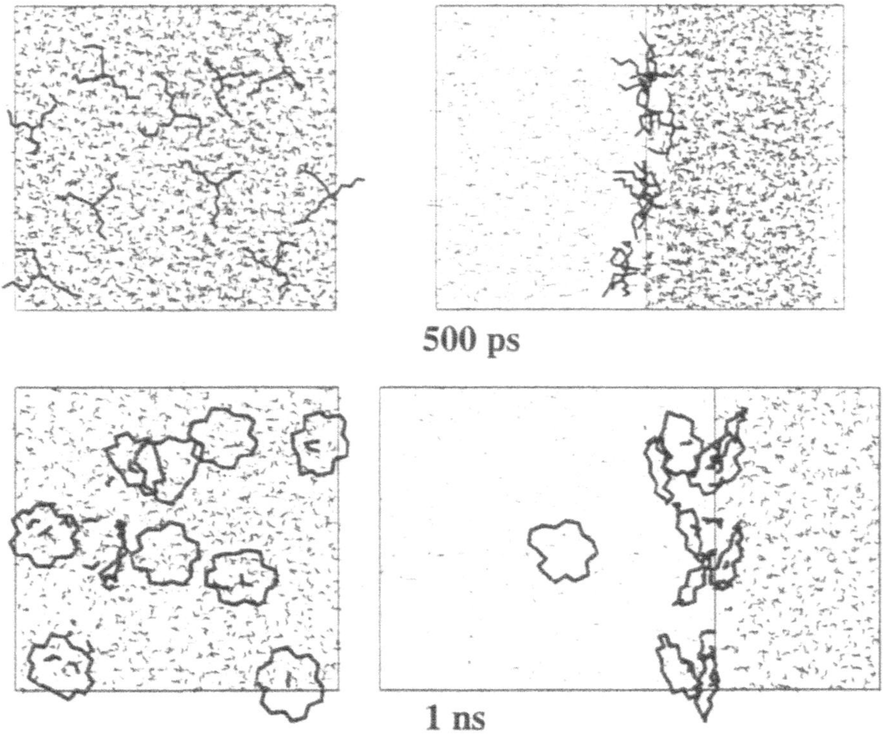

Figure 6: Snapshot of the $(TBP)_{10}$ layer (500 ps) and of the "diluted" $(18C6)_{10}$ aggregate (1 ns) at the interface. Orthogonal views.

A different behaviour has been found with *18C6* as solute [77]. A simulation starting with the $(18C6)_{10}$ decamer right at the interface leads, after 1 ns, to a complex situation (Figure 6). Most of the crowns remain at the interface, hydrogen bonded to water molecules. Others have moved to chloroform, changing their conformation from D_{3d} to C_i, accompanied by a few water molecules ("water dragging effect" [104,105]). The concentration of *18C6* is thus higher at the interface than in the bulk organic phase, without however forming a monolayer. It is noticeable that the *18C6* molecules at the interface retain their D_{3d} form, of zero dipole moment, due to specific hydrogen bonding patterns [25]. Thus, as pointed out above, the *surface activity does not necessarily relate to the asymmetry of the solute.* It is markedly determined by *microsolvation* (*i. e.* formation of "supermolecules" involving some firmly bound solvent molecules) and by the cavitation energies [106] of the two adjacent liquids.

An important concept in host-guest recognition chemistry concerns the preorganization for guest binding [107]. Interestingly, *at the interface, the macrocyclic hosts are preorganized for ion capture*: the D_{3d} form of *18C6*, like the "K" form of cryptand *222* (D_3 symmetry), which are more surface active than the other conformers, display the same conformation as in their complexes with this K^+, the cation which they selectively extract.

3.2 Assemblies of complexed ionophores at the interface: towards the extraction process ?

Simulations show that *increasing the concentration of the ionophores modifies their interfacial behaviour.* Indeed, a monomeric $(18C6K^+Pic^-)_1$ complex, simulated for 1 ns, remains of inclusive type, and anchored at the interface (Figure 5). This contrasts with a simulation on $(18C6K^+Pic^-)_6$ which also starts at the interface, but leads to an equilibrium between some *18C6*K^+ complexes at the interface, some having migrated to chloroform, and others being decomplexed (Figure 7). An important result, in relation with the assisted ion transfer, concerns the *extraction of a 18C6K*$^+$ *complex to chloroform*, seemingly driven by a Pic^- anion (Figure 7). This complex, surrounded by one H_2O molecule at each face, is "dragged" by a Pic^- anion, hydrogen bonded with H_2O. *Counterions thus seem to play a crucial role for assisted cation extraction.* We notice that none of them sits in water, while some K^+ decomplexed cations do. The anions adsorbed at the interface create a negative potential which likely "pushes" some of them to the organic phase.

The $(calixC6Cs^+Pic^-)_9$ system, simulated in similar conditions as $(18C6K^+Pic^-)_6$, behaves differently. All complexes remain of inclusive type and adsorbed with the Pic^- anions at the interface, without migrating to the organic phase (Figure 7). This finding is consistent with the experimental characterization of monolayers of calixarene derivatives at the water/air interface [108-111]. We notice that the *calixC6* molecules adopt multiple orientations with respect to the the interface, and display π-stacking interactions with some Pic^- ions. Whether their concentration is still too low to allow for the Cs^+ extraction to chloroform [112] remains to be further investigated.

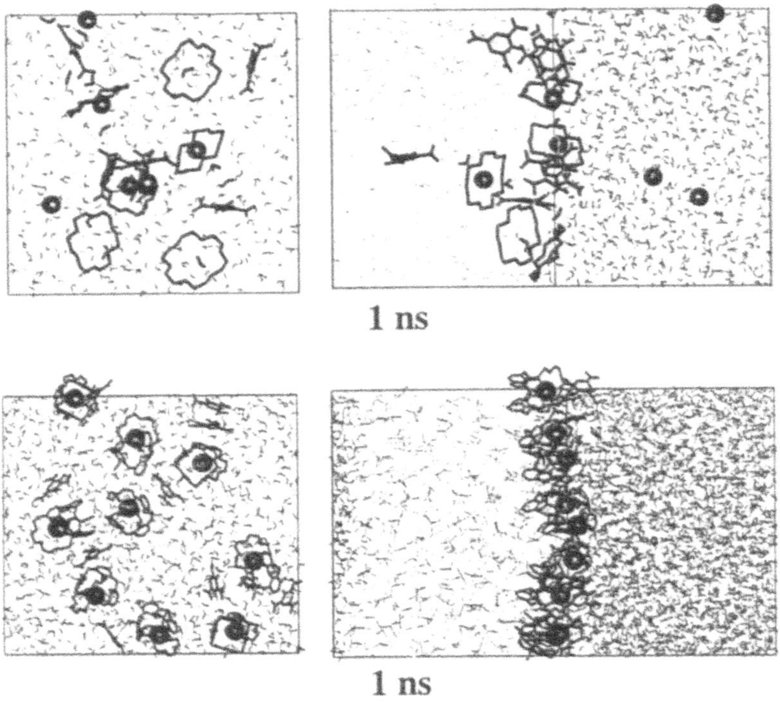

Figure 7: ***Top***: the $(18C6K^+Pic^-)_6$ aggregate, after 1 ns (the simulation started with the solute right at the interface) [77]. ***Bottom***: $(\mathit{calixC6}Cs^+\ Pic^-)_9$ after 1 ns at the water / chloroform interface.

4- From "chaos" to "order". Demixing of binary solutions containing salts or ionophores: phase separation and formation of the liquid-liquid interface.

4.1 Demixing of binary solutions of salts. The above simulations started with the two liquid phases separated by the preformed interface. In relation with the liquid-liquid extraction experiment, the demixing of "perfectly mixed" binary water - chloroform solutions of ions and ionophores has been simulated. Details are given in ref. [113], with free salts (Cs^+Pic^-, $Cs^+NO_3^-$) or complexes of *calixC6* as solutes. Here, we describe solutions containing some of the salts considered above: $(K^+\ Cl^-)_{20}$, $(K^+\ ClO_4^-)_{10}$, $(NMe_4^+\ Cl^-)_{10}$, $(NTMA^+\ Cl^-)_{10}$. It is indeed important to assess to which extent the final situation depends on the initial state, and whether the rate and extent of demixing are determined by the nature of the solute. Figures 8 and 9 show that in all cases, phase separation is achieved in about 1 ns, or less, leading to the two liquid phases separated by a more or less flat interface. Very

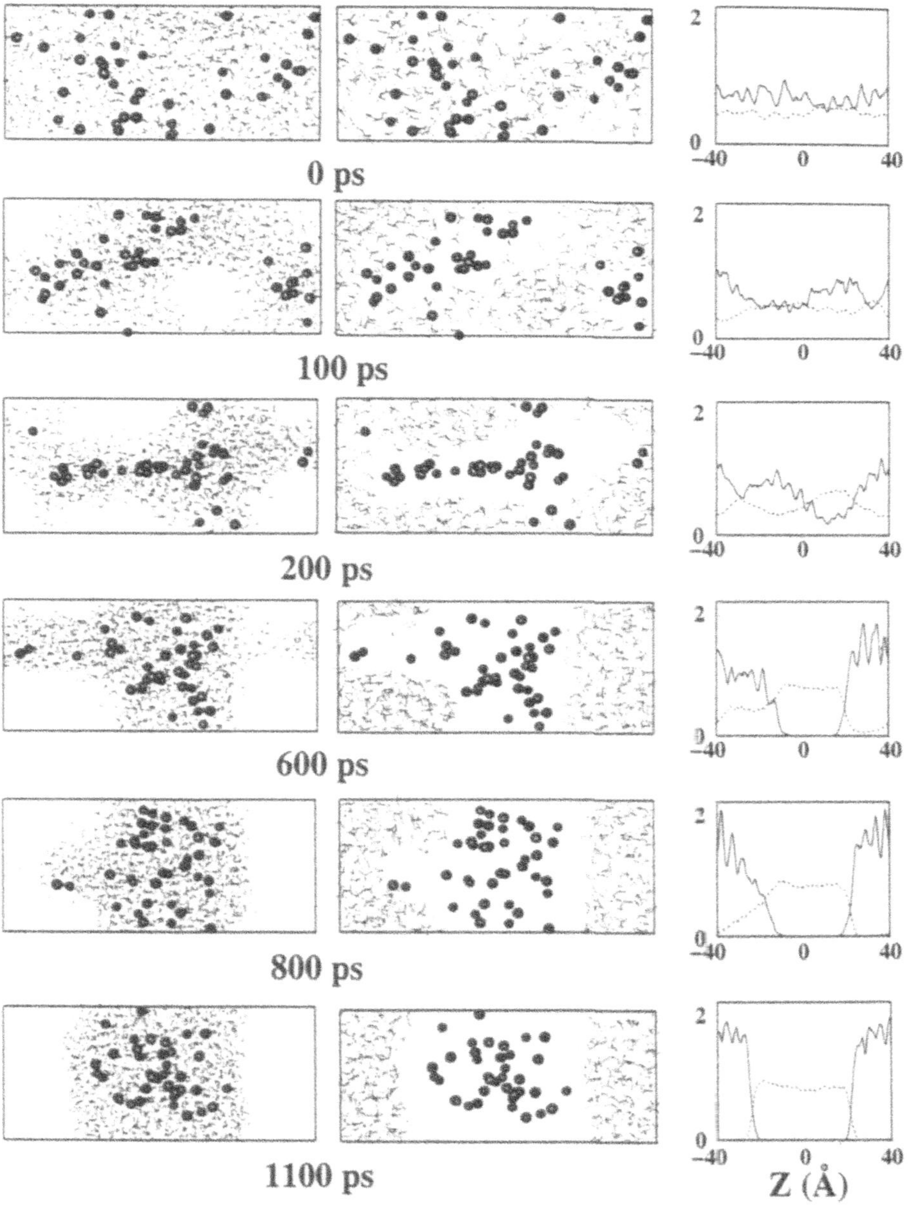

Figure 8: Demixing of a binary water / chloroform mixture containing the $(K^+ Cl^-)_{20}$ salt as solute. Snapshots at 0 ps, 100, 200, 600, 800 and 1100 ps. For clarity, water (*left*) and chloroform (*middle*) are displayed separately. ***Right:*** densities of water (dotted line) and chloroform (solid line) as a function of Z (defined in Chart 2).

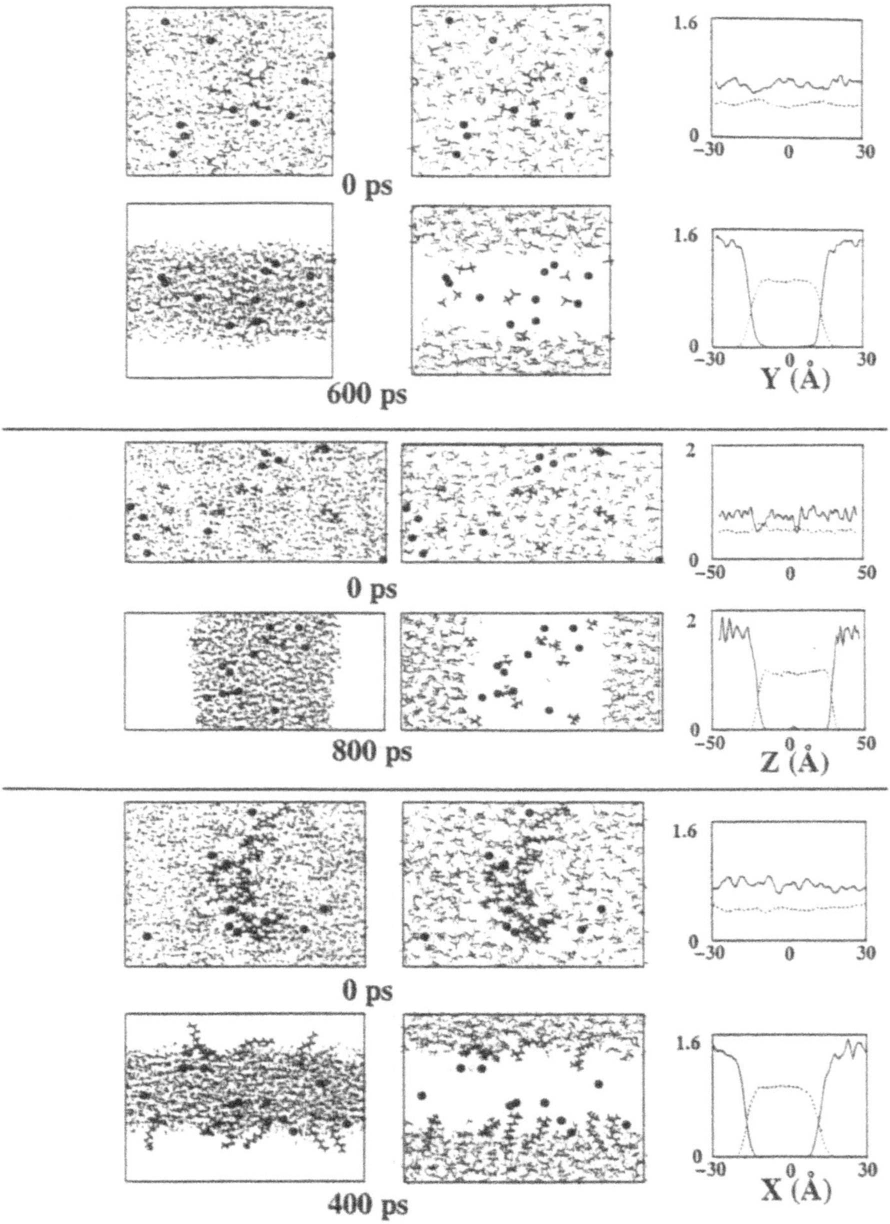

Figure 9: Demixing of binary water / chloroform mixtures containing respectively $(K^+ ClO_4^-)_{10}$ at 0 and 600 ps, $(NMe_4^+ Cl^-)_{10}$ at 0 and 800 ps, $(NTMA^+ Cl^-)_{10}$ at 0 and 400 ps. See Figure 8 for comments.

early (in less than 100 ps), water aggregates around the ions, while chloroform forms "pockets" in the ion-free water area (Figure 8). The most hydrophilic ions (Cl^-, K^+) sit far from the border which gradually forms between the two liquids, while the less hydrophilic ones (NMe_4^+, ClO_4^-) sit in water closer to this border along the demixing process, as well as in the completely demixed state. Concerning the cations ($NTMA^+$), their hydrophilic heads are rapidly surrounded by water and their hydrophobic tails by chloroform molecules. Then these cations seem to catalyze the formation of the water / chloroform interface. After demixing, the organic phase is dry, in agreement with the low water solubility in chloroform.

4.2 Demixing of binary solutions of salts and ionophores. Another "demixing experiment" is illustrated in Figure 10, where the solution contains a mixture of species involved in the extraction of Cs^+ by *calixC6*, in the presence of different counterions: $(Cs^+Pic^-)_4$ and $(Cs^+NO_3^-)_4$ salts, $(calixC6)_2$ uncomplexed and one *calixC6*·Cs^+ complex. The initial state (Figure 10) corresponds to "random" dilution of the solutes. Rapidly, the Cs^+ and NO_3^- ions are surrounded by water, while the amphiphilic *calixC6*, *calixC6*·Cs^+ and Pic^- species remain at the border region between the two phases. As expected, *calixC6* and *calixC6*Cs^+ diffuse more slowly than the solvent molecules. The *calixC6*·Cs^+ complex, more surface active than *calixC6*, adsorbs at the interface earlier than the free ligands which remain for about 500 ps in the chloroform phase, together with a few water molecules (Figure 10). At about 650 ps, the liquids are separated, but the interfaces are not flat: even at 1 ns, adsorption of *calixC6* and *calixC6*·Cs^+ leads to water protuberances. *CalixC6* is anchored at about 6 Å from the interface either by its crown moitey, or by its lower rim, via a relay of hydrogen bonded water molecules ("water finger" [59]). We notice that the latter situation is unfavorable for cation capture, while the first one would easily allow for cation capture via a least motion pathway. There is thus no spontaneous complexation of Cs^+ by *calixC6* free, nor extraction of the *calixC6*·Cs^+ complex to the organic phase. We again notice the surface activity of Pic^- anions which attract some cations close to the interface.

The final situation is nearly identical to the one obtained from simulations which started with the two adjacent liquid phases [70, 76, 81], indicating that the simulated adsorption is not an artefact caused by the choice of the initial state and corresponds to a thermodynamic equilibrium. Concerning the mechanism of ion extraction, these results suggest that *ion capture and recognition take place at the interface* of microdroplets which form on the nanosecond timescale. In reality, the solvents may be less homogeneously mixed than they are in the simulations and display a large interfacial area.

The "driving forces" for phase separation have been analyzed in typical systems [113]. The rate of demixing decays in an exponential mode, and is weakly dependent of the solute. This is because, at the concentrations simulated, the changes in potential energy upon demixing mostly come from the solvent - solute interactions, which are much smaller than those corresponding to the solute - solvent interactions[113] .

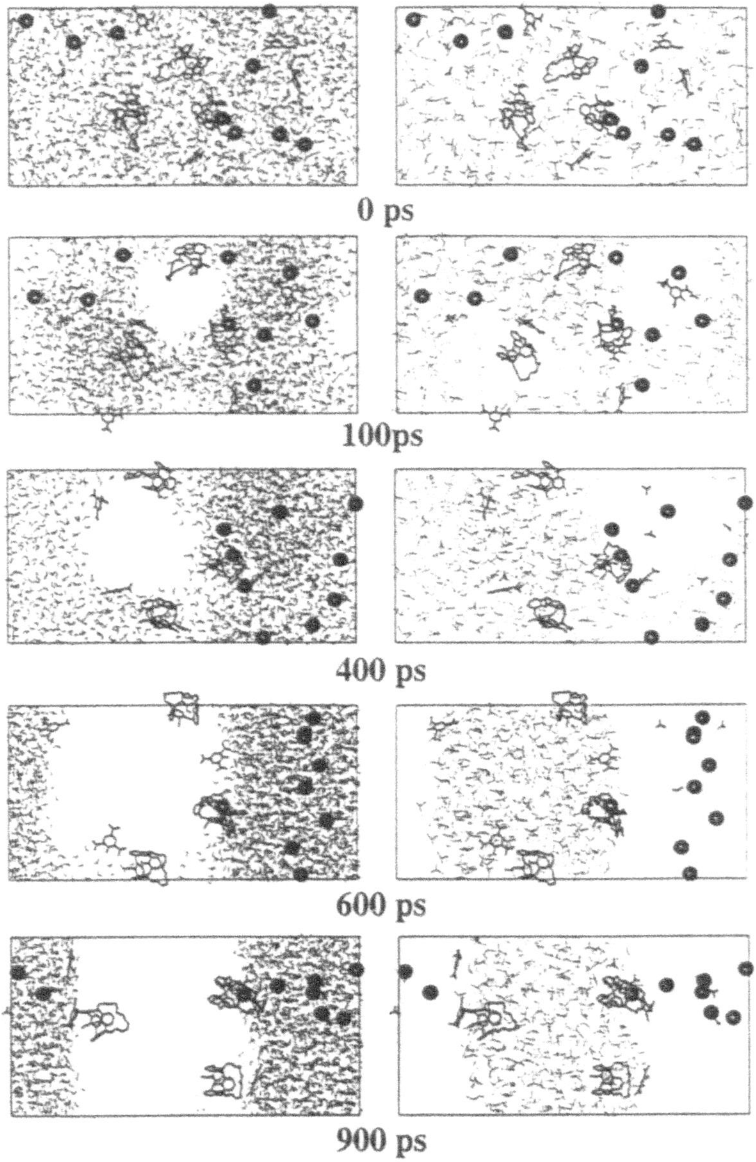

Figure 10: Demixing of a binary water / chloroform mixture containing species involved in the extraction of Cs^+ by *calixC6*: the *CalixC6*.Cs^+Pic^- complex, 2 *CalixC6*, 4(Cs^+ Pic^-), 4(Cs^+ NO_3^-) ions as solute. Snapshots at 0 ps, 100, 400, 600 and 900 ps. For clarity, we display separately the water (*left*) and chloroform (*right*) solvents.

4.3 Demixing of a ternary solution: water / chloroform / TBP. Assisted extraction of the $UO_2(NO_3)_2$ salt. The demixing of more concentrated solutions has also been simulated, to investigate what happens beyond the first layer formed at the interface. A first "computer demixing experiment" of 1 ns has been performed with a chloroform / water binary mixture containing 30 TBP molecules as solute. In relation with the PUREX treatment of nuclear wastes [103,114], a second simulation of 5.1 ns has been repeated with $[UO_2(NO_3)_2]_5$ as solute. Indeed, in this process, a nitric acid solution of the cations is contacted with a solution of TBP in kerosene into which $UO_2(NO_3)_2(TBP)_2$ is extracted [114, 115]. It was thus challenging to test whether such complexes form and are extracted to the organic phase during the demixing simulation.

Demixing with $(TBP)_{30}$ as solute is slower than with purely ionic solutions, as expected from the surfactant behaviour of TBP. For instance, at 200 ps, the water and chloroform liquids are less separated with $(TBP)_{30}$ than with $(K^+Cl^-)_{20}$ (compare Figures 8 and 11). At 500 ps, chloroform forms a pocket, still containing TBP molecules, while others sit at the periphery between the two liquids. At 700 ps, the two liquids are separated, but still linked via TBP molecules. At the end (1 ns), most of the TBP's are spread on the two water / chloroform interfaces, without however forming regular monolayers. This pattern is presumably favoured, compared with formation of a single tick layer, as it decreases the interfacial tension. The TBP's which are more remote from water are disordered, tangled with the first layer, some of them being hydrogen bonded to water molecules. These patterns markedly differ from those found with a monolayer of $(TBP)_{10}$ which is more ordered and flat (Figure 6).

Demixing of the solution containing $(TBP)_{30}.[UO_2(NO_3)_2]_5$ as solute has been achieved with constraints on the U···N distances, to avoid dissociation of $UO_2(NO_3)_2$ salts. Water aggregates form more rapidly than above, because of the "catalytic role" of the $UO_2(NO_3)_2$ salt (compare snapshots at 200 ps in Figures 11 and 12). At 500 ps, the two liquids are almost separated and the TBP's sit at the interface which is "horizontal". This is however metastable, and the whole system reorganizes (between 1 and 2 ns) to form two "vertical interfaces" of smaller interfacial area. No major changes occur with the solvents phases between 2 and 5 ns (Figure 12). However, an important event concerns the *spontaneous formation of TBP complexes with $UO_2(NO_3)_2$*. At 1 ns, three uranyles are complexed with one or two TBP's. From 2 to 5 ns, two uranyles are free, and three form complexes which evolve from a 2:1 to 3:1 stoichiometry. After 5 ns, a layer of TBP molecules is adsorbed at one interface (left hand side; Figure 12), while on the water side of the other interface (right hand side) one finds three $UO_2(NO_3)_2(TBP)_3$ complexes and two $UO_2(NO_3)_2$ complexes near the center of the water slab. *There is thus spontaneous complexation of uranyle by TBP.*

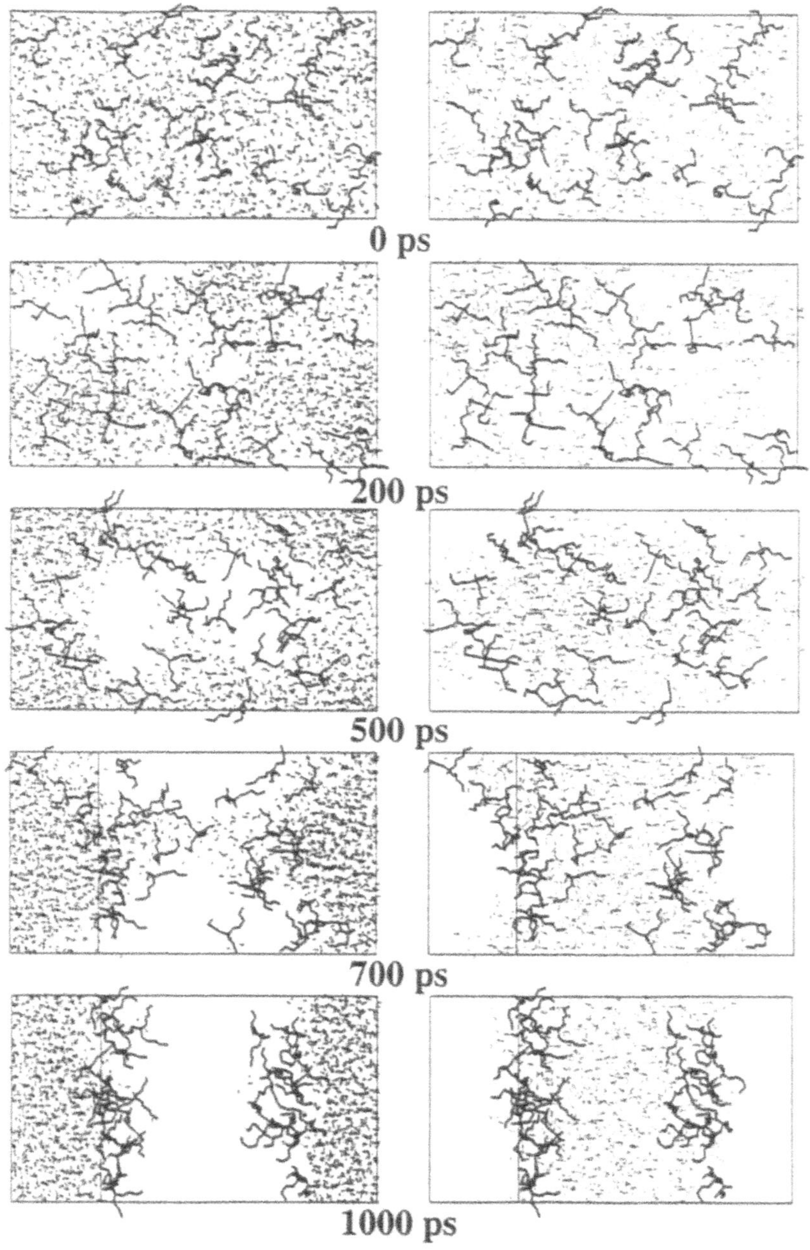

Figure 11: Demixing of a binary water / chloroform mixture containing $(TBP)_{30}$ as solute. Snapshots at 0 ps, 200, 500, 700 and 1000 ps. For clarity, the water (*left*) and chloroform (*middle*) solvents are displayed separately.

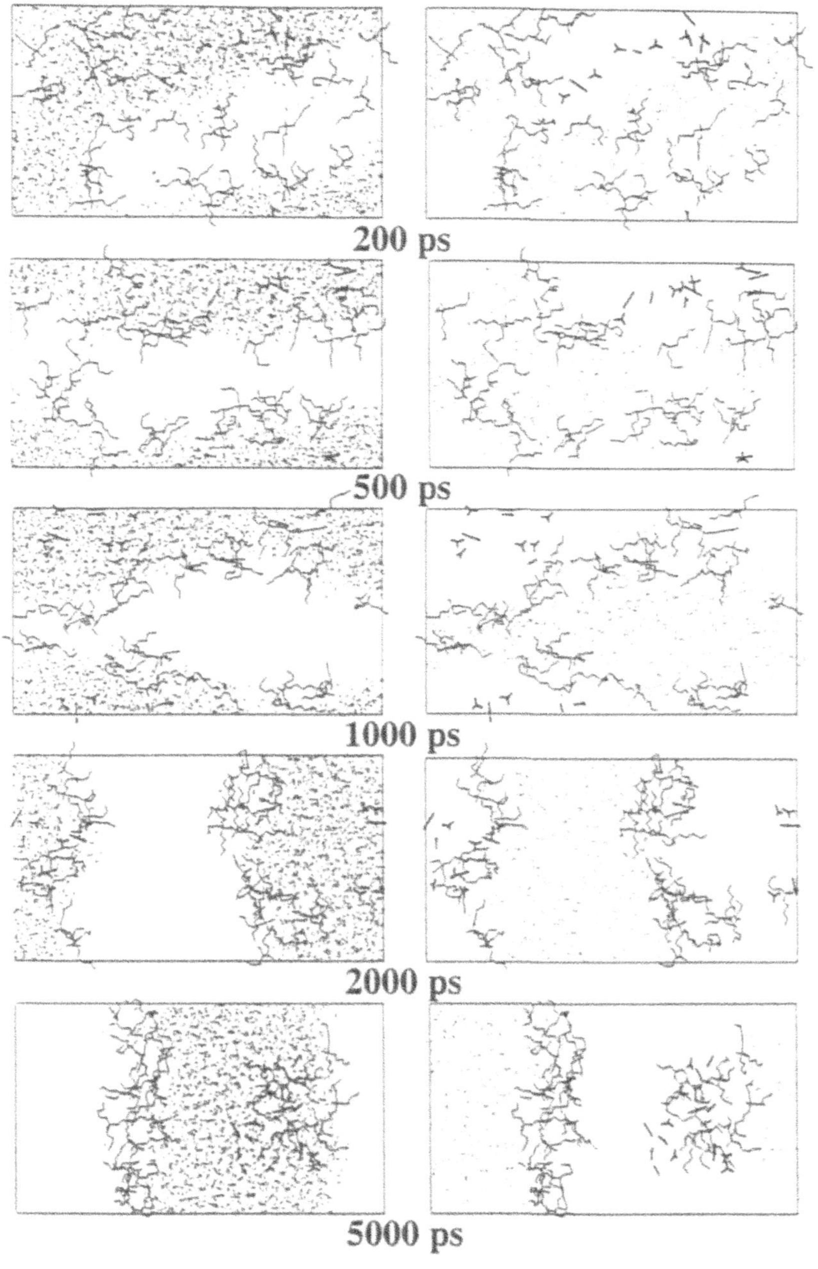

Figure 12: Demixing of a binary water / chloroform mixture containing $(TBP)_{30}$.$[UO_2(NO_3)_2]_5$ as solute. Snapshots at 0 ps, 200, 500, 1000, 2000 and 5000 ps. For clarity, the water (*left*) and chloroform (*middle*) solvents are displayed separately.

The complexes formed, however, are too hydrophilic to leave the aqueous phase, in contrast to what was observed with the other macrocylic complexes. A close examination reveals that this may be related to deficiences in the force field representation of the $UO_2(NO_3)_2$ salts, as the structure of $UO_2(NO_3)_2(TBP)_2$ is far from the D_{2h} one, more hydrophobic, characterized in the solid state [116].

5- Molecular recognition at the interface

Molecular recognition is a central theme of supramolecular chemistry and biology [1]. As far as the mechanism of ion extraction or transport is concerned, it becomes clear that the surface of water, where extractant molecules accumulate, becomes a key border where ions can be captured. On the computational side, it is possible to address the question of ion recognition at the interface, using free energy perturbation calculations [117]. This is described in ref. [113], for the Na^+ / Cs^+ discrimination by the *calixC6* ligand ("**L**"). For this purpose, we defined a thermodynamic cycle

$$\begin{array}{ccc} M_1^+(Pic^-)_{aq} + L_{itf} & \xrightarrow{\Delta G_1} & LM_1^+(Pic^-)_{itf} \\ \Delta G_{3\_aq} \downarrow & & \downarrow \Delta G_{4\_itf} \\ M_2^+(Pic^-)_{aq} + L_{itf} & \xrightarrow[\Delta G_2]{} & LM_2^+(Pic^-)_{itf} \end{array}$$

where the free M^+ cations sit in water and their complexes sit *at the interface* "itf". According to this cycle, the ion recognition at the interface, measured by $\Delta\Delta G = \Delta G_1 - \Delta G_2$, is equal to $\Delta G_{3\text{-}aq} - \Delta G_{4\text{-}itf}$, obtained by the "alchemical route" [118]. $\Delta G_{4\text{-}itf}$ has been calculated first with a Pic^- counterion close to LM^+ at the interface, and then without counterion. In all cases, Na^+ is found to interact best with **L**. However, taking into account the difference in dehydration energies $\Delta G_{3\text{-}aq}$ of the ions leads to positive values of $\Delta\Delta G$, i.e. to the *selective binding of Cs^+ at the interface*. The calculated order of binding affinities at the interface ($Na^+ < K^+ < Rb^+ < Cs^+$) is the same as the experimental order of free energies of extraction by **L** from water to chloroform [112]. This finding is fully consistent with related data on ion recognition by monolayers of calixarenes analogues [108-111] and with the K^+ uptake by valinomycin at the air - water interface [119].

CONCLUSION

We have presented molecular dynamics experiments related to the question of assisted ion extraction, a complex recognition process which results from the interplay between non-covalent interactions between hosts, guests, accompanying ions, and two immiscible solvents. As discussed previously [70-82], the microscopic results obtained at the aqueous interface are in qualitative agreement with conclusions drawn from "macroscopic experiments".

Physical questions concerning the size and complexity of the simulated systems, concentration, pH, salting out and synergistic effects, are challenging topics for future simulations. The computational results deserve words of caution concerning the energy representation of the system and the treatment of short range and long range electrostatics, especially when dealing with highly charged solutes. The precise interactions between unlike solvents have to be accounted for more precisely, which may require explicit account of polarization effects [120]. The behaviour at longer timescales has to be explored.

It becomes clear however, that the problem of ion extraction, revisited from the point of view of the interfacial phenomena, raises the question of *supramolecular organization and recognition at the interface*, actual themes in the context of early beginnings of life [120]. The simulations have also bearing on the question of crystal nucleation, i.e. solvent induced ordering of charged and neutral species. Like amphiphilic ions, the free and complexed extractant molecules are *concentrated* and *ordered* at the interface. The resulting organization at the interface, different from the one in pure solvents (and *a fortiori* in the solid state) may relate to special functions like ion recognition, signal transduction, or photophysical processes. The liquid interface is thus not only a selective border for ions and ionophores. It is a unique, inherently asymmetrical, "supramolecular meeting area", where the interplay between non covalent interactions leads to *order* and *function* [121]. Computer simulations like those performed on membranes or membrane mimetics [64, 95, 96, 99, 120-130], micelles [96, 131-136], solvent mixtures [137-142], should play an increasing role to provide dynamical microscopic models, whose status may range from stimulating mental pictures, to representations of "a complex reality".

Acknowledgements. The authors are grateful to CNRS IDRIS for allocation of computer time, to PRACTIS for support, and to E. Engler for software developments. LT and GW thank EEC (F14W-CT0022 contract) for a grant. FB and NM thank the French Ministry of Research for a grant.

References

(1) J.-M. Lehn, *Supramolecular Chemistry. Concepts and Perspectives*; **1995**, Ed.; VCH: Weinheim.
(2) D. J. Cram, *Science* **1988**, *240*, 760-767.
(3) C. J. Pedersen, *J. Am. Chem. Soc.* **1967**, *89*, 7017.
(4) *The World of Quantum Chemistry*; **1973**, R. Daudel B. Pullman Ed.; Reidel: Dordrecht.
(5) E. Clementi *Determination of Liquid Water Structure, Coordination Numbers for Ions and Solvation for Biological Molecules.*; **1976**, Ed.; Springer Verlag, Berlin.
(6) P. Kollman, *J. Am. Chem. Soc.* **1977**, *99*, 4875.
(7) K. Morokuma, *Acc. Chem. Res.* **1977**, *10*, 294.
(8) N. L. Allinger, *J. Am. Chem. Soc.* **1977**, *99*, 8127.
(9) *Computational Approaches in Supramolecular Chemistry*; **1994**, G. Wipff Ed.; Kluwer Acad. Pub.: Dordrecht.
(10) D. Feller, *J. Phys. Chem. A* **1997**, *101*, 2723-2731.
(11) E. D. Glendening; D. Feller; M. A. Thompson, *J. Am. Chem. Soc.* **1994**, *116*, 10657-10669.
(12) M. A. Thompson; E. D. Glendening; D. Feller, *J. Am. Chem. Soc.* **1994**, *116*, 10465-10476.
(13) U. Cosentino; G. Moro; D. Pitea; A. Villa; P. C. Fantucci; A. Maiocchi; F. Uggerin, *J. Phys. Chem. A* **1998**, *102*, 4606-4614. W. Liu, M. Dolg, P. Fulde, *Inorg. Chem.* **1998**, *37*, 1067.
(14) F. Hutschka; L. Troxler; A. Dedieu; G. Wipff, *J. Phys. Chem. A* **1998**, *102*, 3773-3781.
(15) I.-H. Chu; H. Zhang; D. V. Dearden, *J. Am. Chem. Soc.* **1993**, *115*, 5736-5744.
(16) B. P. Hay; L. Yang; J.-H. Lii; N. L. Allinger, *J. Mol. Struct. (THEOCHEM)* **1998**, *428*, 203-219.
(17) M. P. Allen; D. J. Tildesley *Computer Simulation of Liquids.*; **1987**, W. F. van Gunsteren P. K. Weiner Ed.; Clarendon press: Oxford.
(18) W. F. van Gunsteren; H. J. C. Berendesen, *Angew. Chem. Int. Ed. Engl.* **1990**, 992-1023.
(19) M. Karplus; J. Wiorkiewiczkuzera; A. D. Mackerell, *J. Am. Chem. Soc.* **1995**, *117*, 11946-11975.
(20) P. A. Kollman, *Acc. Chem. Res.* **1996**, *29*, 461-469.
(21) M. L. Lam; V. L. Jorgensen, *Current Opinion in Chemical Biology* **1997**, *1*, 449-457.
(22) G. Wipff; P. Weiner; P. A. Kollman, *J. Am. Chem. Soc.* **1982**, *104*, 3249-3258.
(23) Y. Sun; P. A. Kollman, *J. Comput. Chem.* **1992**, *12*, 33-40.
(24) T. P. Straatsma; J. A. McCammon, *J. Chem. Phys.* **1989**, *91*, 3631-3637.
(25) G. Ranghino; S. Romano; J. M. Lehn; G. Wipff, *J. Am. Chem. Soc.* **1985**, *107*, 7873-7877.
(26) L. Troxler; G. Wipff, *J. Am. Chem. Soc.* **1994**, *116*, 1468-1480.
(27) Y. L. Ha; A. K. Chakraborty, *J. Phys. Chem.* **1994**, *98*, 11193-11203.
(28) T. Kowall; A. Geiger, *J. Phys. Chem.* **1994**, *98*, 6216-6224.
(29) J. V. van Eerden; S. Harkema; D. Feil, *J. Phys. Chem.* **1988**, *92*, 5076-5079.
(30) M. H. Mazor; J. A. Mc Cammon; T. P. Lybrand, *J. Am. Chem. Soc.* **1989**, *111*, 55-56.
(31) Y. L. Ha; A. K. Chakraborty, *J. Phys. Chem.* **1993**, *97*, 11291-11299.
(32) P. A. Kollman; Y. X. Sun, *J. Am. Chem. Soc.* **1995**, *117*, 3599-3604.
(33) L. X. Dang, *J. Am. Chem. Soc.* **1995**, *117*, 6954-6960.
(34) T. J. Marrone; D. S. Hartsough; K. M. Merz, *J. Phys. Chem.* **1994**, *98*, 1341-1343.
(35) F. T. H. Leuwerink; W. J. Briels, *J. Phys. Chem. B* **1997**, *10*.

(36) S. R. Izatt. R. T. Hawkins; J. J. Christensen, R. M. Izatt, *J. Am. Chem. Soc.* **1985**, *107*, 63-66.
(37) J. P. Behr; M. Kirch; J. M. Lehn, *J. Am. Chem. Soc.* **1985**, *107*, 241-246.
(38) H. C. Visser; D. N. Reinhoudt; F. de Jong, *Chem. Soc. Rev.* **1994**, 75-81.
(39) P. R. Danesi, *Separation Science Technology* **1985**, *19*, 857-894.
(40) P. R. Danesi; R. Chirizia; C. F. Coleman, "The Kinetics of Metal Solvent Extraction" in *Critical Reviews in Analytical Chemistry*; **1980**, B. Campbell Ed.; CRC Press, Boca Raton, Florida pp 1.
(41) B. A. Moyer, "Basic Principles of Extraction and Liquid-Liquid Systems Employing Crown Ethers and Related Metal-Ion Receptors" in *Molecular Recognition: Receptors for Cationic Guests*; **1996**, J. L. Atwood; J. E. D. Davies; D. D. McNicol; F. Vögtle J.-M. Lehn Ed.; Pergamon, pp 325-365.
(42) P. R. Danesi, "Solvent Extraction Kinetics" in *Principles and Practices of Solvent Extraction*; **1992**, J. Rydberg; C. Musikas G. R. Chopin Ed.; M. Dekker, Inc., New York, pp 157-207.
(43) V. V. Tarasov; G. A. Yagodin; A. A. Pichugin *Kinetics of the Extraction of Inorganic Compounds (Russ).*; **1984**, Ed.; VINITI: Moscow.
(44) H. Watarai, *Trends in Analytical Chemistry* **1993**, *12*, 313-318.
(45) N. H. Sagert; W. Lee; M. J. Quinn, *Can. J. Chem.* **1979**, *57*, 1218.
(46) G. F. Vandegrift; E. P. Horwitz, *J. Inorg. Nucl. Chem.* **1977**, *39*, 1425-1432.
(47) P. R. Danesi; R. Chiarizia; M. Pizzichini; A. Saltelli, *J. Inorg. Nucl. Chem.* **1978**, *40*, 1119-1123.
(48) A. W. Adamson *Physical Chemistry of Surfaces. 5th ed.*; **1990**, Ed.; Wiley.
(49) W. Lasek; W. Makosza, *J. Phys. Org. Chem.* **1993**, *6*, 412-420.
(50) C. M. Starks; C. L. Liotta; M. Halpern *Phase Transfer Catalysis*; **1994**, C. M. Starks Ed.; Chapmann § Hall: New York.
(51) H. H. Girault; D. J. Schiffrin, "Electrochemistry of Liquid-Liquid Interfaces" in *Electroanalytical Chemistry*; **1989**, A. J. Bard Ed.; Dekker, New York pp 1-141.
(52) P. Vanysek, *Electrochim. Acta* **1995**, *40*, 2841-2847.
(53) J. Koryta, *Ion Selective Electrode Rev.* **1983**, *5*, 131-164.
(54) K. B. Eisenthal, *J. Phys. Chem.* **1996**, *100*, 12997-13006.
(55) H. Watarai; Y. Saitoh, *Chemistry Letters* **1995**, 283-284.
(56) T. M. Fyles, *J. Chem. Soc., Faraday Trans. 1* **1986**, *82*, 617-633.
(57) P. Linse, *J. Chem. Phys.* **1987**, *86*, 4177-4187. J. Gao, W. L. Jorgensen, *J. Phys. Chem.* **1988**, *92*, 5813-5822. I. L. Carpenter, W. J. Hehre, *J. Phys. Chem.* **1990**, *94*, 531-536. D. Michael, I. Benjamin, *J. Phys. Chem.* **1995**, *99*, 1530-1536. A. R. van Buuren, S.-J. Marrink, H. J. C. Berendsen, *J. Phys. Chem.* **1993**, *97*, 9206-9212.
(58) G. M. Torrie; J. P. Valleau, *J. Chem. Phys.* **1980**, *73*, 5807-5816.
(59) I. Benjamin, *J. Chem. Phys* **1992**, *96*, 577-585. K. J. Schweighofer; I. Benjamin, *J. Phys. Chem.* **1995**, *99*, 9974-9985.
(60) I. Benjamin, *Chem. Rev.* **1996**, *96*, 1449-1475.
(61) I. Benjamin, *J. Chem. Phys.* **1991**, *95*, 3698-3709.
(62) D. Michael; I. Benjamin, *J. Phys. Chem.* **1998**, *102*, 5145-5151.
(63) I. Benjamin, *Chem. Phys. Let.* **1998**, 480-486.
(64) W. Guba; H. Kessler, *J. Phys. Chem.* **1994**, *98*, 23-27.
(65) W. Guba; R. Haessner; G. Breipohl; S. Henke; J. Knolle; V. Sandagada; H. Kessler, *J. Am. Chem. Soc.* **1994**, *116*, 7532-7540.
(66) C. Chipot; A. Pohorille, *J. Mol. Struct. (THEOCHEM)* **1997**, *398-399*, 529-535.
(67) C. Chipot; M. A. Wilson; A. Pohorille, *J. Phys. Chem. B* **1997**, *101*, 782-791.
(68) V. P. Sokhan; D. J. Tildesley, *Faraday Discuss.* **1996**, *104*, 193-208.
(69) I. Benjamin, *Acc. Chem. Res.* **1995**, *28*, 233-239.
(70) G. Wipff; E. Engler; P. Guilbaud; M. Lauterbach; L. Troxler; A. Varnek, *New J. Chem.* **1996**, *20*, 403-417.
(71) P. Guilbaud; G. Wipff, *New J. Chem.* **1996**, *20*, 631-642.
(72) A. Varnek; C. Sirlin; G. Wipff, "Interaction of the t-butyl-calix[4]arene Anion with Alkali Cations *in vacuo*, in Water, in Acetonitrile and in Chloroform: Host-Guest Complexes

or *exo* Counterions ? MD and FEP Computer Investigations on the Cs^+ Binding Affinity" in *Crystallography of Supramolecular Compounds*; **1995**, G. Tsoucaris Ed.; Kluwer Acad. Pub., Dordrecht pp 67-100.
(73) A. Varnek; G. Wipff, *J. Comput. Chem.* **1996**, *17*, 1520-1531.
(74) A. Varnek; L. Troxler; G. Wipff, *Chem. Eur. J.* **1997**, *3*, 552-560.
(75) M. Lauterbach; A. Mark; W. van Gunsteren; G. Wipff, *Gazzetta Chim. Ital.* **1997**, *127*, 699-709.
(76) M. Lauterbach; E. Engler; N. Muzet; L. Troxler; G. Wipff, *J. Phys. Chem.* **1998**, *102*, 225-256.
(77) L. Troxler; G. Wipff, *Analytical Sciences* **1998**, *14*, 43-56.
(78) P. Beudaert; V. Lamare; J.-F. Dozol; L. Troxler; G. Wipff, *Solv. Extract. Ion Exch.* **1998**, *16*, 597-618.
(79) G. Wipff, *Lettre des Sciences Chimiques du CNRS* **1998**, *66*, 22-26.
(80) F. Berny; N. Muzet; R. Schurhammer; L. Troxler; G. Wipff, "MD Simulations on Ions and Ionophores at a Liquid-Liquid Interface: from Adsorption to Recognition" in *Current Challenges in Supramolecular Assemblies, NATO ARW Athens*; **1998**, G. Tsoucaris Ed.; Kluwer Acad. Pub., Dordrecht pp 221-248.
(81) G. Wipff; M. Lauterbach, *Supramol. Chem.* **1995**, *6*, 187-207.
(82) M. Lauterbach; G. Wipff, "Liquid-Liquid Extraction of Alkali Cations by Calix[4]crown Ionophores: Conformation and Solvent Dependent Na^+ / Cs^+ Binding Selectivity: A MD FEP study in pure Chloroform and MD Simulations at the Water / Chloroform Interface" in *Physical Supramolecular Chemistry*; **1996**, L. Echegoyen A. Kaifer Ed.; Kluwer Acad. Pub., Dordrecht pp 65-102.
(83) D. A. Pearlman; D. A. Case; J. C. Caldwell; W. S. Ross; T. E. Cheatham III; D. M. Ferguson; G. L. Seibel; U. C. Singh; P. Weiner; P. A. Kollman, *AMBER4.1* **1995**.
(84) W. D. Cornell; P. Cieplak; C. I. Bayly; I. R. Gould; K. M. Merz; D. M. Ferguson; D. C. Spellmeyer; T. Fox; J. W. Caldwell; P. A. Kollman, *J. Am. Chem. Soc.* **1995**, *117*, 5179-5197.
(85) J. Åqvist, *J. Phys. Chem.* **1990**, *94*, 8021-8024.
(86) W. L. Jorgensen; J. Chandrasekhar; J. D. Madura, *J. Chem. Phys.* **1983**, *79*, 926-936.
(87) W. L. Jorgensen; J. M. Briggs; M. L. Contreras, *J. Phys. Chem.* **1990**, *94*, 1683-1686.
(88) D. G. Kalina; G. W. Mason; E. P. Horwitz, *J. Inorg. nucl. Chem.* **1981**, *43*, 579-582.
(89) J. E. Randles, *Discuss. Faraday Soc.* **1957**, *24*, 194-199.
(90) S. Boudon; G. Wipff; B. Maigret, *J. Phys. Chem.* **1990**, *94*, 6056-6061.
(91) J.-C. Soetens; C. Millot; C. Chipot; G. Jansen; J. G. Angyan; B. Maigret, *J. Phys. Chem. B* **1997**, *101*, 10910-10917.
(92) L. Troxler; J. M. Harrowfield; G. Wipff, *J. Phys. Chem.* **1998**, *102*, 6821-6830.
(93) G. Wipff; L. Troxler, "MD Simulations on Synthetic Ionophores and on their Cation Complexes: Comparison of Aqueous / non-Aqueous Solvents." in *Computational Approaches in Supramolecular Chemistry*; **1994**, G. Wipff Ed.; Kluwer, Dordrecht pp 319-348.
(94) A. N. Popov, "Counterions and Adsorption of Ion-Exchange Extractants at the Water/Oil Interface" in *The Interface Structure and Electrochemical Processes at the Boundary Between two Immiscible Liquids*; **1987**, V. E. Kazarinov Ed.; Springer Verlag, Berlin pp 179-205.
(95) K. A. Sharp; B. Madam, *J. Phys. Chem. B* **1997**, *101*, 4343-4348.
(96) J. Böcker; M. Schlenkrich; P. Bopp; J. Brickmann, *J. Phys. Chem.* **1992**, *96*, 9915-9922.
(97) S. Karaborni; K. Esselink; P. A. J. Hilbers; B. Smit; J. Karthäuser; N. M. van Os; R. Zana, *Science* **1994**, *266*, 254-256.
(98) B. Smit; P. A. J. Hilbers; K. Esselind, *Tenside Surf. Det. 30* **1993**, *4*, 287-294.
(99) P. van der Ploeg; H. J. C. Berendsen, *J. Chem. Phys.* **1982**, *76*, 3271.
(100) P. Ahlström; J. C. Berendsen, *J. chem. Phys.* **1993**, *97*, 13691-13702.

(101) A. R. van Buuren; S.-J. Marrink; J. C. Berendsen, *Colloids and Surfaces* **1995**, *102,* 143-157.
(102) M. Lauterbach, *Simulations sur Ordinateur de l'Extraction Liquide-Liquide des Cations Alcalins par des Ionophores de type Calixarène-couronne et Valinomycine.*; Thesis, Université Louis Pasteur Strasbourg, **1997**.
(103) G. Jarvinen *et al*; "Selective Extraction of Trivalent Actinides from Lanthanides with Dithiophosphinic Acids and Tributylphosphate." in *Separations of* f *Elements*; **1994**, K. L. Nash G. R. Choppin Ed.; Plenum Press, New York pp 43-62.
(104) D. W. Tedder, "Water Extraction" in *Science and Technology of Tributyl Phosphate,* **1991**, W. W. Schultz, I. D. Navratil, A. S. Kertes Ed.; CRC Press pp 35-70.
(105) W. Fan; R.-S. Tsai; N. El Tayar; P.-A. Carrupt; B. Testa, *J. Phys. Chem.* **1994**, *98,* 329-333.
(106) R. A. Pierotti, *Chem. Rev.* **1976**, *76,* 717-726.
(107) J.-M. Lehn, *Science* **1993**, *260,* 1762-1763.
(108) Y. Ishikawa; T. Kunitake; T. Matsuda; T. Otsuka; S. Shinkai, *J. Chem. Soc. Chem. Commun.* **1989**, 736-737.
(109) L. Dei; A. Casnati; P. L. Nostro; P. Baglioni, *Langmuir* **1995**, *11,* 1268-1272.
(110) P. L. Nostro; A. Casnati; L. Bossoletti; L. Dei; P. Baglioni, *Colloids and Surfaces A* **1996**, *116,* 203-209.
(111) L. Dei; A. Casnati; P. L. Nostro; A. Pochini; R. Ungaro; P. Baglioni, *Langmuir* **1996**, *12,* 1589-1593.
(112) A. Casnati; A. Pochini; R. Ungaro; F. Ugozzoli; F. Arnaud; S. Fanni; M.-J. Schwing-Weil; R. J. M. Egberink; D. N. Reinhoudt, *J. Am. Chem. Soc.* **1995**, *117,* 2767-2777.
(113) N. Muzet; E. Engler; G. Wipff, *J. Phys. Chem.* **1998**, *102,* in press.
(114) G. R. Choppin; K. L. Nash, *Radiochimica Acta* **1995**, *70/71,* 225-236.
(115) A. M. Rozen, *J. Radioanal. Nucl. Chem., Articles* **1990**, *143,* 337-355.
(116) U. Casellato; P. A. Vigato; M. Vidali, *Coord. Chem.* **1981**, *36,* 183-265.
(117) W. L. Jorgensen, *Acc. Chem. Res.* **1989**, *22,* 184-189.
(118) T. P. Straatsma; J. A. McCammon, *Annu. Rev. Phys. Chem.* **1992**, *43,* 407-435.
(119) H. Rapaport; I. Kuzmenko; K. Kjaer; P. Howes; W. Bouwman; J. Als-Nielsen; L. Leiserowitz; M. Lahav, *J. Am. Chem. Soc.* **1997**, *119,* 11211-11216.
(120) A. Pohorille; M. A. Wilson, *Origins of Life and Evolution of the Biosphere* **1995**, *25,* 21-46 and references cited therein.
(121) J. H. Fuhrhop; J. Köning *Membranes and Molecular Assemblies: the Synkinetic Approach*; **1994**, J. F. Stoddart Ed.; The Royal Society of Chemistry: Cambridge. H. Ringsdorf; B. Schlarb; J. Venzmer, *Angew. Chem. Int. Ed. Engl.* **1988**, *27,* 113-158.
(122) I. Benjamin, *Science* **1993**, *261,* 1558-1560.
(123) S.-J. Marrink; H. J.-C. Berendsen, *J. Phys. Chem.* **1994**, *98,* 4155-4168.
(124) A. Pohorille; M. A. Wilson, *J. Chem. Phys.* **1996**, *104,* 3760-3774.
(125) A. Pohorille; P. Cieplak; M. A. Wilson, *Chem. Phys.* **1996**, *204,* 337-345.
(126) M. Schlenkrich; K. Nicklas; J. Brickmann, *Ber. Bunsenges. Phys. Chem.* **1994**, *94,* 133-145.
(127) T. R. Stouch, *Molecular Simulation* **1993**, *10,* 317-345.
(128) Q. Sheng; K. Schulten, *J. Phys. Chem.* **1995**, *99,* 11018-11027.
(129) R. M. Venable; Y. Zhang; B. J. Hardy, *Science* **1993**, *262,* 223-225.
(130) M. A. Wilson; A. Pohorille, *J. Am. Chem. Soc.* **1996**, *118,* 6580-6587.
(131) B. Jönsson; O. Edholm; O. Teleman, *J. Chem. Phys.* **1986**, *85,* 2259-2271.
(132) S. Karaborni; J. P. O'Connell, *J. Phys. Chem.* **1990**, *94,* 2624-2631.
(133) J. Shelley; K. Watanabe; M. L. Klein, *Inter. J. Quantum Chem. Quantum Biology Symposium* **1990**, *17,* 103-117.
(134) B. Smit; P. A. J. Hilbers; K. Esselink; L. A. M. Rupert; N. M. van Os, *J. Phys. Chem.* **1991**, *95,* 6361-6368.
(135) T. Wallin; P. Linse, *J. Phys. Chem.* **1996**, *100,* 17873-17880.
(136) M. C. Woods, *J. Phys. Chem.* **1986**, 1875-1885.

(137) M. Ferrario; M. Haughney; I. R. McDonald; M. L. Klein, *J. Chem. Phys.* **1990**, *93*, 5156-5166.
(138) H. Kovacs; J. Kowalewski; A. Laaksonen, *J. Phys. Chem.* **1990**, *94*, 7378-7385.
(139) A. Laaksonen; P. G. Kusalik; I. M. Svishchev, *J. Phys. Chem. A* **1997**, *101*, 5910-5918.
(140) A. Wallqvist, *J. Chem. Phys.* **1992**, *96 N° 2*, 1655-1656.
(141) M. S. Skaf; B. M. Ladanyi, *J. Phys. Chem.* **1996**, *100*, 18258-18268.
(142) S. E. DeBolt; P. A. Kollman, *J. Am. Chem. Soc.* **1995**, *117*, 5316-5340.

SYNTHETIC RECEPTORS FOR ANIONIC AND NEUTRAL SUBSTRATES

A. P. DAVIS
Trinity College Dublin
Dublin 2, Ireland

1. Introduction

The study of synthetic receptors has been of great historical importance to, and continues as a major theme within, the area of supramolecular chemistry. It is significant on a number of levels. Firstly, there are practical applications for receptors in sensing, in separation technology and in phase-transfer catalysis, and potentially in pharmaceuticals and in more sophisticated catalytic systems ("artificial enzymes"). Secondly, the behaviour of synthetic receptors can throw light on biomolecular recognition, and thus answer fundamental questions concerning the functioning of living systems. Thirdly, if the ultimate goal of supramolecular chemistry is to establish engineering through non-covalent interactions, the design and synthesis of receptors provides challenges which stimulate interest and focus efforts, impelling the subject in much the same way that the structures of natural products have encouraged progress in conventional organic synthesis.

Over the past few years our group has addressed three specific problems in molecular recognition; (i) the binding of inorganic anions by neutral organic molecules, (ii) the enantioselective recognition of carboxylates, and (iii) the biomimetic recognition of carbohydrates. The principle (but not exclusive) tool for this work has been the inexpensive, highly functionalised steroid cholic acid. This article surveys our progress in all three areas, focussing especially on our most recent results.

2. General Considerations

When designing a synthetic receptor, it is necessary to take account of (at least) 4 major considerations:

R. Ungaro and E. Dalcanale (eds.), Supramolecular Science: Where It Is and Where It Is Going, 125–146.

(i) The nature, positioning and strength of the binding moieties, which must be complementary to those in the substrate.

(ii) The prospects for self-quenching of binding moieties, making them unavailable to act on a substrate.

(iii) The prospects for self-association of the receptor in a manner which might prevent binding.

(iv) The likely solubility of the receptor in the intended solvent

Point (i) is the most obvious, and is the one on which we normally prefer to focus. However, if any of the other three are neglected, there is a danger that much synthetic effort may be wasted.

The question of rigidity impinges on all of (i) - (iv). From the viewpoint of (i), a highly flexible receptor may be able to adopt a conformation complementary to its substrate, but only at the cost of a good measure of entropy. From the viewpoints of (ii) and (iii), a flexible receptor is highly vulnerable to collapse with self-quenching, and cannot be prevented from assembling into higher aggregates (as can a properly designed rigid system). On the other hand rigidity generally correlates with insolubility, so that it should be possible to incorporate flexible components somewhere in the design.[1]

Some of the issues raised above did not apply in the early days of supramolecular chemistry, when the focus was on crown ethers and cryptands which do not have self-complementary functionality. In contrast, our own work has required the disposition of neutral H-bond donor groups, which almost always contain H-bond acceptors. The separation of these groups by rigid spacers has therefore been a strict requirement. The traditional device for this purpose has been the aromatic ring, to which we have turned in some of our most recent work (Section 5). However, we have developed a special interest in an alternative, the steroid nucleus as present in cholic acid (**1**).[2] This

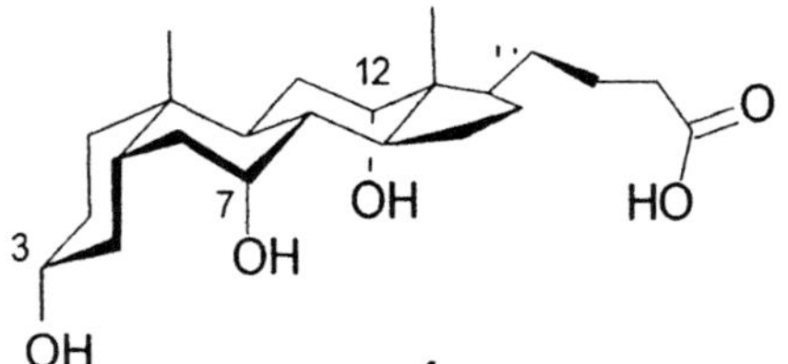

1

inexpensive starting material (currently £55 for 250g from Aldrich) possesses an extended rigid skeleton functionalised at four well-separated positions. These centres can be addressed individually to give a wide range of molecules, monomeric and oligomeric, with co-directed or convergent functional groups. It has served us well in our studies on anion and carbohydrate recognition, and has been exploited by many others, for example in receptors for peptides,[3] alkaloids,[4] DNA,[5] and aromatic compounds.[6]

3. Neutral Organic Receptors for Inorganic Anions

The crown ethers and cryptands, which inaugurated supramolecular chemistry, were neutral organic (and organic-soluble) receptors for cations. Once cation recognition was established, anions became obvious targets.[7] However, most of the early work in this area employed positively charged species, studied mainly in aqueous environments. Some years ago we became interested in the notion of a true "anion-equivalent" of a crown ether or cryptand. This would be a *neutral lipophilic* molecule capable of binding anions in organic solvents. Such a species could find applications in sensing devices such as ion-selective electrodes, and might possibly act as a novel phase transfer catalyst, solubilising inorganic salts by binding the anion as opposed to the cation. As organic chemists we preferred to consider organic, rather than inorganic, solutions to the problem; receptors based on Lewis acidic metals might be less stable, and, depending on the metal, undesirably toxic for practical applications.[8]

Cholic acid appeared to be a good basis for our efforts, in view of its lipophilic β-face and co-directed nuclear hydroxyl groups. Acting as H-bond donors, these hydroxyls could act as binding units without further elaboration. Our first design was the "cryptand" **2**, the cyclodimer of a monomer unit derived by replacing the side-chain in **1** with an α-directed amino group, and the 3-OH with α-carboxymethyl and β-pentyloxy substituents. The pentyloxy groups were intended to promote solubility in organic solvents, while the largely rigid core provided a binding site apparently suitable for the smaller halide anions. Addition of tetrabutylammonium halides to **2** in $CDCl_3$ caused NMR spectral changes consistent with 1:1 complex formation. As shown in Table 1, the binding constants (K_a) ranged up to 3220 (for fluoride ions), with selectivities favouring the smaller (but also, of course, more basic) anions.[9]

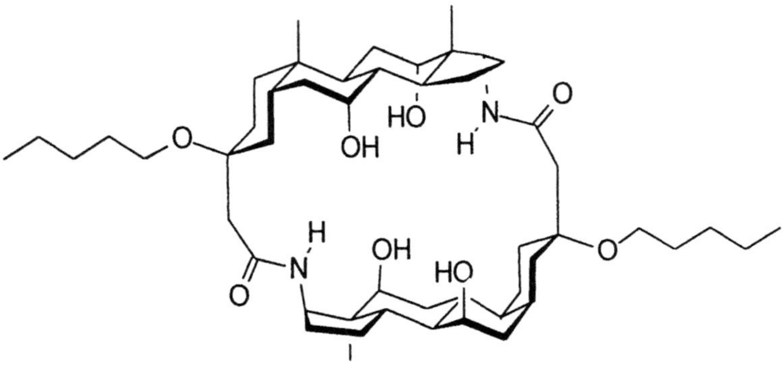

2

Table 1. Results from ^{1}H NMR titrations of **2** with tetrabutylammonium halides in $CDCl_3$ at 298 K.

Halide	K_a [M^{-1}]	$-\Delta G^0$ [$KJmol^{-1}$]	$\Delta\delta$ N*H* [p.p.m.][a]
F^-	3220 ±350	20.01	1.569
Cl^-	990 ±80	17.09	0.611
Br^-	250 ±20	13.68	0.393

[a] Movement of N*H* signal extrapolated to 100% saturation.

While the performance of **2** was pleasing, this receptor design undoubtedly suffered from disadvantages. The most obvious was its relative inaccessibility. To synthesize **2** from **1** had required *ca.* 15 steps, several of which proved quite troublesome. In many respects the challenge was an interesting one, but from a pragmatic viewpoint there were two negative consequences. The first was the time taken; between design and synthesis of **2** a number of alternative neutral organic anionophores were reported,[10] so that our "cryptand" was less novel than we might have hoped. The second was the lack of material, which restricted the depth to which we could study its properties. A further problem was the reliance on secondary hydroxyl groups as H-bond donors. Their modest acidity limits their potency, and cannot easily be increased.

A solution to these problems could be found in the podand architecture **3**,[11] based on a single cholic acid nucleus and employing NHZ groups (tunable through Z) as H-bond donors. Reinhoudt and co-workers had already found that simple podands such as

4 were quite effective anionophores,[10a] and the preorganisation imposed by the steroid seemed likely to yield improvements.

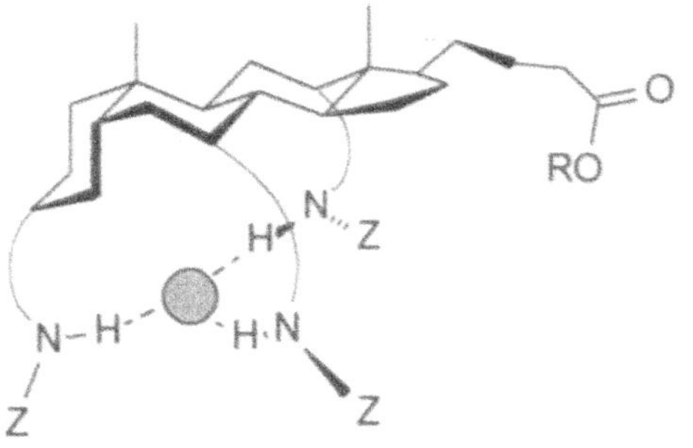

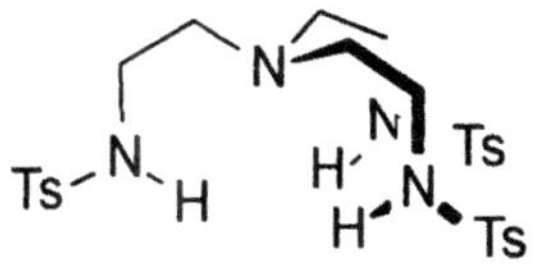

4

3 Z = electron-withdrawing group
shaded sphere = anionic substrate
curved lines = various linkages

Of the many possible variations on **3**, we decided to study first the sulfonamido bis-carbamate **7**.[11] This molecule proved quite accessible, especially once we discovered that the Mitsunobu reaction could be accomplished with methanesulfonate anion as nucleophile.[12] As shown in Scheme 1, this reaction is selective for the equatorial 3-OH in methyl cholate (**5**), and allows the 2-step generation of 3α-azide **6**. Three straightforward transformations thereafter lead to **7**. Molecular modelling on **7** suggested that intramolecular hydrogen bonds were limited to S=O···HN interactions (expected to be weak), and that placing the carbamate groups in unstrained, planar conformations gave a binding pocket with potential NH···X^- distances of *ca.* 2.7 Å. ^{1}H NMR titrations of **7** against tetrabutylammonium salts showed that this receptor did indeed bind anions, with the association constants shown in Table 2. The figures are significantly higher than those achieved with **2**, and there are interesting changes in selectivity. In particular, bromide is bound as strongly as chloride despite its lower basicity, suggesting geometric complementarity between **7** and the larger halide.

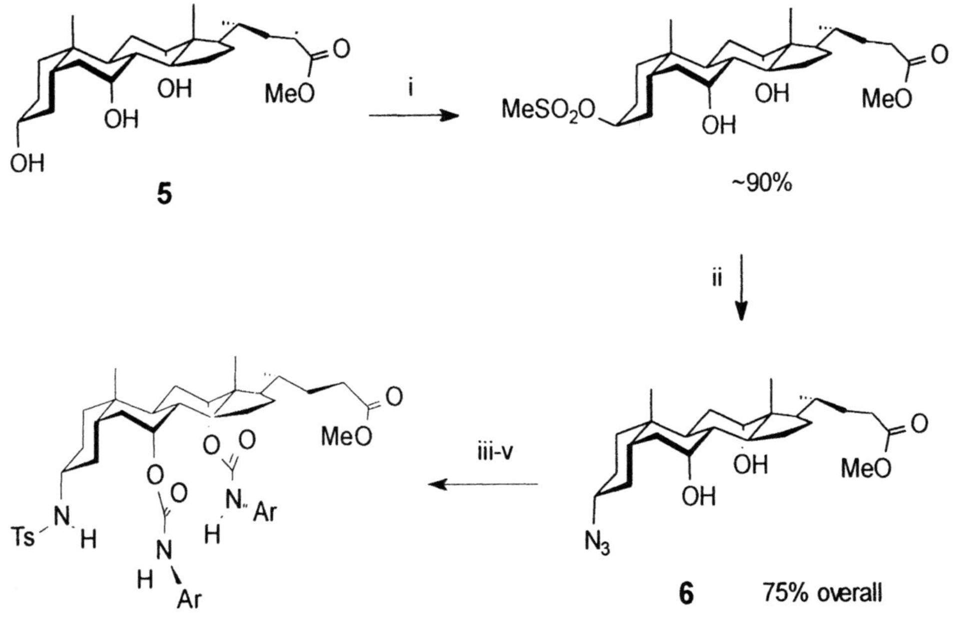

7 Ar = 3,5-dimethylphenyl

Scheme 1: i, Ph_3P, DEAD, THF, $MeSO_3H$; ii, NaN_3, DMPU; iii, ArNCO, TMSCl cat.; iv, Zn, AcOH; v, TsCl, Et_3N

Table 2. Results from ^1H NMR titrations of **7** with tetrabutylammonium salts in $CDCl_3$ at 298 K.

Anion	K_a [M^{-1}]	$-\Delta G^0$ [$KJmol^{-1}$]
F^-	15,400 (±1500)	-23.89
Cl^-	7,200 (±660)	-22.01
Br^-	7,200 (±760)	-22.01
I^-	930 (±70)	-16.94
TsO^-	865 (±120)	-16.76

An alternative version of **3**, the tris-sulfonamide **8**, held extra attraction from an architectural point of view. In this case, the binding units at positions 7 and 12 are based on nitrogen atoms directly attached in *axial* orientations to the steroid nucleus.

Prevented by steric hindrance from rotation about the steroidal C-N bonds (Eq. 1), these groups are forced to direct their polar hydrogens inwards against the α-face of the steroid nucleus. The reduced flexibility also keeps the sulfonamide groups apart, so that there is essentially no possibility of intramolecular hydrogen bonding. Though less accessible than **7**, **8** was finally synthesized on foot of a somewhat long-winded preparation of triamine **9**[13] (itself an interesting "facial amphiphile"). Our efforts were rewarded by the binding results shown in Table 3.[11] Appearing to possess a smaller binding site than **7**, receptor **8** shows an affinity for chloride which is roughly two orders of magnitude greater than that of our original cryptand **2**.

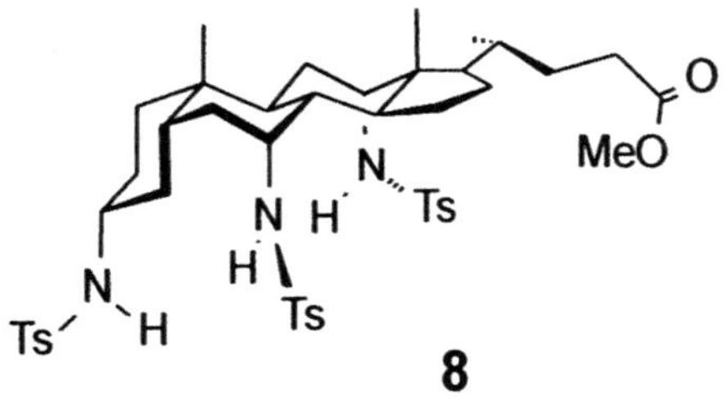

8

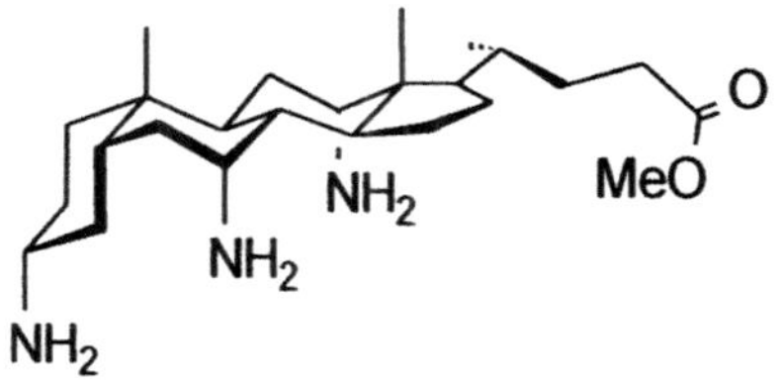

9

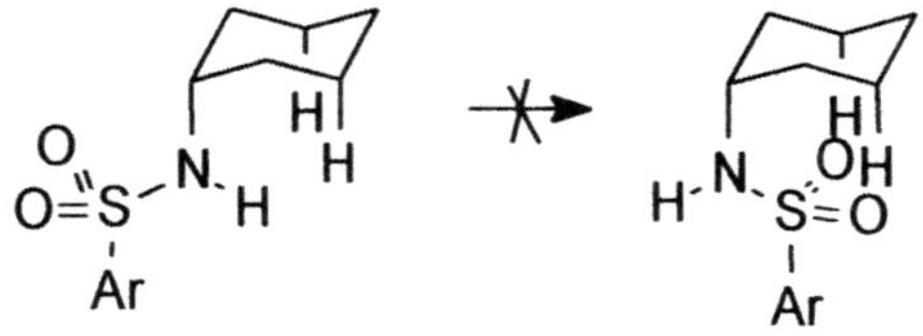

(1)

Table 3. Results from ^{1}H NMR titrations of **8** with tetrabutylammonium salts in $CDCl_3$ at 298 K.

Anion	K_a [M^{-1}]	$-\Delta G^0$ [$KJmol^{-1}$]
Cl^-	92,000 (±28,000)^c	-28.32
Br^-	9,200 (±700)	-22.61
I^-	525 (±45)	-15.52
TsO^-	950 (±80)	-16.99

Receptors **7** and **8** illustrate well the potential of podands **3** for anion recognition. Although this is a fast moving field,[14] the preorganisation imparted by the steroid gives the design a distinctive advantage which we hope to exploit for some time to come.

4. The Enantioselective Recognition of Carboxylates

A striking feature of the previous section is the contrast in symmetry properties between host and guest. The steroid-based receptors are sophisticated, asymmetrical structures, while their major substrates, the halides, are featureless spheres. It almost seems wasteful to apply such complex structures to such simple targets, which cannot benefit from, for example, the enantiodiscriminatory potential of the steroidal frameworks.

Steroidal podands similar to **3** are, in fact, quite well-adapted for enantioselective recognition. The three functionalised sites on the steroid nucleus may be modified to give asymmetrical derivatives of general form **10**, capable in principle of the three-point contact required for the classical model of enantioselectivity.[15] Whether or not this idealised situation is ever achieved in practice, the chiral, functionalised steroidal α-face appears to provide an excellent environment for enantioselection.

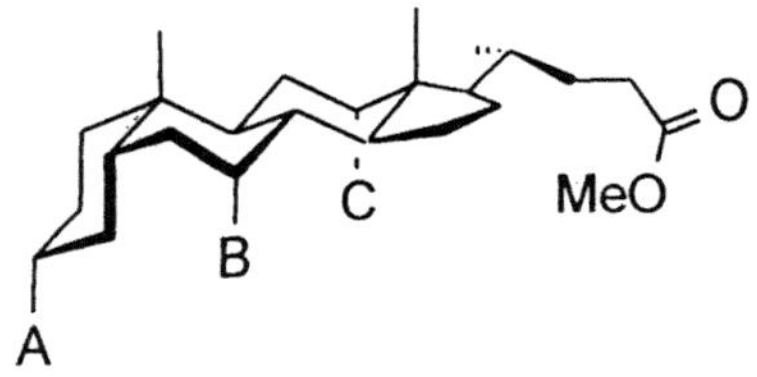

10

In seeking to realise this notion, we chose carboxylates as substrates for two reasons. Firstly, we could avail of the well-known carboxylate-guanidinium recognition motif, as in **11**, to establish and dominate complex formation.[16] Secondly, the amphiphilic nature of carboxylates allowed us to aim for enantioselective phase transfer, an easy phenomenon to detect and potentially very useful for the large scale resolution of racemates.[17]

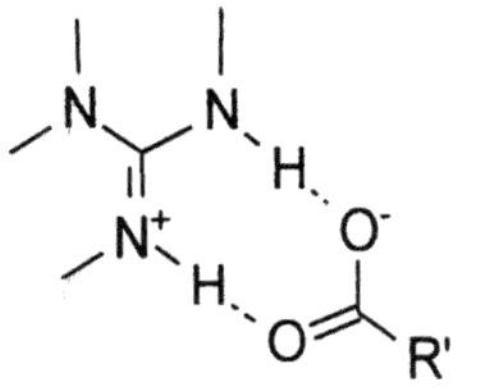

11

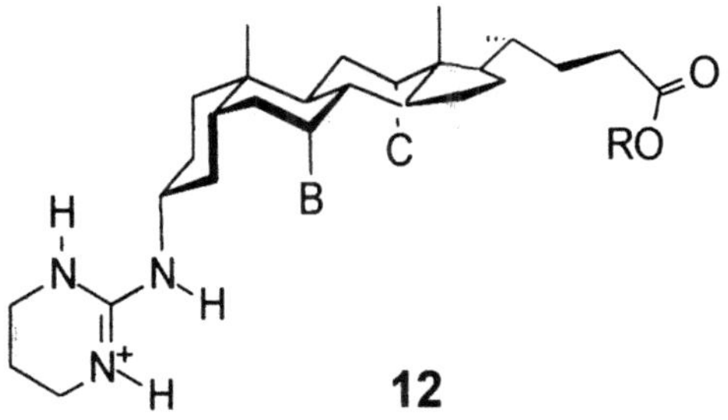

12

We have thus undertaken the study of receptors of general form **12**, and analogues with guanidinium groups in positions 7 and 12, as phase-transfer resolving agents for carboxylates. Though still young, this programme has yielded encouraging preliminary results. In early explorations of the synthetic chemistry required for **12**, we have prepared the simple guanidinium bis-dicarbamate **13** as shown in Scheme 2, and also the asymmetrically-substituted analogue **14**.

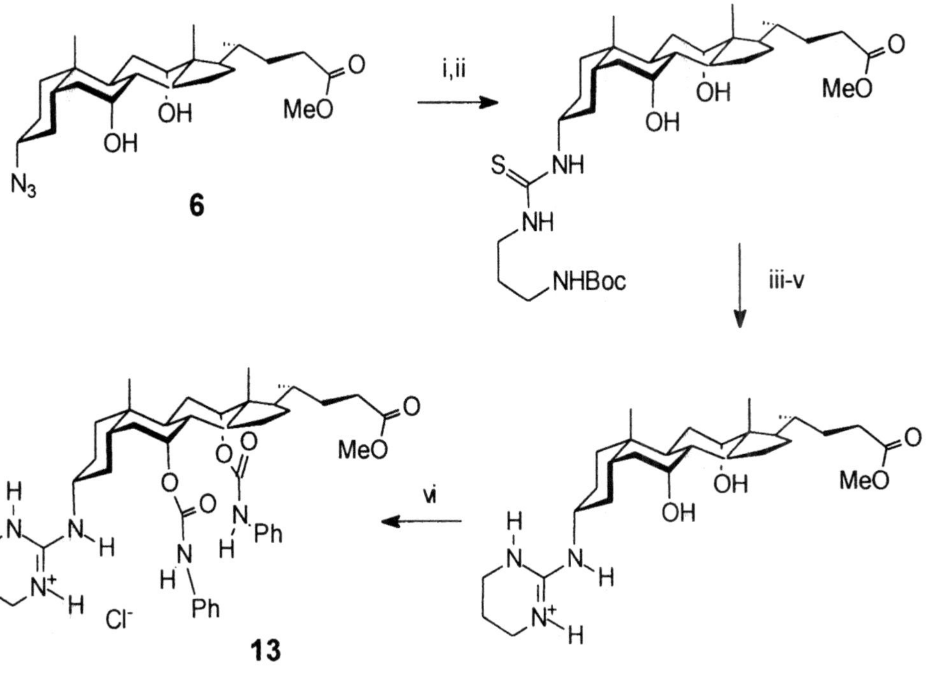

Scheme 2: i, Zn, AcOH; ii, $SCN(CH_2)_3NHBoc$, $^iPr_2NEt$; iii, MeI, MeOH; iv, TFA, CH_2Cl_2; v, $^iPr_2NEt$, MeOH; vi, PhNCO, 11.65 M HCl (cat.), CH_2ClCH_2Cl.

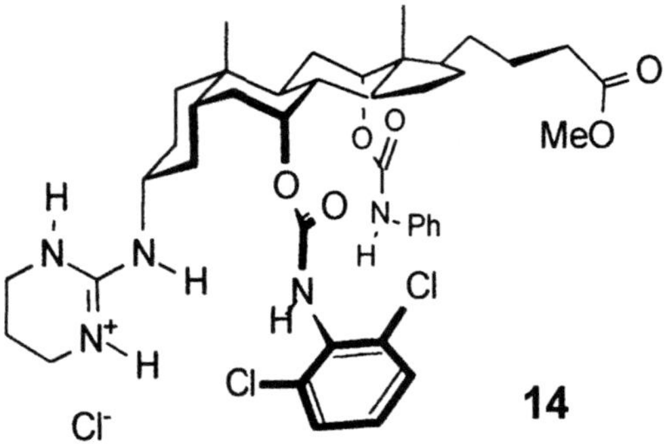

14

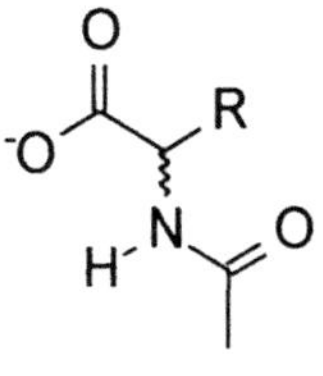

15

As expected for lipophilic guanidinium cations,[16,17] both **13** and **14** extract carboxylates from aqueous into organic phases, presumably exchanging for chloride. As hoped, this can happen with significant enantioselectivity for amino acid derivatives. Table 4 summarises the results from a series of experiments employing *N*-acetyl α-amino carboxylates **15**. The selectivities are not exceptional, but are close to the best previously reported for extractions of amino acid derivatives.[18] Considering the variations possible on the theme of **12**, we can reasonably hope for improved performance in future, more refined versions of the system.

Table 4. Enantioselective extractions by **13** and **14** of racemic *N*-acetyl α-amino acids from aqueous buffer (pH 7.4) into $CHCl_3$.

Substrate	Enantioselectivity (L:D)[a]	
	13	14
N-Ac-alanine	7:1	6:1
N-Ac-phenylalanine	7:1	9:1
N-Ac-tryptophan	7:1	6:1
N-Ac-valine	7:1	9:1
N-Ac-*tert*-leucine	*b*	5:2
N-Ac-methionine	*b*	9:1
N-Ac-proline	*b*	4:1

[a] Measured by integration of ^{1}H NMR signals due to substrate protons in diastereomeric complexes. [b] Not determined.

5. (Towards) The Biomimetic Recognition of Carbohydrates

Carbohydrate nuclei represent important and interesting targets for supramolecular chemists, but also especially challenging ones. Their significance derives from the newly-appreciated rôle played by oligosaccharides in the control of biological processes, notably of cell-cell recognition and related phenomena such as infection by pathogens.[19] The challenge arises because (a) carbohydrates have complex 3-dimensional structures, which must be straddled for fully specific recognition, and (b) the individual functional groups in carbohydrates, typically secondary hydroxyl groups, are not especially "sticky".[20] Although capable of H-bond donation and acceptance, alkanols are uncharged and only mildly acidic or basic, thus forming relatively weak interactions with functionality in a receptor. The difficulty is multiplied if one attempts fully biomimetic carbohydrate recognition, i.e. the binding of carbohydrates through non-covalent interactions[21] in aqueous solution. The distinction between a carbohydrate and an aggregate of water molecules is not an easy one to make. Indeed, the means by which it is achieved in Nature is not entirely clear,[22] adding zest to the interest felt by many supramolecular chemists in the area.[23]

Our own involvement in carbohydrate recognition grew out of our interest in cholic acid as a building block for supramolecular chemistry. Having designed and synthesized the first "cholaphane" **16**,[24] we were looking for an application. Prompted by the work of Aoyama on calixarene **17**,[25] the first synthetic carbohydrate receptor, we realised that the cavity of **16** was nicely complementary to a monosaccharide. ^{1}H NMR

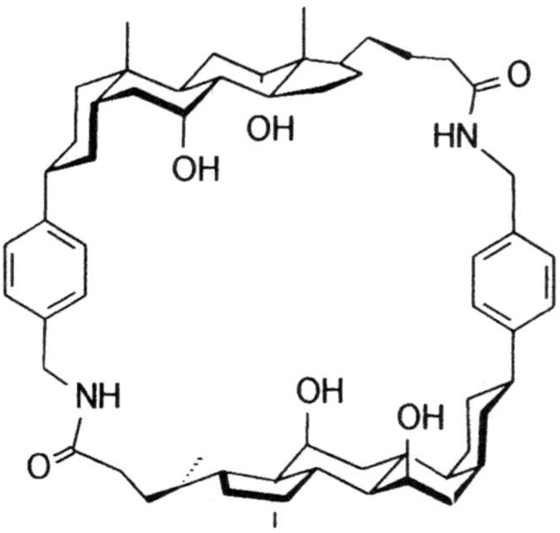

16

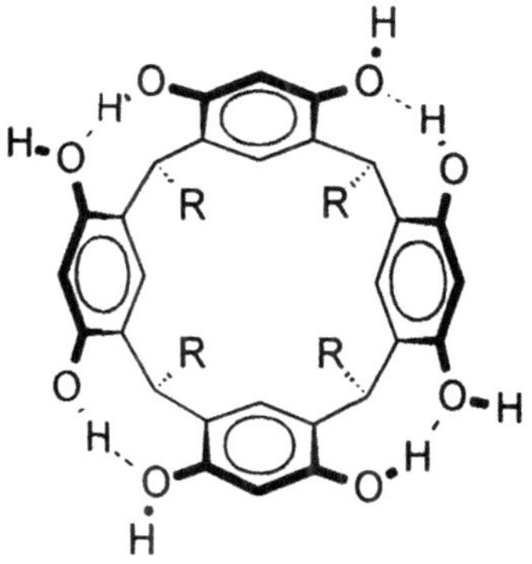

17 R = n-$C_{11}H_{23}$

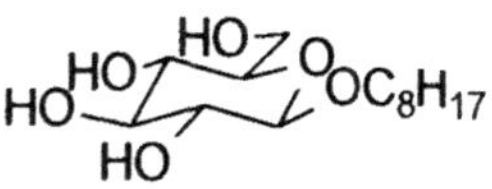

18

titration experiments with alkyl glucosides in $CDCl_3$ indicated that 1:1 complexation did indeed take place, with association constants of up to 3000 M^{-1} [for the case of octyl β-D-glucoside (**18**)] and significant enantioselectivity (3:1 for **18** *vs*. the L enantiomer).[26]

Further work, over a number of years, led to second-generation receptors such as cholaphane **19**[27] and "cyclocholamide" **20**.[28] Unfortunately, the improved preorganisation of these new designs did not appear to be reflected in their performance; **19** and **20** were found to bind **18** with K_a's of just 1560 and 750 M^{-1} respectively. Moreover, the appeal of the steroid-based architectures was lessened by their hydrophobic exteriors. The recognition of carbohydrates through hydrogen bonding in organic media was, by now, a well-established phenomenon.[23a-e] It seemed necessary to address the more difficult challenge, that of carbohydrate recognition in water.

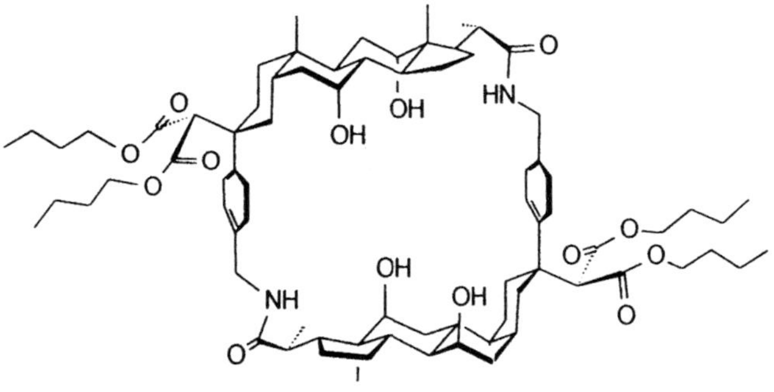

19

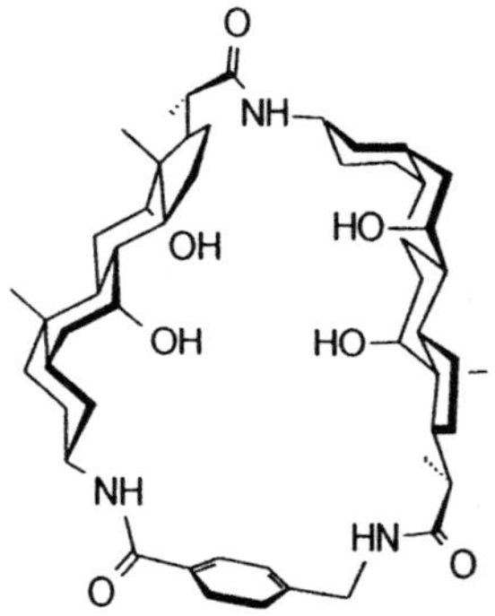

20

As mentioned above, the driving force for natural carbohydrate recognition is a mysterious and somewhat controversial issue. Some authors highlight the rôle of hydrogen bonding, clearly the strongest intermolecular force in most protein-carbohydrate complexes,[29] while others place emphasis on solvent reorganisation (as in hydrophobic binding) as an impetus for recognition.[22] Whichever view is taken, common sense (backed by protein crystallography)[19b,29] suggests that a receptor designed for operation in aqueous solution should aim to satisfy both the polar and apolar "valencies" of its substrate. The crystal structures of carbohydrate-protein complexes further suggest that the apolar moieties of carbohydrates (CH groups) are best matched by aromatic surfaces, perhaps implying a specific CH-π effect.

Considering glucose (**21**), as an interesting and exceptionally challenging substrate,[30] a crude "low-resolution" structural analysis yields the picture shown in Figure 1. The general requirement for glucose recognition would therefore seem to be a pair of roughly parallel hydrophobic (preferably aromatic) surfaces, held apart by bridges containing polar, hydrogen-bonding functional groups. As discussed in section 2, a rigid framework is probably required to prevent, at least, self-quenching of the hydrophobic surfaces. Externally-directed functionality is also desirable so that solubility can be controlled independently of the effective core.

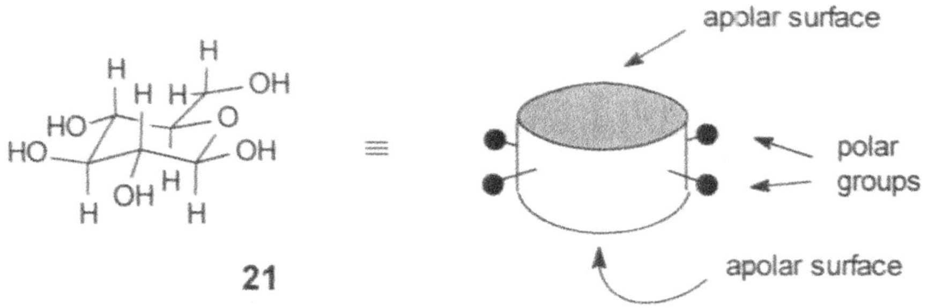

Figure 1: β-D-glucopyranose (**21**) interpreted as polar and apolar binding regions.

After many trials, molecular modelling revealed a tricyclic structure **22** which seemed to satisfy the above criteria. As tetra-ester **22a** it could be synthesized and studied in organic media, then as tetra-anion **22b** it could be investigated in water. The biphenyl surfaces and isophthaloyl spacers create a cavity large enough to encapsulate a pyranose (Figure 2), and lined with 8 secondary amide groups capable of H-bond donation or acceptance.

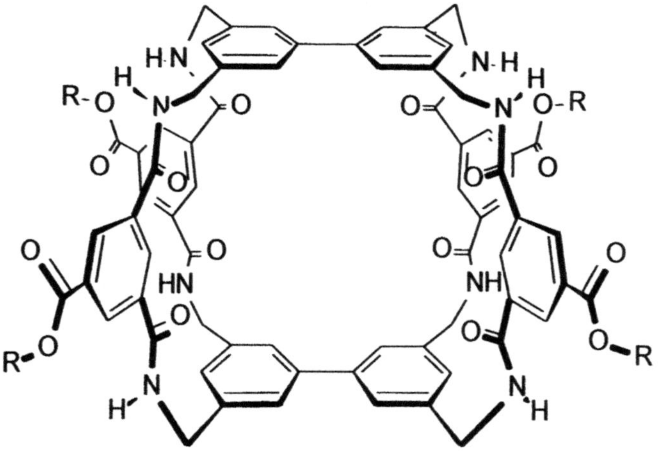

22a R = C_5H_{11}

22b R = $^-$

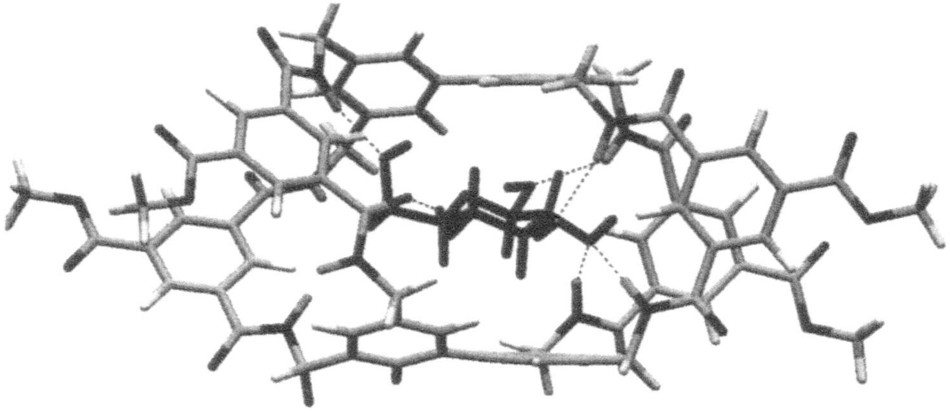

Figure 2: Receptor **22** , modelled as the tetramethyl ester, binding β-D-glucopyranose (**21**). The configuration shown is the baseline conformation from a 1000 step Monte Carlo molecular mechanics study employing MacroModel V5.5 (MM2* force field, $CHCl_3$ GB/SA solvation). Intermolecular hydrogen bonds are shown as broken lines. Several carbohydrate CH groups are positioned to take part in CH-π interactions with the receptor. A similar study on free receptor revealed collapsed low-energy conformations, but only at the expense of *syn*-amide linkages.

At time of writing, we have not yet synthesized **22b**, and have therefore not been able to investigate truly biomimetic carbohydrate recognition in our laboratory. However, we have succeeded in the preparation of **22a**, using the route shown in Scheme 3.[31] Critical to success were the two [2 + 2] macrocyclisations involving bis-pentafluorophenyl ester **23**. Modelling suggested that the [1 + 1] product would be too strained to form easily, and that the [2 + 2] combination would therefore be favoured at high dilution. This proved to be the case, allowing us to bypass a complex and prohibitively lengthy stepwise synthesis.

As hoped, **22a** is freely soluble in solvents such as chloroform, and can therefore be used to investigate the receptor framework in organic media. The ^{1}H NMR spectra in $CDCl_3$ were broadened, and could not be used for binding studies, but the spectra run in $CDCl_3$-CD_3OH (92:8) were well resolved. Despite the presence of a hydroxylic co-solvent, binding studies in this system gave some interesting results. As shown in Table 5, quite substantial binding constants were recorded for β-glycosides **18** and **25**, but a very low figure for the α anomer **24**. This extreme selectivity, if nothing else, suggests a tight, specific association between glycoside and receptor.

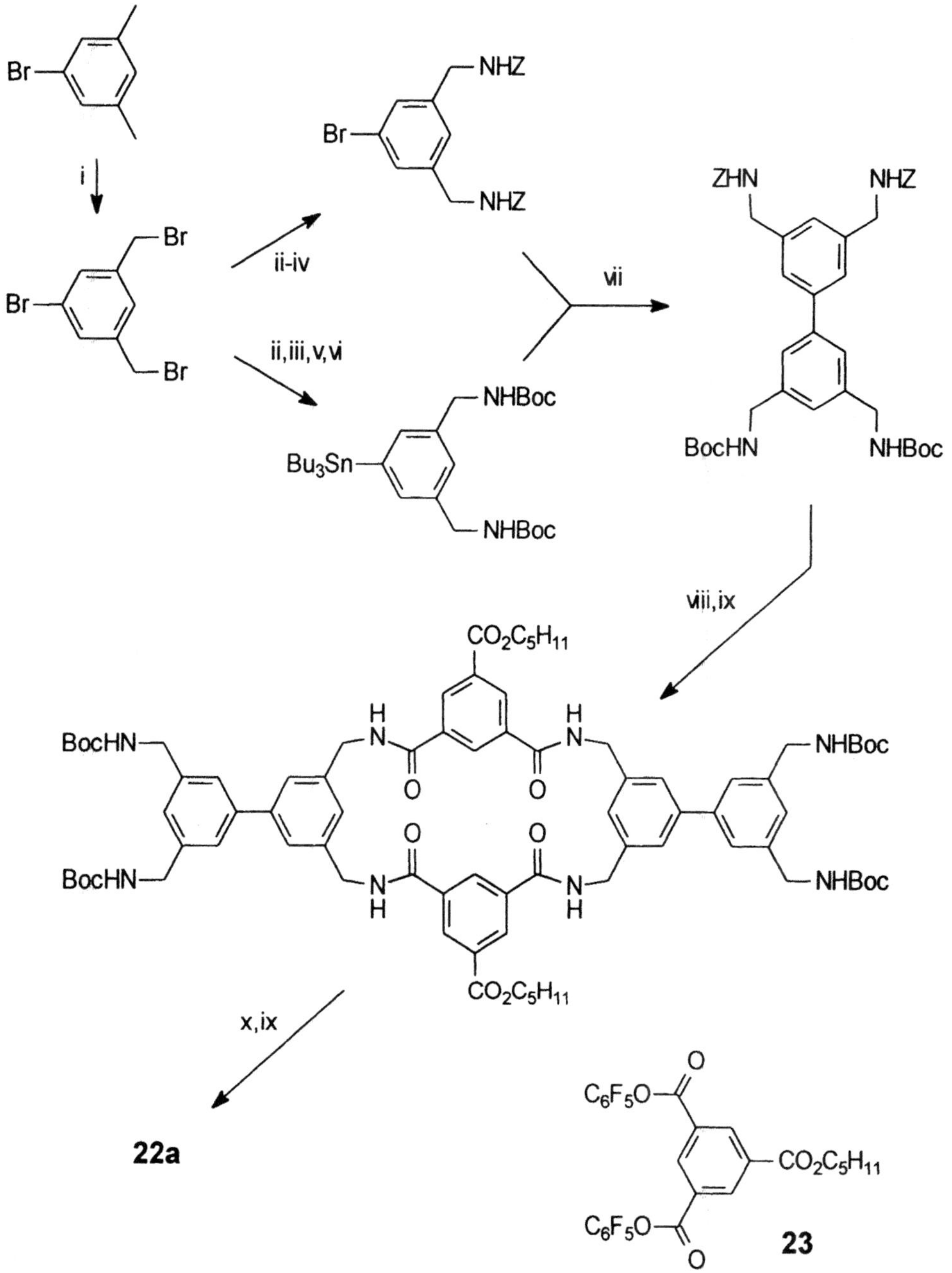

Scheme 3: i, NBS, AIBN, HCOOMe, hν; ii, Potassium phthalimide, DMF; iii, $NH_2NH_2.H_2O$, CH_3OH-CH_2Cl_2, then HCl aq.; iv, $PhCH_2O(CO)Cl$, THF, *i*Pr_2NEt; v, $(Boc)_2O$, THF, *i*Pr_2NEt; vi, $(SnBu_3)_2$, $Pd(OAc)_2$, PPh_3, toluene; vii, $Pd_2(dba)_3$, tri-2-furylphosphine, NMP; viii, H_2, 10% Pd/C, CH_3OH-CH_2Cl_2; ix, **4**, *i*Pr_2NEt, THF-DMF, high dilution; x, TFA, CH_2Cl_2.

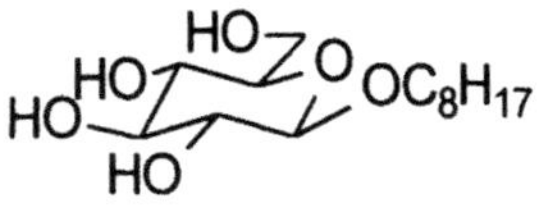
18

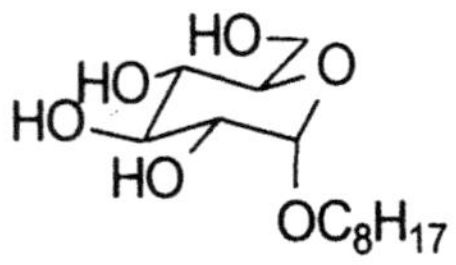
24

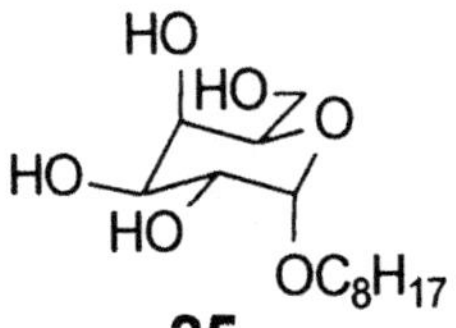
25

Table 5. Results from ^{1}H NMR titrations of **22a** with octyl glycosides in $CDCl_3$-CD_3OH (92:8) at 303 K.

pyranoside	K_a [M^{-1}]	$-\Delta G^0$ [$KJmol^{-1}$]
n-octyl *β*-D-glucopyranoside (**18**)	980 (± 2%)	17.4
n-octyl *α*-D-glucopyranoside (**24**)	20 (± 23%)	7.8
n-octyl *β*-D-galactopyranoside (**25**)	220 (± 5%)	13.6

Returning to chloroform, the solvent which would best allow comparison of **22a** with other systems, we discovered that the fluorescence spectrum of the receptor was affected by the presence of the glycosides. The changes were consistent with 1:1 complex formation, and could be analysed to give binding constants. As shown in Table 6, they confirmed the selectivity of **22a** for *β*-pyranosides, and clearly revealed its unusual potency. The new system binds **18** more strongly by two orders of magnitude than **16**, the best of our steroid-based receptors.

Table 6. Results from fluorescence titrations of **22a** with octyl glycosides in $CHCl_3$

pyranoside	K_a [M^{-1}]	$-\Delta G^0$ [$KJmol^{-1}$]
n-octyl *β*-D-glucopyranoside (**18**)	300,000 (± 6%)	30.7
n-octyl *α*-D-glucopyranoside (**24**)	13,000 (± 8%)	23.1
n-octyl *β*-D-galactopyranoside (**25**)	110,000 (± 12%)	28.3

The exact mode of complexation remains uncertain, but we can be fairly sure that the complete structure of **22a** is needed for effectiveness. A ^{1}H NMR study of the monocyclic control **26** with **18** in $CDCl_3$ suggested a binding constant of just 850 M^{-1}, nearly three orders of magnitude smaller than that of the full tricyclic system.

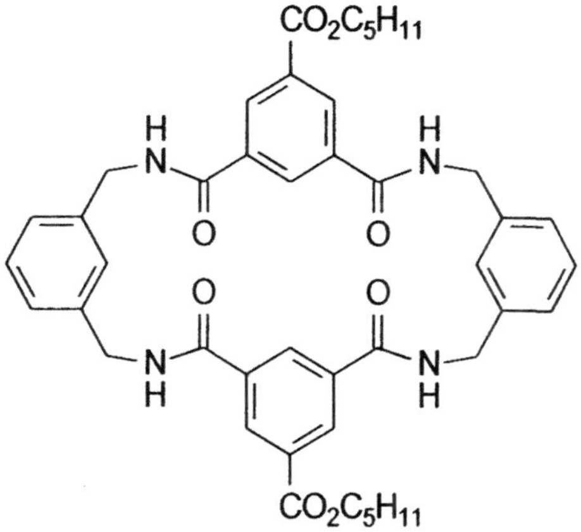

26

Finally, **22a** has one property which is unprecedented for a preorganised receptor operating through non-covalent interactions. Although previously-reported systems have been able to promote the solution of certain other carbohydrates in non-polar solvents, glucose has always defeated them.[32] However, stirring **22a** with glucose in $CHCl_3$ gives a homogeneous solution which, after microfiltration and analysis, is found to contain 0.8-0.9 equivalents of the carbohydrate. Interestingly, the ^{1}H NMR spectrum of the mixture [after evaporation and resolution in $(CD_3)_2SO$] implied an anomeric ratio of 7:1 (β:α), very different from the equilibrium ratios in $(CD_3)_2SO$ (55:45) and D_2O (65:35).[33]

Our next move, of course, must be to hydrolyse the ester groups of **22a**, giving the water-soluble **22b** for studies in aqueous solution. It may or may not prove to be genuinely biomimetic carbohydrate receptor, but this is unquestionably "where we are going".

The author must thank the co-workers listed in the references for the intellectual and practical achievements which have made this article possible. Also acknowledged is the financial support of Forbairt/Enterprise Ireland, Bioresearch Ireland, Materials Ireland, Schering Plough (Avondale) Ltd. and the EU "Human Capital and Mobility" and "Training and Mobility for Researchers" Programmes. Collaborations with David Williams (Imperial College, London) and Peter Ashton (University of Birmingham) have been vital to our progress, as has the generous supply of cholic acid provided by Freedom Chemical Diamalt GmbH.

References and Footnotes

1. It should be said that extreme rigidity is often undesirable, because it may prevent access to the binding site and slow binding kinetics. Moreover, it may not be possible to design a completely rigid receptor with precise complementarity to the target. A limited degree of flexibility allows the receptor to settle around its substrate, and is often present in effective systems.
2. For general accounts, see: Davis, A. P. (1993) Cholaphanes *et al.*; Steroids as Structural Components in Molecular Engineering, *Chem. Soc. Rev.* **22**, 243-253; Davis, A. P., Bonar-Law, R. P. and Sanders, J. K. M. (1996) Receptors based on cholic acid, in *Comprehensive Supramolecular Chemistry*, vol. 4 (Supramolecular Reactivity and Transport: Bioorganic Systems), Y. Murakami (ed.), Pergamon, Oxford, pp. 257-286. Other relevant reviews: Walliman, P., Marti, T., Fürer, A. and Diederich, F. (1997) Steroids in Molecular Recognition, *Chem. Rev.* **97**, 1567; Li, Y. X. and Dias, J. R. (1997) Dimeric and oligomeric steroids, *Chem. Rev.* **97**, 283-304.
3. Boyce, R., Li, G., Nestler, H. P., Suenaga, T. and Still, W. C. (1994) Peptidosteroidal receptors for opioid peptides. Sequence-selective binding using a synthetic receptor library, *J. Am. Chem. Soc.* **116**, 7955; Cheng, Y. A., Suenaga, T. and Still, W. C. (1996) Sequence-selective peptide binding with a peptido-A,B-trans-steroidal receptor selected from an encoded combinatorial receptor library, *J. Am. Chem. Soc.* **118**, 1813-1814.
4. Bonar-Law, R. P., Mackay, L. G. and Sanders, J. K. M. (1993) Morphine recognition by a porphyrin-cyclocholate molecular bowl, *J. Chem. Soc., Chem. Commun.*, 456.
5. Hsieh, H.-P., Muller, J. G. and Burrows, C. J. (1994) Structural effects in novel steroidal polyamine-DNA binding, *J. Am. Chem. Soc.* **116**, 12077.
6. D'Souza, L. J. and Maitra, U. (1996) Design, synthesis, and evaluation of bile acid-based molecular tweezers, *J. Org. Chem.* **61**, 9494-9502.
7. Reviews: Dietrich, B. (1993) Design of anion receptors: Applications, *Pure and Appl. Chem.* **65**, 1457; Bianchi, A., Bowman-James, K. and Garcia-Espana, E. (1997) *Supramolecular Chemistry of Anions*, Wiley-VCH, New York; Schmidtchen, F. P. and Berger, M. (1997) Artificial Organic Host Molecules for Anions, *Chem. Rev.* **97**, 1609-1646.
8. For a leading reference to such "inorganic" solutions, see: Zinn, A. A., Zheng, Z. P., Knobler, C. B. and Hawthorne, M. F. (1996) A hexamethyl derivative of

[9]mercuracarborand-3: Synthesis, characterization, and host-guest chemistry, *J. Am. Chem. Soc.* **118**, 70-74.

9. Davis, A. P., Gilmer, J. F. and Perry, J. J. (1996) A steroid-based cryptand for halide anions, *Angew. Chem., Int. Ed. Engl.* **35**, 1312-1315.

10. Representative examples: (a) Valiyaveettil, S., Engbersen, J. F. J., Verboom, W. and Reinhoudt, D. N. (1993) Synthesis and complexation studies of neutral anion receptors, *Angew. Chem., Int. Ed. Engl.* **32**, 900; (b) Beer, P. D., Gale, P. A. and Husek, D. (1995) A neutral upper to lower rim linked bis-calix[4]arene receptor that recognises anionic guest species, *Tetrahedron Lett.* **36**, 767; (c) Worm, K. and Schmidtchen, F. P. (1995) Molecular Recognition of Anions by Zwitterionic Host Molecules in Water, *Angew. Chem., Int. Ed. Engl.* **34**, 65-66; (d) Savage, P. B., Holmgren, S. K. and Gellman, S. H. (1994) Anion and Ion Pair Complexation by a Macrocyclic Phosphine Oxide Disulfoxide, *J. Am. Chem. Soc.* **116**, 4069-4070.

11. Davis, A. P., Perry, J. J. and Williams, R. P. (1997) Anion Recognition by Tripodal Receptors Derived from Cholic Acid, *J. Am. Chem. Soc.* **119**, 1793-1794.

12. Davis, A. P., Dresen, S. and Lawless, L. J. (1997) Mitsunobu reactions with methanesulfonic acid; The replacement of equatorial hydroxyl groups by azide with net retention of configuration, *Tetrahedron Lett.* **38**, 4305-4308.

13. Broderick, S., Davis, A. P. and Williams, R. P. (1998) The "Triamino-analogue" of Methyl Cholate; A Facial Amphiphile and Scaffold with Potential for Combinatorial and Molecular Recognition Chemistry, *Tetrahedron Lett.* **39**, 6083-6086.

14. For recently published neutral organic halide receptors, see: Kavallieratos, K., deGala, S. R., Austin, D. J. and Crabtree, R. H. (1997) A readily available non-preorganized neutral acyclic halide receptor with an unusual nonplanar binding conformation, *Journal of the American Chemical Society* **119**, 2325-2326; Jagessar, R. C. and Burns, D. H. (1997) (cis)5,10,15,20-Tetrakis [2(arylurea) phenyl] porphyrins: novel neutral ligands for remarkably selective and exceptionally strong chloride anion complexation in $(CD_3)_2SO$, *Chem. Commun.*, 1685-1686; Boerrigter, H., Grave, L., Nissink, J. W. M., Chrisstoffels, L. A. J., vanderMaas, J. H., Verboom, W., deJong, F. and Reinhoudt, D. N. (1998) (Thio)urea resorcinarene cavitands. Complexation and membrane transport of halide anions, *J. Org. Chem.* **63**, 4174-4180.

15. Dalgliesh, C. (1952) The optical resolution of aromatic amino acids on paper chromatograms, *J. Chem. Soc.* , 3940-3942.

16. Hannon, C. L. and Anslyn, E. V. (1993) The guanidinium group: its biological role and synthetic analogs, *Bioorg. Chem. Frontiers* **3**, 193.

17. Selected relevant examples: (a) Galán, A., Andreu, D., Echavarren, A. M., Prados, P. and de Mendoza, J. (1992) A Receptor for the Enantioselective Recognition of Phenylalanine and Tryptophan under Neutral Conditions, *J. Am. Chem. Soc.* **114**, 1511-1512; (b) Konishi, K., Yahara, K., Toshishige, H., Aida, T. and Inoue, S. (1994) A Novel Anion-Binding Chiral Receptor Based on a Metalloporphyrin with Molecular Asymmetry. Highly Enantioselective Recognition of Amino Acid Derivatives, *J. Am. Chem. Soc.* **116**, 1337-1344; (c) Sessler, J. L. and Andrievsky, A. (1998) Efficient transport of aromatic amino acids by sapphyrin-lasalocid conjugates, *Chem. Eur. J.* **4**, 159-167.

18. See ref. 15b and: de Mendoza, J. and Gago, F. (1994) Molecular recognition of dinucleotides and amino-acids by artifical receptors containing a bicyclic guanidinium unit, in *Computational approaches in supramolecular chemistry*, G. Wipff (ed.), Kluwer Academic Publishers, pp. 79-99.

19. Leading references: (a) Sharon, N. and Lis, H. (1995) Lectins - proteins with a sweet tooth: functions in cell recognition, *Essays in Biochemistry* **30**, 59-75; (b) Weis, W. I. and Drickhamer, K. (1996) Structural basis of lectin-carbohydrate recognition, *Ann. Rev. Biochem.* **65**, 441-473; (c) Lee, Y. C. and Lee, R. T. (1995) Carbohydrate-protein interactions: basis of glycobiology, *Acc. Chem. Res.* **28**, 321-327.

20. The irony of this point will not be lost on readers with small children.

21. It should be noted that carbohydrate recognition through formation of boronates has been remarkably successful. However, it involves covalent bonds and cannot be seen as biomimetic. For reviews, see: James, T. D., Sandanayake, K. R. A. S. and Shinkai, S. (1996) Saccharide sensing with molecular receptors based on boronic acid, *Angew. Chem., Int. Ed. Engl.* **35**, 1911-1922; Smith, B. D. (1996) Liquid membrane transport using boronic acid carriers, *Supramolecular Chemistry* **7**, 55-60.

22. Toone, E. J. (1994) Structure and energetics of protein-carbohydrate complexes, *Current Opinion in Structural Biology* **4**, 719-728; Lemieux, R. U. (1996) How water provides the impetus for molecular recognition in aqueous solution, *Acc. Chem. Res.* **29**, 373-380.

23. Selected leading references: (a) Anderson, S., Neidlein, U., Gramlich, V. and Diederich, F. (1995) A new family of chiral binaphthyl-derived cyclophane receptors: complexation of pyranosides, *Angew. Chem., Int. Ed. Engl.* **34**, 1596. (b) Bonar-Law, R. P. and Sanders, J. K. M. (1995) Polyol recognition by a steroid-capped porphyrin. Enhancement and modulation of misfit guest binding by added water or methanol, *J. Am. Chem. Soc.* **117**, 259. (c) Mizutani, T., Kurahashi, T., Murakami, T., Matsumi, N. and Ogoshi, H. (1997) Molecular recognition of carbohydrates by zinc porphyrins: Lewis acid Lewis base combinations as a dominant factor for their selectivity, *J. Am. Chem. Soc.* **119**, 8991-9001. (d) Das, G. and Hamilton, A. D. (1997) Carbohydrate recognition: Enantioselective spirobifluorene diphosphonate receptors, *Tetrahedron Lett.* **38**, 3675-3678; (e) Huang, C.-Y., Cabell, L. A. and Anslyn, E. V. (1994) Molecular recognition of cyclitols by neutral polyaza-hydrogen-bonding receptors: the strength and influence of intramolecular hydrogen bonds between vicinal alcohols, *J. Am. Chem. Soc.* **116**, 2778; (f) Jiménez-Barbero, J., Junquera, E., Martin-Pastor, M., Sharma, S., Vicent, C. and Penadés, S. (1995) Molecular Recognition Of Carbohydrates Using a Synthetic Receptor - a Model System to Understand the Stereoselectivity Of a Carbohydrate-Carbohydrate Interaction In Water, *J. Am. Chem. Soc.* **117**, 11198-11204.
24. Bonar-Law, R. P. and Davis, A. P. (1989) Synthesis of Steroidal Cyclodimers from Cholic Acid; A Molecular Framework with Potential for Recognition and Catalysis, *J. Chem. Soc., Chem. Commun.*, 1050.
25. Aoyama, Y., Tanaka, Y., Toi, H. and Ogoshi, H. (1988) Polar host-guest interaction. Binding of non-ionic polar compounds with a resorcinol-aldehyde cyclooligomer as a lipophilic polar host, *J. Am. Chem. Soc.* **110**, 634.
26. Bonar-Law, R. P., Davis, A. P. and Murray, B. A. (1990) Artificial Receptors for Carbohydrate Derivatives, *Angew. Chem., Int. Ed. Engl.* **29**, 1407; Bhattarai, K. M., Bonar-Law, R. P., Davis, A. P. and Murray, B. A. (1992) Diastereo- and Enantio-selective Binding of Octyl Glucosides by a Tetrahydroxycholaphane, *J. Chem. Soc., Chem. Commun.*, 752.
27. Bhattarai, K. M., Davis, A. P., Perry, J. J., Walter, C. J., Menzer, S. and Williams, D. J. (1997) A New Generation of "Cholaphanes": Steroid-Derived Macrocyclic Hosts with Enhanced Solubility and Controlled Flexibility, *J. Org. Chem.* **62**, 8463.

28. Davis, A. P., Menzer, S., Walsh, J. J. and Williams, D. J. (1996) Steroid-based receptors with tunable cavities; A series of polyhydroxylated macrocycles of varying size and flexibility, *Chem. Commun.* , 453-455.
29. Quiocho, F. A. (1989) Protein-Carbohydrate interactions: basic molecular features, *Pure & Appl. Chem.* **61**, 1293-1306.
30. Taking account of its presence in starch and cellulose, glucose is the world's most plentiful biomolecule. It is also among the most hydrophilic of carbohydrates. See: Miyajima, K., Machida, K., Taga, T., Homaksu, H. and Nakagaki, M. (1988) Correlation Between the Hydrophobic Nature of Monosaccharides and Cholates, and their Hydrophobic Indices, *Journal of the Chemical Society Faraday Transactions 1* **84**, 2537-2544.
31. Davis, A. P. and Wareham, R. S. (1998) A Tricyclic Polyamide Receptor for Carbohydrates in Organic Media, *Angew. Chem., Int. Ed. Engl.* **37**, 2270.
32. Two earlier systems have been able to dissolve glucose, but both have been essentially micellar in nature. See: Greenspoon, N. and Wachtel, E. (1991) Reverse Micelles as a Model System for Carbohydrate Binding, *J. Am. Chem. Soc.* **113**, 7233-7236; Kobayashi, K., Ikeuchi, F., Inaba, S. and Aoyama, Y. (1992) Accommodation of Polar Guests in Unimolecular Polyamine-Polyhydroxy Cores: Solubilization of Sugars in Apolar Organic Media via Intramolecular Polar Microsolvation, *J. Am. Chem. Soc.* **114**, 1105-1107.
33. Franks, F., Lillford, P. J. and Robinson, G. (1989) Isomeric equilibria of monosaccharides in solution; influence of solvent and temperature, *J. Chem. Soc., Faraday Trans. 1* **85**, 2417.

COMPUTATIONAL STUDIES OF THE MOLECULAR RECOGNITION OF HALIDE ANIONS BY CALIX[4]AROMATICS

NORA A. McDONALD,[&] WILLEM P. van HOORN,
ERIN M. DUFFY[‡] and WILLIAM L. JORGENSEN

Department of Chemistry, Yale University
New Haven, Connecticut 06520-8107, USA

Abstract. Monte Carlo statistical mechanics simulations have been carried out to elucidate the complexation of halide anions by a bis(phenylurea) calix[4]arene and by a calix[4]pyrrole. Optimized structures for the complexes in the gas phase are found to be maintained in chlorocabon solvents, though there are reorganization penalties of 20-30 kcal/mol for the hosts to achieve the binding geometries. All complexes feature coordination of the anions through four hydrogen bonds. The binding affinities decrease with increasing size of the anion. It is demonstrated that fluoride affinities are underestimated experimentally owing to difficulties in obtaining anhydrous tetraalkylammonium fluoride salts.

1. Introduction

Selective binding in solution requires steric and electrostatic complementarity and limited attraction to the solvent. Development of an understanding of the balance of these factors is at the heart of research on molecular recognition. Associated predictive abilities for the structures and binding strengths of complexes have important applications ranging from the design of molecular devices and supramolecular systems to the development of effective enzyme inhibitors.[1] In our work, Monte Carlo simulations with full atomic detail are used to probe molecular recognition in solution for both prototypical systems and specific applications. This course allows direct connections to be made between observed and predicted free energies of binding and provides detailed structural information that can clarify variations in binding affinities.

Though early work in the molecular recognition field centered on selective hosts for cations, interest in anion complexation has been increasing.[2] Anion binding is generally achieved in aprotic solvents with receptors incorporating ammonium, guanidinium, or Lewis-acid fragments. The present study was stimulated by recent experimental reports on the use of urea-derivatized calix[4]arenes and of calix[4]pyrroles as neutral hosts for

Current affiliations: [‡] Pfizer Inc., Groton, CT. [&] SmithKline Beecham, Inc., King of Prussia, PA

R. Ungaro and E. Dalcanale (eds.), Supramolecular Science: Where It Is and Where It Is Going, 147–156.

anion binding.[3,4] Strong binding for halide ions was obtained in chlorocarbon solvents with the bis(phenylurea) *p-tert*-butylcalix[4]arene (**1**) and the octamethylcalix[4]pyrrole (**2**) shown in Figure 1. These hosts are unusual in that they can bind anions solely through hydrogen bonding rather than via ion pairing. Such hosts may facilitate the transport of anions into non-polar media or through membranes. The studies are of interest not only owing to the unique qualities of the hosts, but also because they illustrate the importance of environmental details in binding studies. Although the observed order of binding affinities was determined to be $Cl^- > Br^- > I^-$, it was reported that for the calix[4]arenes "none of the urea derivatives show complexation with F^-".[3] Given that the observed binding affinity decreases with increasing ionic size, it could have been expected that fluoride ion would show the greatest affinity. Furthermore, with the calix[4]pyrrole, though the strongest binding was reported for fluoride ion, the fluoride/chloride selectivity was the same as the chloride/bromide selectivity.[4] This is surprising in view of the far stronger intrinsic interaction energies that are found for fluoride ion with hydrogen-bond donors.[5] The present study explores the structures and complexation energetics of the halide ions with the hosts and with prototypical Brønsted acids, water, pyrrole, imidazole, N-methylacetamide (NMA) and urea. The anomalous observations for fluoride ion binding are attributed to the inadvertent addition of water with the hygroscopic fluoride salts.

2. Methodology

Results are presented here for halide ion – molecule complexes in the gas phase and for complexation of halide ions in solution by **1** and **2**. The gas-phase interaction energies were obtained from energy minimizations and the relative binding affinities in solution were calculated with the free-energy perturbation (fep) method using Monte Carlo sampling.[1c,d] All calculations were carried out with the BOSS program.[6]

Force Field and Sampling. A classical force field is used in which the energy expression consists of harmonic terms for bond stretching and angle bending, a Fourier series for each torsional angle, and Coulomb and Lennard-Jones interactions between atoms separated by three or more bonds (eqs 1-4). The latter "non-bonded" interactions are also evaluated between intermolecular atom pairs, and they are reduced by a factor of 2 for intramolecular 1,4-interactions. An all-atom model is applied for the solutes except

$$E_{bond} = \Sigma_i\, k_{b,i}\,(r_i - r_{0,i})^2 \tag{1}$$

$$E_{bend} = \Sigma_i\, k_{\vartheta,i}\,(\vartheta_i - \vartheta_{0,i})^2 \tag{2}$$

$$E_{torsion} = \Sigma_i\, \{V_{1,i}(1 + \cos\varphi_i)/2 \;+\; V_{2,i}(1 - \cos 2\varphi_i)/2 \;+\; V_{3,i}(1 + \cos 3\varphi_i)/2\} \tag{3}$$

$$E_{nb} = \Sigma_i\, \Sigma_j\, \{q_i q_j e^2/r_{ij} + 4\,\varepsilon_{ij}\,[(\sigma_{ij}/r_{ij})^{12} - (\sigma_{ij}/r_{ij})^6]\} \tag{4}$$

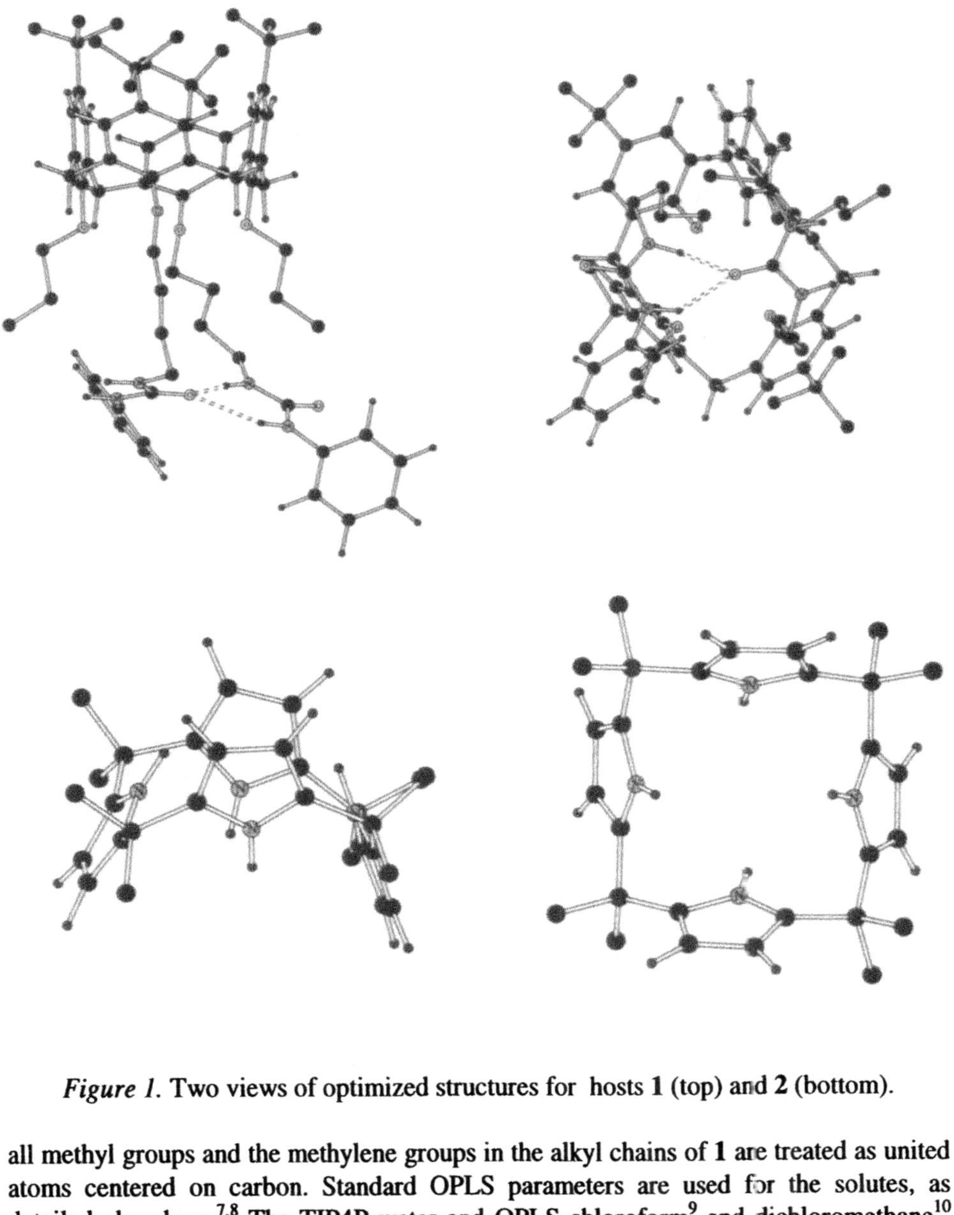

Figure 1. Two views of optimized structures for hosts **1** (top) and **2** (bottom).

all methyl groups and the methylene groups in the alkyl chains of **1** are treated as united atoms centered on carbon. Standard OPLS parameters are used for the solutes, as detailed elsewhere.[7,8] The TIP4P water and OPLS chloroform[9] and dichloromethane[10] molecules are treated as rigid bodies that only translate and rotate. The solutes are represented in internal coordinates with sampling for translations, rotations, bond angle variations, and torsional motion. Bond lengths are generally kept fixed except for ring-closure bonds. Attempted MC moves included variations of up to fifteen randomly selected internal variables of each type.

<u>Free Energy Methods.</u> The fep calculations use the Zwanzig expression (eq 5) to compute the free energy change between a reference system X and the perturbed system Y.[1c,d] The average is taken for sampling configurations of the reference system.

$$\Delta G(X \rightarrow Y) = -\, k_BT \ln \langle \exp [-(E_Y - E_X)/k_BT] \rangle_X \qquad (5)$$

The perturbations are made to convert one anion to another using the thermodynamic cycle shown below. Two series of mutations are performed to convert X to Y unbound in solution and complexed to the host, which yield ΔG_u and ΔG_c. In practice, the conversions are made in a series of steps or "windows" using a coupling parameter, λ.

$$\begin{array}{ccccc} H & + & X & \xrightarrow{\Delta G_X} & HX \\ & \Delta G_u & \downarrow & \Delta G_c & \downarrow \\ H & + & Y & \xrightarrow{\Delta G_Y} & PY \end{array}$$

Thus, one ion is gradually mutated to the other by changing the force field parameters as λ goes from 0 (X) to 1 (Y). The total free-energy change is then the sum of the incremental values. Combination yields the difference in free energies of binding for the ions X and Y via eq 6. In both cases, five windows were used with double-wide sampling such that $\Delta\lambda = 0.10$.

$$\Delta\Delta G_b = \Delta G_X - \Delta G_Y = \Delta G_u - \Delta G_c \qquad (6)$$

<u>System Setup and MC Details.</u> To correspond to the experimental conditions, the fep calculations were performed in the isothermal, isobaric (NPT) ensemble at 25 °C and 1 atm. Furthermore, the solvents were chloroform for the calixarene and dichloromethane for the calixpyrrole.[3,4] The calculations employed periodic boundary conditions and 12-Å spherical cutoffs based on the C-C distance for solvent pairs and on the distance between the solvent molecule's carbon atom and a well distributed set of solute atoms. If any of these interatomic distances is below the cutoff, the entire solute-solvent interaction is included. For the calixarene, the unbound and bound calculations used 626 and 127 chloroform molecules, respectively, and the corresponding numbers were 327 and 127 dichloromethane molecules for the smaller calixpyrrole. A rectangular cell, ca. 39 x 46 x 48 Å, contained the calixarene, ion and chloroform, while a cubic box ca. 33 Å on a side was used for the calixpyrrole complexes. Each window involved ca. 2 x 10^6 configurations of equilibration followed by 2 x 10^6 configurations of averaging for the unbound mutations and ca. 6 x 10^6 configurations of averaging for the interconversions of the complexes.

3. Results for Gas-Phase Structures and Interaction Energies

Results of gas-phase optimizations with the OPLS-AA force field for complexes of the halide ions with several Brønsted acids are summarized in Table 1. The optimal structures feature the expected hydrogen bonds, as illustrated for urea and pyrrole in Figure 2. The interaction energies for the ions with a TIP4P water molecule in Table 1 are essentially identical to the corresponding enthalpy changes from the mass spectrometry measurements of Kebarle and co-workers, namely, -23.3, -13.1, -12.6, and -10.2 kcal/mol for F^-, Cl^-, Br^-, and I^-.[11] Proper comparison requires conversion of

TABLE 1. Gas-phase interaction energies (kcal/mol) for halide ions with molecules.

Halide ion	Water	Pyrrole	NMA	Imidazole	Urea
F^-	-23.2	-25.7	-26.4	-30.1	-31.5
Cl^-	-13.2	-15.1	-16.1	-18.4	-21.2
Br^-	-12.6	-14.6	-15.5	-17.8	-20.5
I^-	-10.4	-12.2	-13.2	-15.0	-17.6

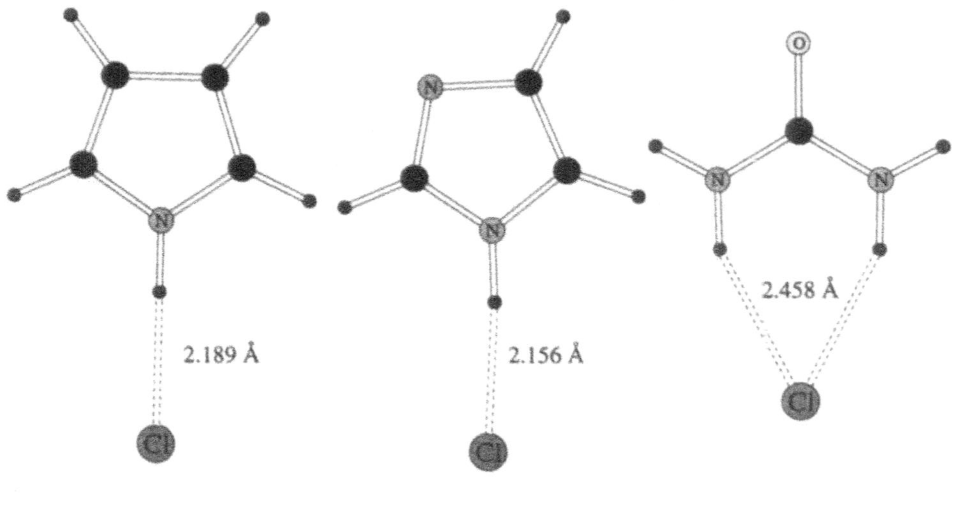

Figure 2. Optimized structures for complexes of Cl^- with pyrrole, imidazole, and urea.

the computed interaction energies to enthalpies at 298 K, which adds ca. 0.5 kcal/mol to the values in Table 1.

There are several points to note. Fluoride ion is by far the strongest base, while the differences between chloride and bromide are small. The interaction energies are most favorable with urea. This is not surprising in view of urea's ability to form two hydrogen bonds with the ions. However, this advantage is diminished since the hydrogen bonds are bent and longer than for the other donors. Imidazole is predicted here to have notably high halide affinities since it is a stronger gas-phase acid than pyrrole.[12] The results suggest that anion hosts, which incorporate imidazole units, should be considered.

The lowest-energy structures that were found for the isolated hosts are shown in Figure 1. The two urea groups of the calixarene **1** form intramolecular hydrogen bonds with NH --- O lengths of 1.78 and 2.29 Å. The conformation draws the NH that is closer to the cone near the face of the phenyl ring on the opposite side chain. This gas-phase structure agrees well with the NMR picture for the host in chloroform solution. Hydrogen bonding of the urea groups is indicated and the 0.45 ppm upfield shift for the proximal NH proton is attributed to shielding by the opposing phenyl ring.[3] For the uncomplexed calix[4]pyrrole **2**, the only low-energy form that was found has the 1,3-alternate structure (Figure 1); other energy minima are 7-16 kcal/mol higher in energy. The 1,3-alternate is also found in the crystal structure for **2**.[4a] Other alternatives, especially the cone conformer, are destabilized by the repulsions between the amino hydrogens.

Structures for the complexes of the hosts with the halide ions were then optimized in the gas phase. The net interaction energy, ΔE, is listed for each ion in Table 2 along with its components, the total ion-host interaction energy, ΔE_{inter}, and the penalty, ΔE_{deform}, for deformation of the host from the lowest-energy unbound structure (Figure 1) to the geometry in the complex. The optimized structures for the complexes with chloride ion are shown in Figure 3. For the calixarene, the urea moieties are canted to yield a distorted tetrahedral geometry with four hydrogen bonds to the ions with NH --- Cl^- distances of ca. 2.4 Å. The optimized structures of the bromide and iodide complexes are similar with average NH --- X^- separations of 2.4 and 2.7 Å. The **1**-fluoride ion complex has the ion positioned closer to the calixarene cone with NH --- F^- distances of 1.7-1.8 Å. For the calixpyrrole, the only low-energy complexes have the cone conformation with four linear hydrogen bonds with the anions (Figure 3); the average NH --- X^- distances increase from 1.6 Å for fluoride ion to 2.7 Å for iodide. The crystal structure for the chloride complex agrees closely with the present results. The cone is observed with N --- Cl^- distances of 3.26 - 3.33 Å,[4a] while the computed average distance here is 3.31 Å.

TABLE 2. Gas-phase interaction energies (kcal/mol) for halide ions with the hosts.

	Host **1**			Host **2**		
Halide ion	ΔE_{inter}	ΔE_{deform}	ΔE	ΔE_{inter}	ΔE_{deform}	ΔE
F^-	-83.9	25.0	-58.9	-90.5	31.4	-59.1
Cl^-	-51.3	20.8	-30.5	-50.3	27.0	-23.3
Br^-	-49.3	21.1	-28.2	-47.8	26.9	-20.9
I^-	-41.5	21.3	-20.2	-37.9	26.4	-11.5

For both sets of complexes, there are substantial deformation penalties, 20-30 kcal/mol, for the hosts to distort to the binding geometries. This follows from disruption of the urea - urea hydrogen bonding for the calixarene and from flipping from the 1,3-alternate to the cone for the calixpyrrole. Clearly, neither host is well preorganized for anion binding. The raw ion-host interaction energies, ΔE_{inter}, are similar for both hosts.

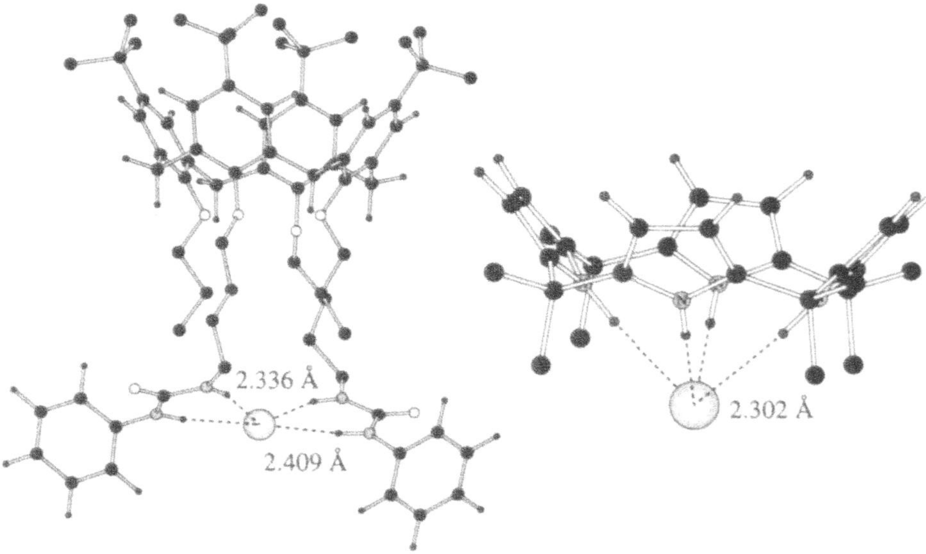

Figure 3. Optimized structures for the complexes of host **1** (left) and host **2** (right) with chloride ion in the gas phase.

However, the distortion penalties are worse for the calixpyrrole, so that its net interaction energies are ca. 8 kcal/mol less favorable except for fluoride ion. Comparisons of the ΔE_{inter} values and the results in Table 1 show that for the calixpyrrole, the electrostatic interactions in the complex are only a little less favorable than for four optimal pyrrole-X^- interactions. Thus, the hydrogen bonds in the complexes are nearly ideal, e.g., the NH --- Cl^- distance has only lengthened by 0.11 Å in going from the pyrrole complex in Figure 2 to **2**-Cl^- in Figure 3. The comparison for the calixarene is even more favorable; the ΔE_{inter} values are more negative than twice the urea-X^- interaction energies. This reflects the nearly ideal hydrogen-bond geometries in the complexes and charge differences, i.e., the phenyl ureas are better hydrogen-bond donors than urea itself, as reflected in the shorter NH --- Cl^- distances in **1**-Cl^- (Figure 3) than in the urea complex in Figure 2.

4. Results for Relative Free Energies of Binding in Solution

Results of the Monte Carlo simulations for the perturbations of chloride ion to the other halides are summarized in Table 3. The resultant $\Delta\Delta G_b$ values can be compared to the experimental data at 25 °C for **1** in chloroform and **2** in dichloromethane.[3,4] The

TABLE 3. Free energy changes (kcal/mol) for binding halide ions by the hosts.

	Host 1				Host 2			
		Calc.		Exptl.[3]		Calc.		Exptl.[4]
Cl^- to	ΔG_u	ΔG_c	$\Delta\Delta G_b$	$\Delta\Delta G_b$	ΔG_u	ΔG_c	$\Delta\Delta G_b$	$\Delta\Delta G_b$
Br^-	1.5	1.9	-0.3	-0.6	1.8	3.1	-1.3	-2.1
I^-	6.5	8.6	-2.1	-1.5	8.6	14.8	-6.2	<-2.1
F^-	-16.1	-26.4	10.4	<0	-21.1	-38.2	17.1	2.3

computed values have statistical uncertainties of 0.1 - 0.2 kcal/mol, as established from subaverages during the simulations and from duplicate simulations with different starting points. Of course, solvation and the thermal averaging diminish the intrinsic gas-phase energy differences, $\Delta\Delta E$, from Table 2. The computations do well in reproducing the differences in free energy of binding for the complexes of Cl^-, Br^-, and I^-. Theory and experiment find the order of binding affinities is $Cl^- > Br^- > I^-$ for both hosts. Furthermore, the computations predict that fluoride ion would bind with tremendous affinity to either host in solution. However, the experimental studies reported no binding of fluoride ion by the calixarene in chloroform and modest selectivity for fluoride over chloride ion with the calixpyrrole.[3,4] On the structural side, the complexes in these non-competitive solvents were found to mimic closely the optimal gas-phase structures, each with the same four ion-host hydrogen bonds.

The origin of the different results for fluoride ion binding has been considered. First, in the calculations, the halide ions are isolated in the pure solvents and in the presence of the hosts. The tetrabutylammonium counterions of the halide salts from the experiments are not included. So, one possibility is that the tetrabutylammonium fluoride may be largely ion-paired in solution and the ion-pair has diminished affinity for the hosts. Another possibility is that there is interference by water in the experiments. Difficulties in obtaining anhydrous fluoride salts are known.[13] In fact, attempts to obtain anhydrous tetrabutylammonium fluoride typically lead to decomposition including production of hydrated fluoride ion and bifluoride ion.[13] While details of the preparation of the fluoride salt were not given in reference 3, the tetrabutylammonium fluoride used with the calixpyrrole was, in fact, the trihydrate.[4] This was pursued with additional Monte Carlo fep calculations by computing the absolute free energies of binding of fluoride ion with host **1** in chloroform vs. fluoride ion with one or two water molecules in chloroform.[7] The key finding is that the free energy change would be unfavorable for a fluoride ion with two water molecules to shed the waters and bind to the calixarene in chloroform. Clearly, it would take little water to seriously distort the binding measurements. Furthermore, if one truly had uncomplexed fluoride ion in chloroform or dichloromethane, it would bind tenaciously to the hosts.

5. Conclusion

The results summarized here illustrate the capabilities of computer simulations to elucidate intermolecular interactions and to quantify binding affinities for organic host-guest complexes in solution. Methodologies based on rigorous free-energy perturbation calculations in conjunction with Monte Carlo sampling show great value as tools for investigating molecular recognition and for ligand design. The structural results can further enrich the quantitative predictions by providing additional insights on the origin of variations in binding affinities. In the present cases, the computations indicated that the gas-phase optimized structures are largely maintained in solution, though there are 20-30 kcal/mol reorganization penalties for the hosts to achieve the binding geometries. Though good agreement was obtained with experiment for the relative free energies of binding of chloride, bromide and iodide ions, the calculations predicted that fluoride ion would be bound by hosts **1** and **2** with far greater affinity than observed. The key contributor to the discrepancy is likely the introduction of water with the fluoride salts in the experiments. The results illustrate the importance of such environmental details in the outcome of binding measurements and the constructive interplay of computation and experiment.

6. Acknowledgments

Gratitude is expressed to the National Institutes of Health and the National Science Foundation for support of this research.

7. References

1. (a) Lybrand, T. P. (1995) *Curr. Opin. Struct. Biol.* **5**, 224. (b) Lamb, M. L. and Jorgensen, W. L. (1998) *Curr. Opin. Chem. Biol.* **1**, 449. (c) Jorgensen, W. L. (1991) *Chemtracts - Org. Chem.* **4**, 91. (d) Kollman, P. A. (1993) *Chem. Rev.* **93**, 2395.
2. Lehn, J.-M. (1995) *Supramolecular Chemistry*, VCH Press, Weinheim. Bianchi, A., Bowman-James, K. and Garcia-Espana, E. (1997) *Supramolecular Chemistry of Anions*, Wiley, New York. Schmidtchen, F. P. and Berger, M. (1997) *Chem. Rev.* **97**, 1609.
3. Scheerder, J.; Fōchi, M.; Engbersen, J. F. J.; Reinhoudt, D. N. (1994) *J. Org. Chem.* **59**, 7815.
4. (a) Gale, P. A., Sessler, J. L., Král, V. and Lynch, V. (1996) *J. Am. Chem. Soc.* **118**, 5140. (b) Gale, P. A., Sessler, J. L. and Král, V. (1998) *J. Chem. Soc., Chem. Commun.*, 1.

5. Gao, J., Garner, D. S. and Jorgensen, W. L. (1986) *J. Am. Chem. Soc.* **108**, 4784.
6. Jorgensen, W. L. (1997) *BOSS, Version 3.8*; Yale University; New Haven, CT.
7. McDonald, N. A., Duffy, E. M. and Jorgensen, W. L. (1998) *J. Am. Chem. Soc.* **120**, 5104.
8. van Hoorn, W. P. and Jorgensen, W. L., submitted for publication.
9. Jorgensen, W. L., Briggs, J. M. and M. L. Contreras (1990) *J. Phys. Chem.* **94**, 1683.
10. Lim, D. C., Hrovat, D. A., Borden, W. T. and Jorgensen, W. L. (1994) *J. Am. Chem. Soc.* **116**, 3494.
11. Arshadi, M., Yamdagni, R. and Kebarle, P. (1970) *J. Phys. Chem.* **74**, 1475.
12. Taft, R. W., Anvia, F., Taagepara, M., Catalán, J. and Elguero, J. (1986) *J. Am. Chem. Soc.* **108**, 3237.
13. Sharma, R. K. and Fry, J. L. (1983) *J. Org. Chem.* **48**, 2112. Christe, K. O., Wilson, W. W., Wilson, R. D., Bau, R. and Feng, J. (1990) *J. Am. Chem. Soc.* **112**, 7619.

SACCHARIDES-FROM SENSING TARGETS TOWARD COMBINATORIAL LIBRARIES AND CHIRAL COMMANDANTS

SEIJI SHINKAI and MASAYUKI TAKEUCHI

Department of Chemistry and Biochemistry, Graduate School of Engineering, Kyushu University, Fukuoka 812-8581, Japan

1. Introduction: Where We Were

Ions and molecules are abundant in nature and the need to measure the concentration of selected ions and small organic molecules both *in vivo* and *in vitro* processes can be critical. This field, namely called "Host-Guest Chemistry" or more in general "Molecular Recognition", has developed in relation to comprehension of the principle in nature, where enzymes, antibodies, and a plethora of other biological macromolecules bind ions and small organic molecules in order to perform their various physiological tasks. The secret of this process is related to molecular recognition.

In 1960's the concept of an "enthalpy-entropy compensation relationship" was proposed by Leffler[1]. This concept was very helpful to obtain insights into a number of thermodynamic data for the association and kinetic processes. However, it was rather prosaic for us because this also implies that high selectivity and high activity cannot appear in one reaction series. Then, *how can we create such an exceptional system with high selectivity and high activity which deviates from the enthalpy-entropy compensation relationship?* We reached one potential breachthrough: that is, as long as one association process or one kinetic process is treated independently, it is still restricted by the relationship whereas if two or more than two systems are linked reversibly, one may find an exceptional process diverted from the relationship through "switching" to and from other conjugated systems.

This original concept has enabled us to create a number of new ion and

R. Ungaro and E. Dalcanale (eds.), Supramolecular Science: Where It Is and Where It Is Going, 157–179.

molecule recognition systems which are combined with switch-functionalized systems and molecular-assembly systems[2-4]. The typical examples are metal cation bindings, allosteric phenomena, temperature switching, light switching, *etc.*

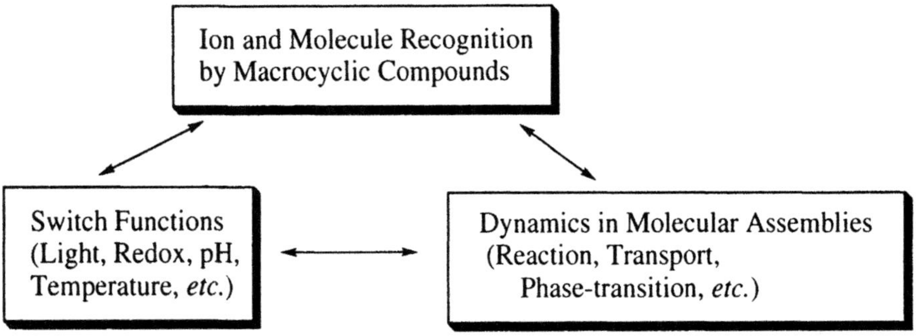

2. Aqueous Sugar Sensing By Boronic-Acid-Based Artificial Receptors: Where We Are Now

The molecular design of artificial receptors which show high affinity and high selectivity comparable with natural systems has long been a dream for scientists and has recently become a very active area of endeavor. The overview of the past literatures teaches us that the hydrogen-bonding interaction is the primary tool used for recognition of guest molecules. We have currently been interested in the sugar recognition as well as the reading-out of the recognition process and in combining these new receptors with various switch-functionalized systems[5-7]. Although hydrogen-bonding interactions are also useful for sugar recognition in several systems, the effect is exerted only in aprotic organic solvents. Hence, the hydrogen-bonding interactions are nearly useless for sugar recognition in water while sugars show the practical solubility only in water. Then, *how can we break this dilemma to "touch" sugars and "recognize" them in water?* As an attempt to solve this we and others have proposed to use a boronic acid which self-associatively forms covalent complexes with a variety of sugar molecules in water[5-13]. Since this covalent-bond formation process is reversible and its reaction rate is much faster than the human time-scale, one can treat this system like noncovalent interactions frequently used for molecular recognition. Although this strategy is quite different from that employed by Nature (using hydrogen-bonding interactions), this is undoubtedly a practical (and probably the sole) way to "*touch*" sugars in water.

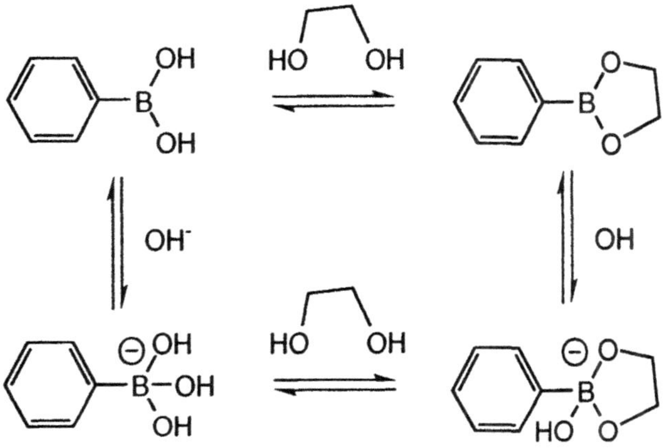

2.1. RECOGNITION OF SACCHARIDES BY MONO- AND DIBORONIC ACIDS

It is known that the acidity of monoboronic acids is intensified when they form covalent complexes with diols[3-7,11-13]. Hence, at constant pH the saccharide addition can change neutral boronic acids to anionic boronate esters. In fluorescent monoboronic acids this change is reflected as a decrease in the fluorescence intensity (I) of the neighboring fluorophore[11-13]. Typical examples are compounds **1**-**5**.

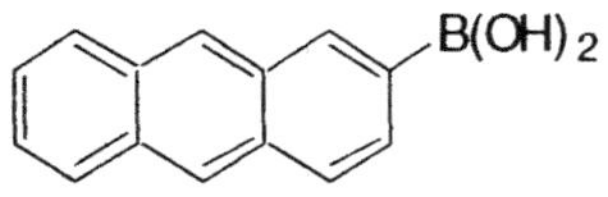

1

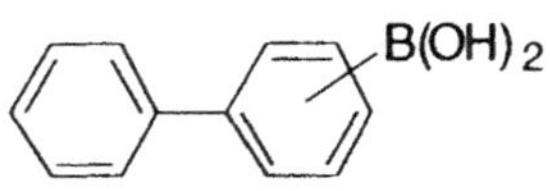

2a : 2-$B(OH)_2$
2b : 3-$B(OH)_2$
2c : 4-$B(OH)_2$

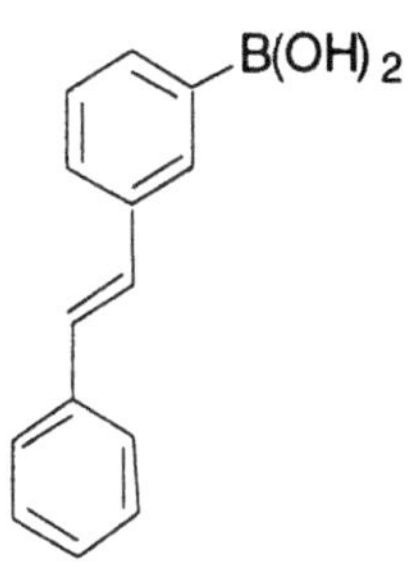

3

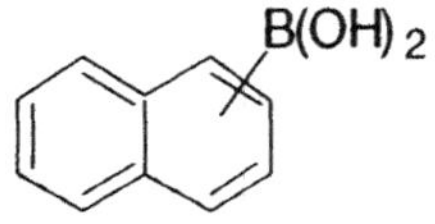

4a : 1-$B(OH)_2$
4b : 2-$B(OH)_2$

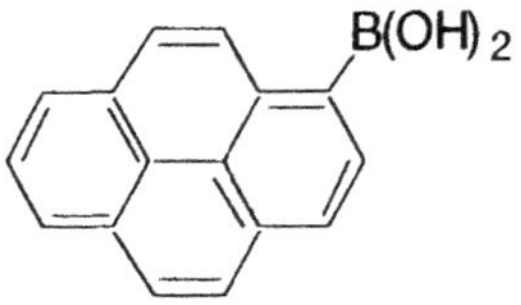

5

Regular monosaccharides have five OH groups. Since a boronic acid can react with a 1,2-diol (HO-C-C-OH) or 1,3-diol (HO-C-C-C-OH), diboronic acids can immobilize four of these five OH groups. We thus expected that diboronic acids would show selectivity toward some selected saccharides, depending on the relative spatial position of two intramolecular boronic acids.

Compound **6** is a flexible diboronic acid, but when it adopts a folded *syn* conformation, the distance between two boronic acids is comparable with that between 1,2-diol and 4,6-diol in monosaccharides (*ca.* 6 Å). It was shown that at 25 °C and pH 11.3, **6** can complex several monosaccharides such as glucose, mannose, galactose and talose to form intramolecular 2 : 1 boron / saccharide complexes (Figure 1)[8]. The highest affinity (K_{ass} = 19000 M^{-1}) was observed for glucose[8]. Thus, this is the first example that boronic acid derivatives show the highest affinity for a saccharide other than fructose.

Compound **7** was designed for disaccharides, since the spacing between the two boronic acid units is similar to the spacing between the 1,2 -diol and 4'-OH and 6'-OH of disaccharides (*ca.* 7.4 Å)[9]. In the presence of D-maltose a distinct CD band appeared. The split CD band which crosses the [θ] =0 line at 210 nm (λ_{max} = 207 nm in the absorption spectrum) is ascribed to exciton coupling. The negative sign for the first Cotton effect (223 nm) and the positive sign for the second Cotton effect (201 nm) indicate that the two dipoles along the phenylboronic acid molecular axis are oriented in a chiral, anti-clockwise direction when they interact in the excited state. These findings reveal that when **7** forms a complex with D-maltose, the two dipoles favorably adopt (S)-chirality. Very interestingly, D-cellobiose induced the positive sign for the first Cotton effect whereas D-lactose induced the negative sign.

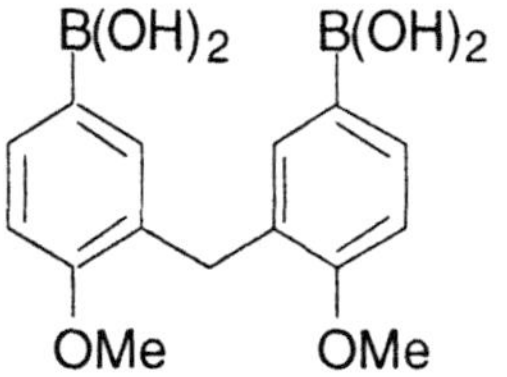

6

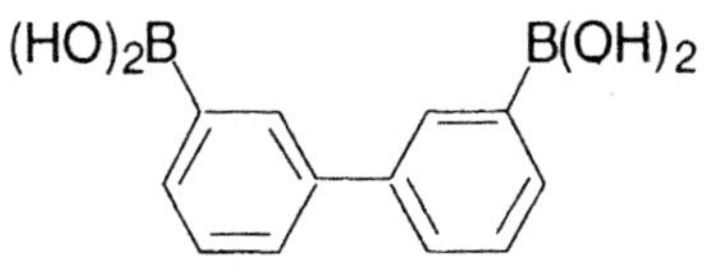

7

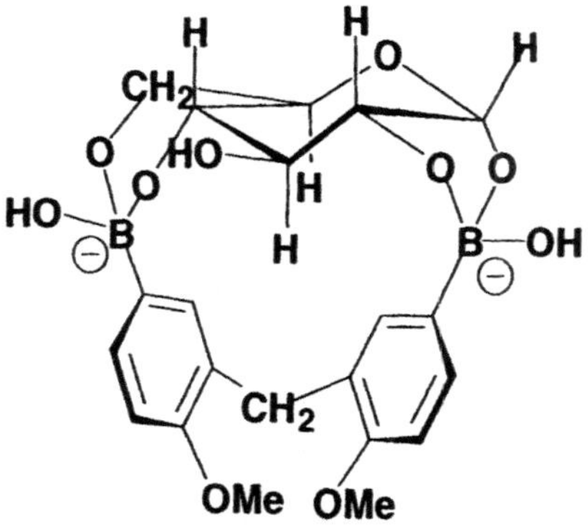

Figure 1. Proposed structure of a **6**·D-glucose complex. Here the pyranose form of D-glucose is given, but the furanose complex cannot be ruled out[14].

This rigidification process can be utilized in the design of spectroscopic sensors. The main path of nonradiative deactivation of the lowest excited singlet state of stilbene is known to be *via* rotation of the ethylenic double bond. Inhibited bond rotation, followed by enhanced fluorescence emission is known for stilbenes in solid matrices, viscous solvents and cyclodextrin inclusion complexes. The fluorescence of stilbene-3,3'-diboronic acid **8** increases upon binding to disaccharides in basic aqueous media[15]. Large fluorescence increases were observed specifically for the disaccharide D(+)-melibiose in basic aqueous media compared to small increases observed for monosaccharides (D(+)-glucose, D(+)-mannose and D(-)-arabinose). This concept, a fluorescence increase caused by rigidification of fluorophores, was extended to a cyanine dye **9** bearing conjugated C = C double bonds[16]. As expected, a large fluorescence increase was observed only when two boronic acids were intramolecularly bridged by saccharides[16].

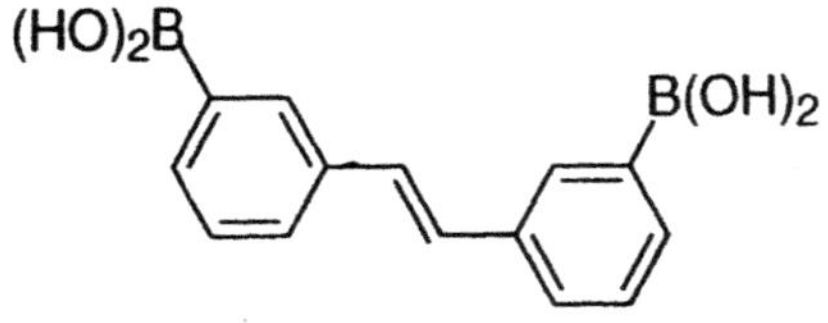

8

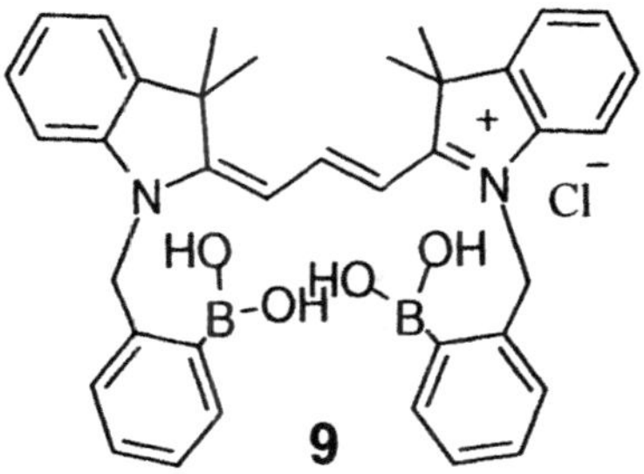

9

2.2. COMBINATION WITH THE CONCEPT OF PET (PHOTOINDUCED ELECTRON TRANSFER) SENSORS

Photoinduced electron transfer (PET) has been wielded as a tool of choice in fluorescent sensor design for protons and metal ions. Design of fluorescent sensors for neutral organic species presents a harsher challenge due to the lack of electronic changes upon inclusion. The design of a fluorescent sensor based on the boronic acid saccharide interaction has been difficult due to the lack of sufficient electronic changes found in either the boronic acid moiety or in the saccharide moiety. Furthermore, facile boronic acid-saccharide complexation occurs only at high pH conditions required to create a boronate anion. It is known that saccharide complexation changes the pK_a of the boronic acid moiety[5-13]. It has been observed that 2- and 9-anthrylboronic acids display enhanced acidity upon binding to saccharides and consequent fluorescent suppression by the boronate anion *via* a photoinduced electron transfer mechanism[11]. However, the photoinduced electron transfer from the boronate anion was not efficient despite the fact that the boronate anion is directly bound to the chromophore (I (in the presence of saccharide) / (I_0 (in the absence of saccharide) = *ca.* 0.7)[11].

In order to overcome the above-mentioned disadvantages of boronic acid-saccharide interactions, we have modified the boronic acid binding site to create a better electron center around the boronic acid moiety[5-7,17]. A given monosaccharide possesses at least two binding sites, which differ from other monosaccharides. Thus, by controlling the spatial disposition of two boronic acids, it should become possible to construct saccharide selective receptors. Our molecular design strategy is depicted in **10**[18]. Also, the "switch-on" factor (ratio of maximum to minimum fluorescence intensity) for **10** is much greater than those for **1~5**. The formation of the large macrocyclic structure upon 1:1 binding of glucose to **10** holds glucose close to the anthracene aromatic face (Figure 2).

Chiral recognition of saccharides by **11** utilizes both steric and electronic factors[19,20]. The asymmetric immobilization of the amine groups relative to the binaphthyl moiety upon 1:1 complexation of saccharides by D- or L-isomers creates a difference in PET. This difference is manifested in the maximum fluorescence intensity of the complex. Steric factors arising from the chiral binaphthyl building block are chiefly represented by the stability constant of the complex. However, the interdependency of electronic and steric factors upon each other is not excluded. This new molecular cleft, with a longer spacer unit compared to the anthracene based diboronic acid **10**, gave the best recognition for fructose. D-fructose was best bound by (R) -**11** with a large fluorescence increase. In this system steric factors and electronic factors bimodally discriminate the chirality of the saccharide.

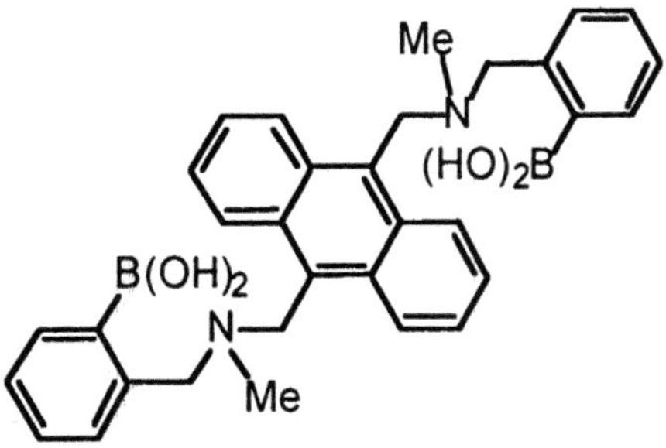

10

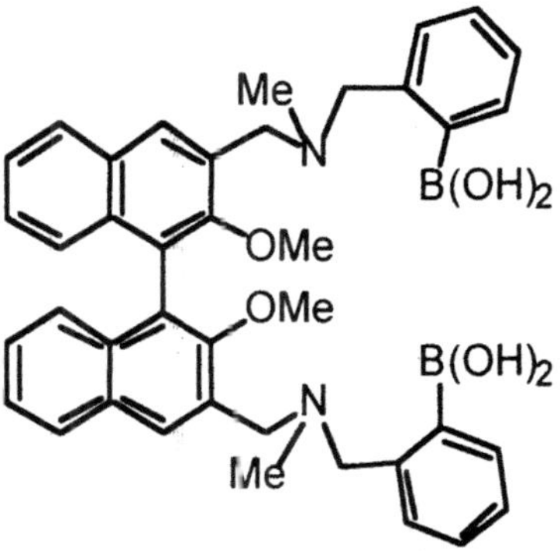

11 (R and S)

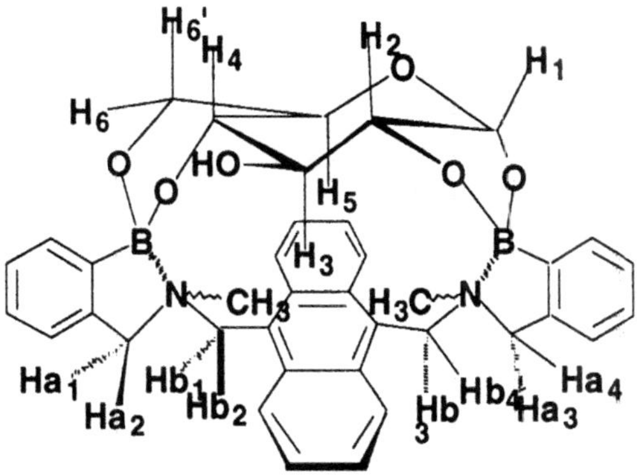

Figure 2. Proposed structure of a **10**·D-glucose complex. The ^{1}H NMR study has established that **10** immobilizes the furanose form of D-glucose[18]. This structure was identified in methanol-d_4. In an aqueous solution the furanose form tends to be immobilized [Norrild, private communication].

2.3. MOLECULAR RECOGNITION ON A PORPHYRIN PLATFORM

Porphyrins are undoubtedly rare supermolecules from viewpoints of spectroscopic properties and catalytic functions. They not only have characteristic absorption and fluorescence bands but also act as unique receptors and catalysts using the coordination ability of central metal cations[21]. We thus considered that combination of the above-mentioned concept with porphyrin chemistry would provide further extension of sugar sensing chemistry[22].

Saccharides are nature's conveyors of energy and therefore indispensable to cell survival. In particular, conversion of D-glucose to D-glucose-6-phosphate mediated by hexokinase in the presence of ATP and interconversion between D-glucose-6-phosphate (G-6-P) and a-D-glucose-1-phosphate (G-1-P) mediated by phosphoglucomutase are essential steps to produce glycogen stored as an energy source. It occurred to us that if we have some appropriate receptors in our hands which selectively bind one of these three D-glucose homologs, the equilibria may be controllable by the co-existence of such artificial receptors. Then, how can we discriminate between G-1-P and G-6-P? After trial-and-errors we reached the molecular design of a porphyrin (**12**) and its metalloporphyrin (**13**) bearing an intramolecular boronic acid[23]. Examination of CPK molecular models reveals that in G-6-P the phosphate group can interact with the central metal when the 1,2-diol is bound to the boronic acid. In G-1-P, on the other hand, the distance between the phosphate group and the 4,6-diol is relatively short and the phosphate group cannot interact with the central metal when the 4,6-diol is bound to the boronic acid. Hence, we expected that **13** would bind G-6-P by the two-points interaction in preference to G-1-P[23].

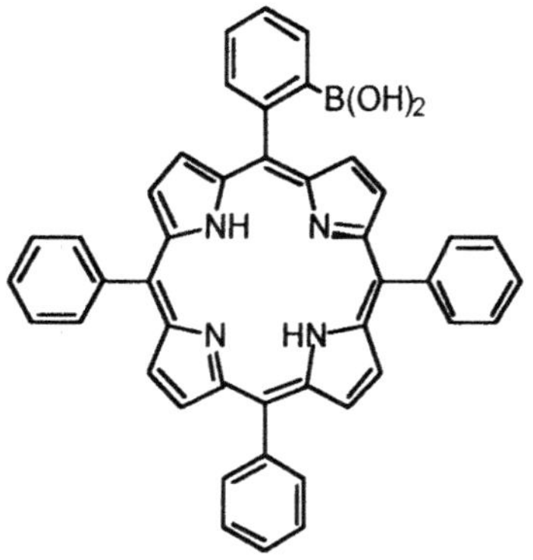

12

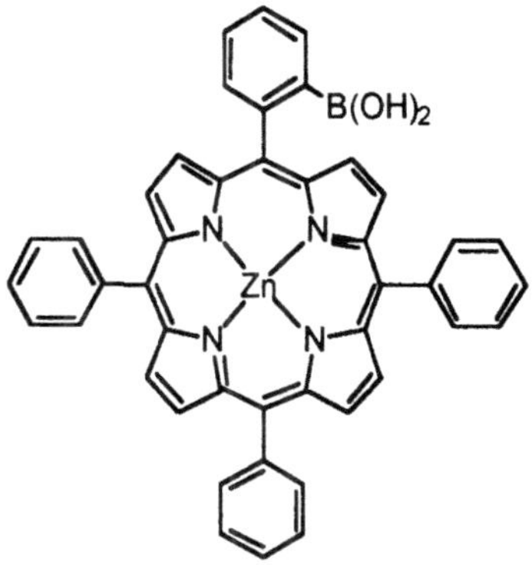

13

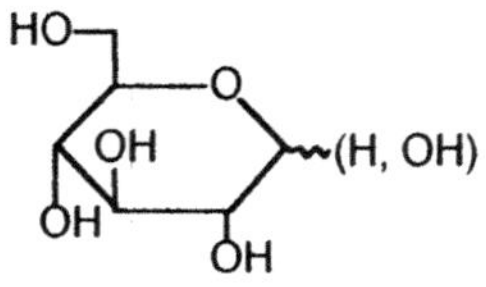

D-Glucose

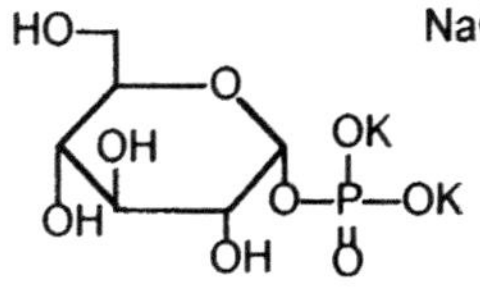

α-D-Glucose-1-phosphate dipotassium salt (G-1-P)

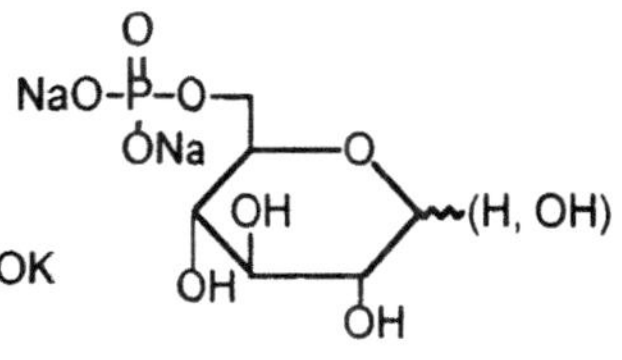

α-D-Glucose-6-phosphate disodium salt (G-6-P)

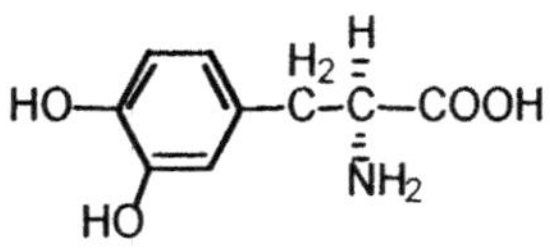

L-DOPA

The spectroscopic data consistently support the view that compound **13** is capable of discriminating G-6-P from G-1-P through the two-points interaction. Figure 3 is a proposed structure for the two-points binding mode between **13** and G-6-P.

It is known that catechol shows the very high affinity (probably the highest among diols) with boronic acids. One can expect, therefore, that guest molecules bearing a catechol moiety and a ligand group within a molecule would be selectively bound to **13** by the two-points interaction if the distance between two functional groups is appropriate for the binding[23]. We noticed that this concept can be extended to recognition of 3,4-dihydroxyphenylalanine (DOPA) and its derivatives because examination with CPK molecular models suggests that when the catechol moiety in

DOPA is bound to the boronic-acid moiety, the amine moiety can intramolecularly coordinate to the central metal. Hence, this system seems useful to discriminate DOPA

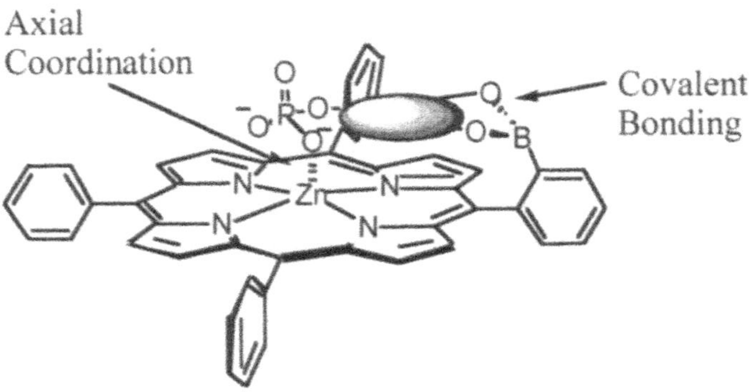

Figure 3. Complex structure proposed for the **13**·G-6-P complex.

and its derivatives which can experience the two-points interaction from analogous amino acids which can enjoy only the one-point interaction[23].

2.4. SUGAR SENSING WITH DIMERIC PORPHYRINS

In general, sugar-binding to the boronic acid group tends to make the complex more hydrophilic and deaggregate the porphyrin stack. Here, it occurred to us that simple dimeric porphyrins may satisfy the basic requirement to generate such exciton-coupling bands (ECB) in the presence of sugars if the boronic acid groups are introduced into the appropriate positions[24].

To attain this purpose one must manipulate two boronic acids in an appropriate spatial position to achieve successful two-point interrogation of a specific saccharide guest. In these systems it is known that only when two boronic acids are intramolecularly bridged by a saccharide, resultant saccharide-containing macrocycles become CD-active. In order to arrange two boronic-acid-appended porphyrins in an appropriate spatial position a μ-oxo dimer (**15**) of porphinatoiron(III) (**14**) would have a great potential: the μ-oxo dimer is stable in basic aqueous solution where the boronic acid-saccharide complex is also stable. Furthermore, the distance between two porphyrin planes (3.8 Å) is comparable with the molecular size of monosaccharides (3.0 Å). We found that **15** can selectively bind glucose and galactose (intramolecularly as in Figure 4) among many monosaccharides and their absolute configuration is readily determined from the sign of the CD spectra[25].

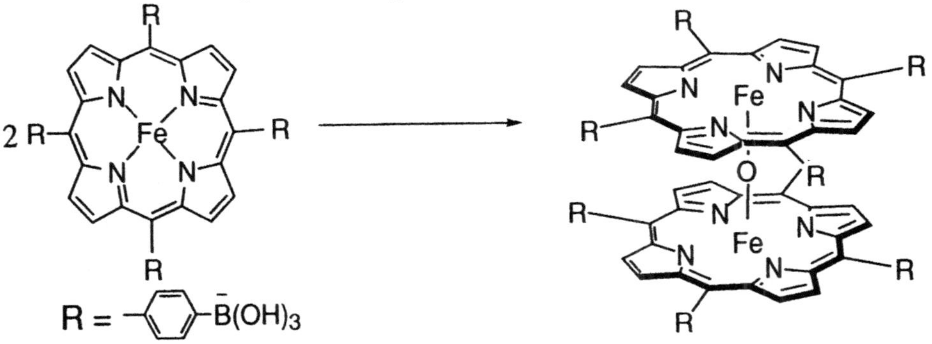

14 **15**

Examination of the absorption spectra of **15** at pH 10.5 in the absence and the presence of saccharides showed that they are scarcely affected by saccharide addition. In CD spectroscopy, only glucose and galactose could gave the strong CD bands at Soret band region among many monosaccharides tested herein[25]. The CD intensity could be detected even for ~10^{-5} M glucose and galactose. Thus, one can conclude that **15** *acts as a highly selective and sensitive "sugar tweezer" for glucose and galactose.*

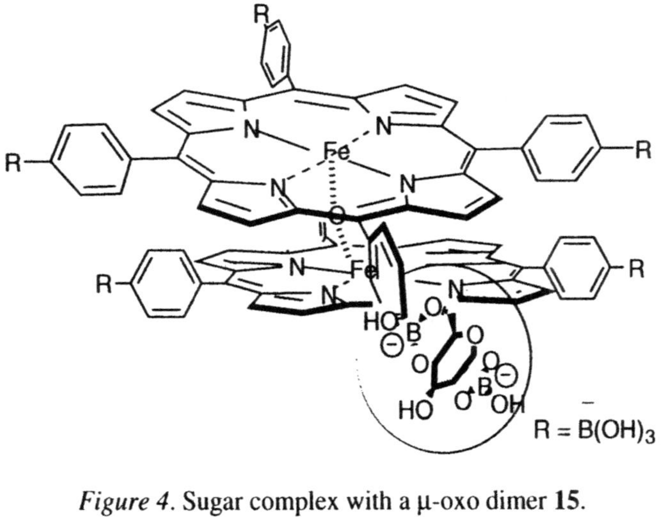

Figure 4. Sugar complex with a μ-oxo dimer **15**.

From plots of CD intensity (θ_{obs} at 380nm) vs. [saccharide] we estimated the association constants to be 1.51 x 10^5 dm^3 mol^{-1} for glucose and 2.43 x 10^4 dm^3 mol^{-1} for galactose. *These values are the largest for artificial saccharide receptors and one to two orders of magnitude greater than those achieved so far.* The results clearly indicate that μ-oxo dimers provide an excellent platform for designing boronic-acid-based saccharide receptors. Furthermore, it was shown that **15** forms only 1:1 complexes with glucose and galactose[25], indicating that only two of eight boronic acid groups are utilized for the saccharide binding. This negative allosterism is due to the inclination of two porphyrin planes which is induced by the binding of the first saccharide guest. Since the distance between the two boronic acid groups is elongated, the residual three clefts are no longer useful as the binding site[25].

2.5. WHAT HAS BECOME CLEAR FROM THE SUGAR SENSING STUDIES

The recognition of saccharides by boronic acid based molecular receptors has shown tremendous growth during the last few years. From inherent saccharide selectivity with monoboronic acids and controlled selectivity with simple diboronic acids through to the chiral recognition of saccharides. The biggest breakthrough in this study was a combination of the PET sensor concept with the boronic-acid sugar-binding, which enabled us to solve two difficult problems at one time, sugar-binding at neutral pH region and reading-out of the sugar-binding process. We believe that such sensors will find many applications in biological systems for both the monitoring and mapping of biologically important saccharides. This relatively new field will further attract many scientists' attention in the years to come.

3. Saccharides for Combinatorial Libraries and Chiral Commandants: Where We Are Going

So far, we have been devoting our research effort toward exploitation of new sugar sensing systems which are useful in aqueous solutions. There, we recognized saccharides as a sensing target: *i. e.* , we always looked at saccharides "subjectively". There remains an opposite viewpoint, however, which is scarcely developed and has escaped attention. It is to utilize them more "objectively" according to their inherent characteristics. The Nature presents a variety of saccharide structures to us which are already utilized as a "recognition tag" or a "supramolecular building-blocks" in the life processes. If we gain some efficient interfaces to control chemical and physical properties of saccharides, it will become possible to mimic this chemistry in artifical systems. As a research target to challenge this new concept, we can now raise three examples which are currently studied in our laboratory.

3.1. AS NEW CHIRAL BUILDING-BLOCKS

The distance and the angle of two porphyrins are the most important controlling factors governing the efficiency of electron and energy transfers to and from porphyrins and are known to play a decisive role in the photosynthetic system as an "*essential pair*". Thus, the reconstitution of the essential pair in an artificial system is of current interest to many photochemists. However, in most cases the molecular design, actual synthesis and fine tuning of such porphyrin dimers with a specific distance and angle are very difficult. It is known that monosaccharides can form saccharide-boronic acid complexes (1:2) and chirally arrange two pendent groups in the specific distance and angle, depending on the absolute configuration of the monosaccharides involved. Judging from the variety and diversity of monosaccharides, one can expect that the fine-tuning of the distance and the angle of the porphyrin dimers may be achieved by simply mixing a boronic acid-appended porphyrin **16** with monosaccharides[26].

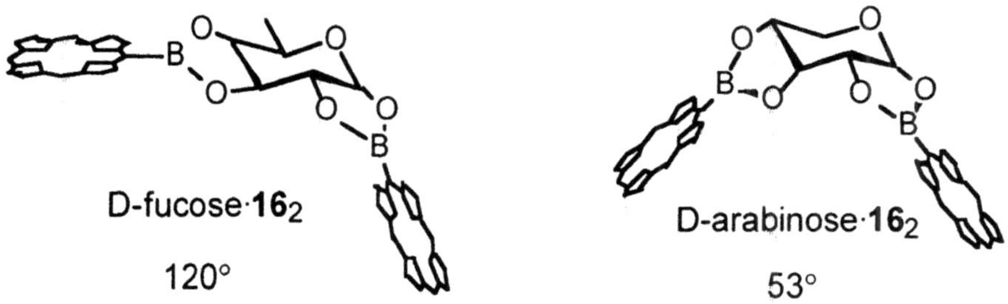

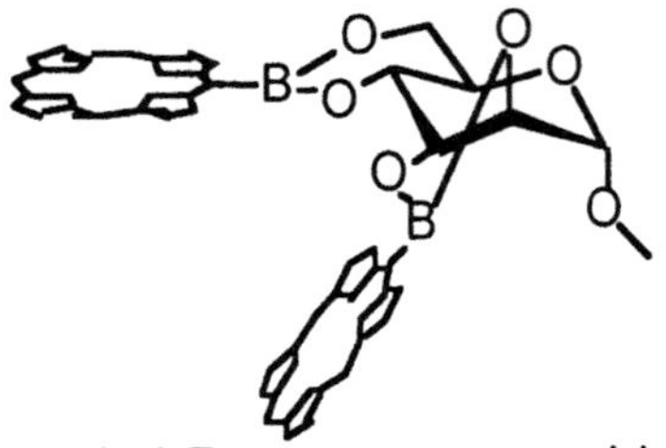

α-methyl-D-mannopyranoside·**16**$_2$

40°

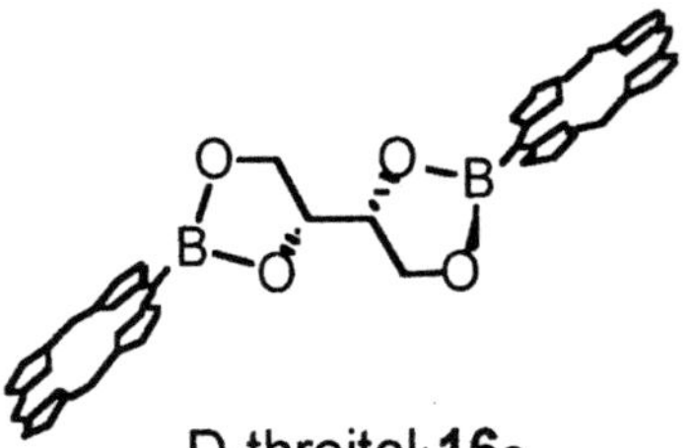

D-threitol·**16**$_2$

12°

The heredity information is transferred to descendants with the aid of DNA. To execute this function DNA has a double helix structure. To twist the polymer into such a helical structure Nature utilizes a saccharide (D-deoxyribose) as a chiral component. This idea in Nature may be applicable to the synthesis of helical polymers in an artificial system: that is, when an appropriate saccharide unit is introduced into the main chain, the resultant polymer would be twisted into a helical structure. Polymers containing sugar units in the main chain were synthesized by polycondensation of diboronic acid **17** and C_2-symmetrical chiral tetrols[27]. Spectroscopic studies and theoretical calculations established that they possess a DNA-like helical structure[27]. In particular polycondensation of diboronic acid with intramolecular amines and C_2-symmetrical chiral tetrols affords the polymers, the most stable conformation of which is regarded to be helical. Their DNA-like functions are currently being investigated in this laboratory.

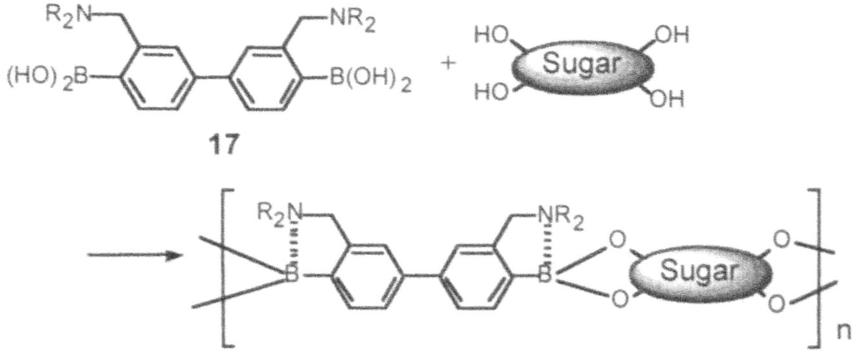

helical polymer

The development of new gelators of organic fluids has recently received much attention. They not only gelatinize various organic fluids but also create novel networks with fibrous superstructures which can be characterized by SEM pictures of the xerogels. The gelators can be classified into two categories according to the difference in the driving force for the molecular aggregation, *viz.* hydrogen-bond-based gelators and nonhydrogen-bond-based gelators. Typical examples of the former group are the aliphatic amide derivatives, whereas those of the latter group are cholesterol derivatives. The superstructures of the organic gels of aliphatic amide derivatives show that they satisfy the complementarity of the intermolecular hydrogen-bonding interactions. This observation stimulated us to use saccharides as a hydrogen-bond-forming segment in the gelators, because one can easily introduce a variety of hydrogen-bond-forming, chiral segments into gelators by selection form a saccharide library. In the literature examples of saccharide-containing gelators are very limited in spite of their high potential[28]. We synthesized glucose-based **18**, galactose-based **19** and mannose-based **20** and studied their gelation abilities[29]. We found that the gelation properties, such as gel stability, superstructure and solvent-dependence, were profoundly related to the saccharide structure[28,29].

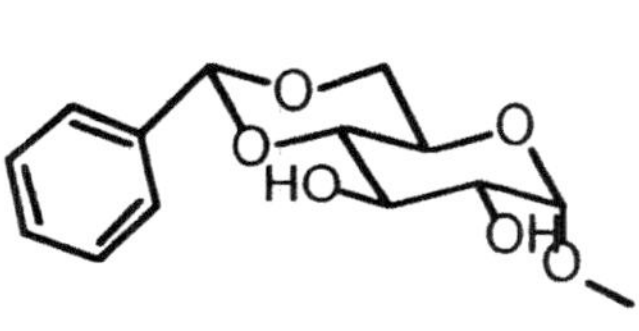

18

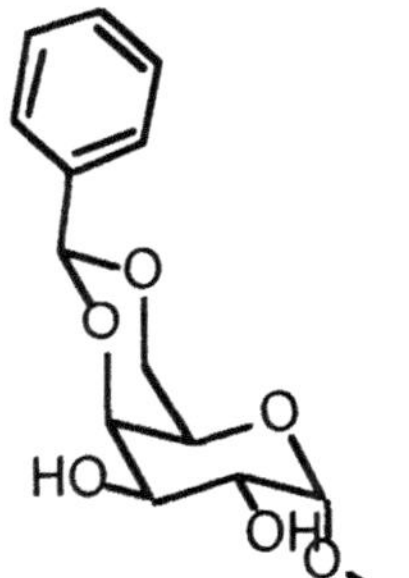

19

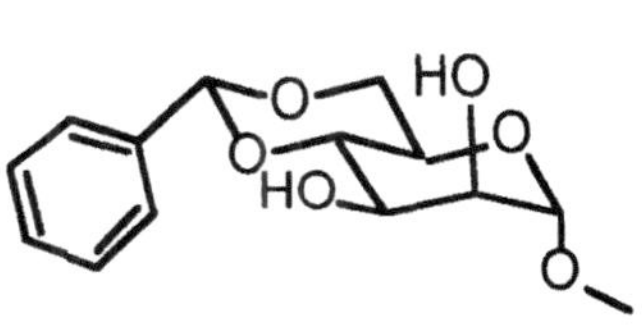

20

3.2. AS NEW CHIRAL COMMANDANTS

Helical complexes have been of great concern as examples of self-assembled supramolecular structures in an artificial system and as a model for the DNA and RNA structures in Nature. Among them, one astonishing example is the chirality control of the twisting direction in the helical structure[30,31]. These systems are created by introduction of chiral substituents into "helicates". Here, it occurred to us that the chirality of the helical metal complexes produced from helicates bearing a terminal boronic acid group may be reversibly controlled by the boronic acid-saccharide interaction. If this working hypothesis is correct, it follows that a saccharide library containing abundant chirality resources is useful to create a variety of helical structures. As the first step to test this intriguing working hypothesis, we designed compound **21** which has a 1,10-phenanthroline moiety to constitute the helical metal complex and an *o*-aminomethylphenyl boronic acid moiety to bind saccharides within a molecule. Interestingly, we have found that added saccharides can influence the equilibrium between plus (P) and minus (M) enantiomers, reflecting their absolute configurational structure[32].

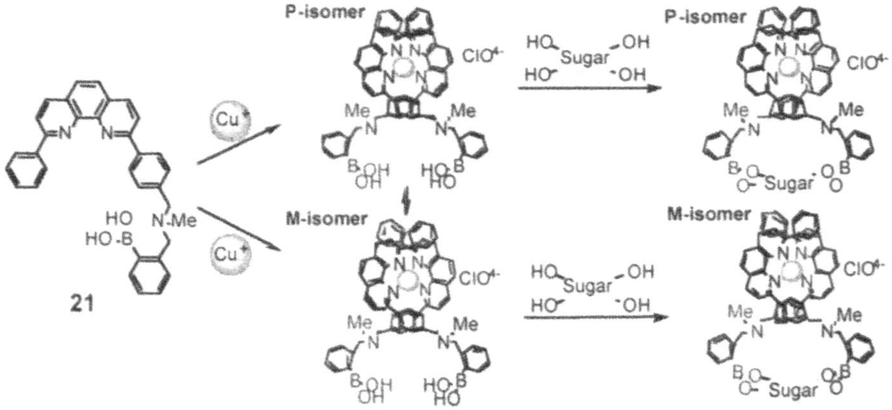

A boronic-acid-appended amphiphile bearing an azobenzene chromophore at the chain center (**22**) was synthesized. In aqueous media, compound **22** formed a scarcely-oriented aggregate in the absence of saccharides, but in the presence of saccharides the boronic acid groups formed saccharide complexes and the resultant amphiphiles formed well-ordered aggregates[33]. The ΔH value in DSC depended on the inherent structure of the added saccharides. Although D-glucose and α-1-*O*-methyl-D-glucopyranoside could only form less ordered aggregates, D-fructose and D-xylose formed the well-ordered aggregates. The saccharide complexes with **22** became CD-active with the appearance of exciton-coupling bands inherent to the saccharides. The CD band intensity was increased with increasing saccharide concentration but decreased by the further increase in the saccharide concentration. These results indicate that the reversible boronic acid-saccharide interaction is useful to induce the chirality in the ordered aggregate structure, where saccharides are used as a trigger for the chiral induction[33].

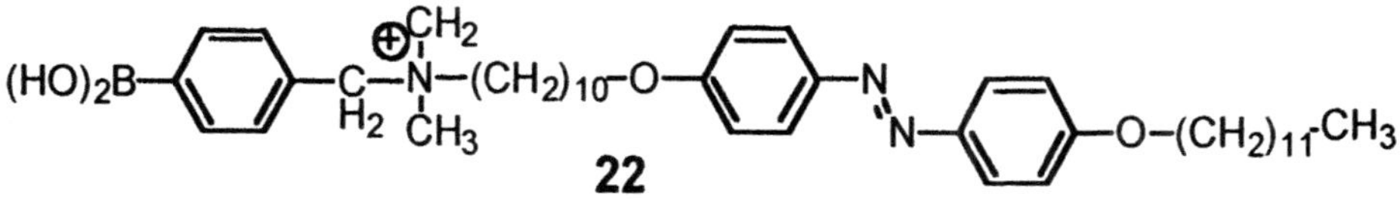

22

3.3. AS NEW CHEMICAL INFORMATION STORAGE SYSTEMS

The molecular imprinting technique attracted considerable attention in the 1970s, but recently has been revived as an active research area. The principle involves copolymerisation of vinyl monomers with divinyl monomers in the presence of guest metals or molecules to produce three-dimensional network polymers. Although this technique has achieved some degree of success, two complex problems have been left unresolved, *i.e.* the evaluation of the imprinting effect is difficult because it can be performed only in a heterogeneous system, and the storage capacity is small because only the particle surface is useful for the re-binding of guests. [60]Fullerene and its homologues are moderately soluble in organic solvents and have plenty of reactive C=C double bonds which are useful for the immobilization of functional groups. It thus occurred to us that they would be useful as a base for the imprinting of functional groups and for use in a homogeneous system. In fact, we have found that, using saccharides as template molecules, two boronic acid groups can be regioselectively introduced into [60]fullerene; the regiospectrum is closely related to the structure of saccharides used as template molecules[34].

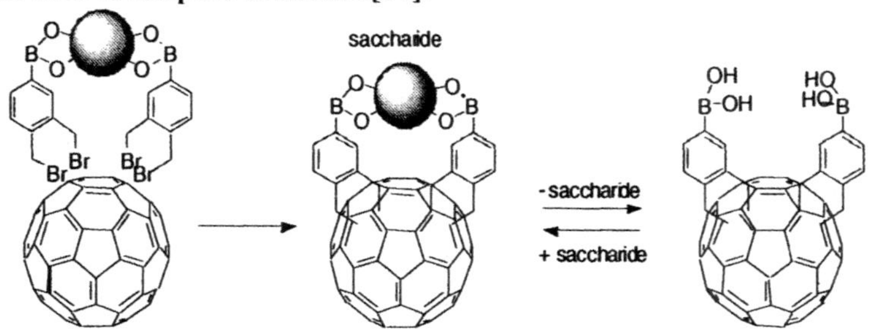

Cholesterol is a potential natural product useful as chiral building-blocks for designing host molecules, liquid crystals, molecular assemblies, *etc.* Of particular interest among those is a family of saccharide complexes with the cholesteryphenylboronic acid (**23**): the 1:2 saccharide/**23** complexes efficiently gelatinize several organic fluids and the gelation properties such as the sol-gel phase-transition temperature, the xerogel fiber structure, the gel stability difference between the D- vs. L-complexes, *etc.* are changeable by the saccharide structure acting as a central building-block of the gelator[28]. From the thermodynamic and spectral studies, it has been established that the gel fibers are more or less crystal-like and are scarcely wetted by solvent molecules. These results have stimulated us to apply the xerogel fibers of **23** to a "host" for chiral saccharide recognition: *i.e.*, the boronic acid groups arranged on the helical xerogel fibers should selectively bind D- or L-saccharides. We found that the saccharide can be extracted out of the fibers without destroying the xerogel structure and the "porous xerogel" can retain a memory for the original saccharide[35].

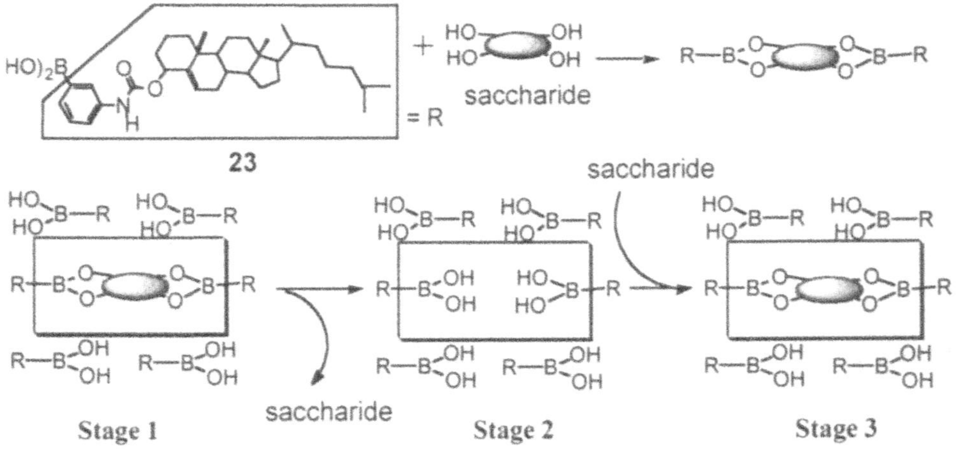

Another possible transcription method is to use chiral metal complexes. Tris[2,2'-bipyridine(bpy)]-metal complexes have inherent Δ *vs.* Λ chirality. It occurred to us that two pyridine rings may be asymmetrically oriented by a boronic acid-saccharide interaction and the metal complexes thus formed may contain an excess amount of either Δ- or Λ- isomer over the other. With this object in mind we synthesized 2,2'-bipyridine-4,4'-diboronic acid (bpyba: **24**). We found that the saccharide-induced chiral orientation of this ligand eventually leads to the enantiomeric excess of either Δ- or Λ-isomer in the resultant metal complexes[36]. Under the optimum conditions in the presence of D-glucose the enantiomeric excess reached 79% e.e.[36].

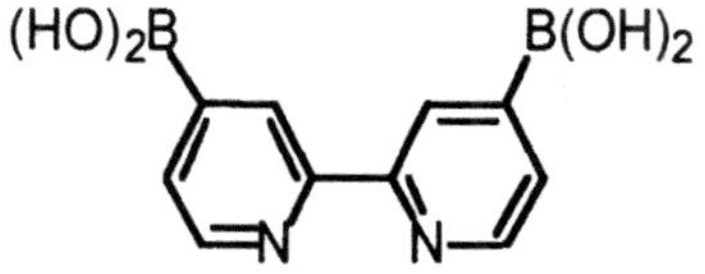

24

References

1. Leffler, J. E. (1963) *Rates and Equilibria of Organic Reactions*, Wiley, New York.
2. Shinkai, S. and Manabe, O. (1984) Photocontrol of Ion extraction and ion transport by photoresponsive crown ethers, *Top. Curr. Chem.*, **121**, 67-104.
3. Shinkai, S. (1987) Switch-functionalized systems in biomimetic chemistry, *Pure Apple. Chem.*, **59**, 425-430.
4. Shinkai, S. (1990) Dinamic Control of Cation Binding, in Y. Inoue and G. W. Gokel (eds.), *Cation Binding by Macrocycles*, Marcel Dekker, New York. pp. 397-428.
5. James, T. D., Sandanayake, K. R. A. S., and Shinkai, S. (1995) Recognition of sugars and related compounds by reading-out-type interfaces, *Supramol. Chem.*, **6**, 141-157.
6. James, T. D., Linnane, P., and Shinkai, S. (1996) Fluorescent saccharide receptors: a sweet solution to the design, assembly and evaluation of boronic acid derived PET sensors, *J. Chem. Soc., Chem. Commun.* (*Feature Article*), 281-288.
7. James, T. D., Sandanayake, K. R. A. S., and Shinkai, S. (1996) Saccharide sensing with molecular receptors based on boronic acid, *Angew. Chem. Int. Ed. Engl.*, **35**, 1910-1922.

8. Tsukagoshi, K. and Shinkai, S. (1991) Specific complexation with mono- and disaccharides that can be detected by circular dichroism, *J. Org. Chem.*, **56**, 4089-4091; Shiomi, Y., Saisho, M., Tsukagoshi, K., and Shinkai, S. (1993) Specific complexation of glucose with a diphenylmethane-3,3'-diboronic acid derivative, *J. Chem. Soc., Perkin Trans. 1*, 2111-2117.

9. Kondo, K., Shiomi, Y., Saisho, M., Harada, T., and Shinkai, S. (1992) Specific complexation of disaccharides with diphenyl-3,3'-diboronic acid that can be detected by circular dichroism, *Tetrahedron*, **48**, 8239-8252.

10. Lorand, J. P. and Edwards, J. O. (1959) Polyol complexes and structure of the benzeneboronate ion, *J. Org. Chem.*, **24**, 769-774.

11. Yoon, J. and Czarnik, A. W. (1992) Fluorescent chemosensors of carbohydrates. A means of chemically communicating the binding of polyols in water based on chelation-enhanced quenching, *J. Am.Chem. Soc.*, **114**, 5874-5875; Mohler, L. K. and Czarnik, A. W. (1993) α-Amino acid chelative complexation by an arylboronic acid, *Ibid.*, **115**, 2998-2999.

12. Nagai, Y., Kobayashi, K., Toi, H., and Aoyama, Y. (1993) Stabilization of sugar-boronic esters of indolylboronic acid in water *via* sugar-indol interaction: A notable selectivity in oligasaccharides, *Bull. Chem. Soc., Jpn.*, **66**, 2965-2971.

13. Suenaga, H., Mikami, M., Sandanayake, K. R. A. S., and Shinkai, S. (1995) Screening of fluorescent boronic acids for sugar sensing which show a large fluorescence change, *Tetrahedron Lett.*, **36**, 4825-4828.

14. Norrild, J. C. and Eggert, H. (1995) Evidence for mono- and bisdentate boronate complexes of glucose in the furanose form, *J. Am. Chem. Soc.*, **117**, 1479-1484.

15. Sandanayake, K. R. A. S., Nakashima, K., and Shinkai, S. (1994) Specific recognition of disaccharides by *trans*-3,3'-stilbenediboronic acid, *J. Chem. Soc., Chem. Commun.*, 1621-1622.

16. Takeuchi, M., Mizuno, T., Shinmori, H., Nakashima, M., and Shinkai, S. (1996) Fluorescence and CD spectroscopic sugar sensing by a cyanine-appended diboronic acid probe, *Tetrahedron*, **52**, 1195-1204.

17. James, T. D., Sandanayake, K. R. A. S., and Shinkai, S. (1994) Novel photoinduced electron-transfer sensor for saccharides based on the interaction of boronic acid and amine, *J. Chem. Soc., Chem. Commun.*, 477-478.
18. James, T. D., Sandanayake, K. R. A. S., and Shinkai, S. (1994) A glucose-selective molecular fluorescence sensor, *Angew. Chem. Int. Ed. Engl.*, **33**, 2207-2209; James, T. D., Sandanayake, K. R. A. S., Iguchi, R., and Shinkai, S. (1995) Novel saccharide-photoinduced electron transfer sensors based on the interaction of boronic acid and amine, *J. Am. Chem. Soc.*, **117**, 8982-8987.
19. James, T. D., Sandanayake, K. R. A. S., and Shinkai, S. (1995) Chiral discrimination of monosaccharides using a fluorescent molecular sensor, *Nature*, **374**, 345-347.
20. Chiral discrimination with 79% e.e. was also attained by a diboronic-acid appended binaphthyl derivative although the PET mechanism was not integrated: Takeuchi, M., Yoda, S., Imada, T., and Shinkai, S. (1997) Chiral sugar recognition by a diboronic-acid-appended binaphthyl derivative through rigidification effect, *Tetrahedron*, **53**, 8335-8348.
21. For recent comprehensive reviews see Collman, J. P., Wagenknecht, P. S. and Hutchison, J. E. (1994) Molecular catalysts for multielectron redox reactions of small molecules, *Angew. Chem. Int. Ed. Engl.*, **33**, 1537-1554; Ogoshi, H. and Mizutani, T. (1998) Multifunctional and chiral porphyrins-model receptors for chiral recognition, *Acc. Chem. Res.*, **31**, 81-89.
22. Shinkai, S. (1997) Aqueous sugar sensing by boronic-acid-based artificial receptors, in J. P. Desvergne and A. W. Czarnik (eds.), *Chemosensors of Ion and Molecule Recognition*, Kluwer Academic Pub., Dordrecht, pp. 37-59.
23. Imada, T., Kijima, H., Takeuchi, M., and Shinkai, S. (1996) Selective binding of glucose-6-phosphate, 3,4-dihydroxyphenylalanine (DOPA) and their analogs with a boronic-acid-appended metalloporphyrin, *Tetrahedron*, **52**, 2817-2826.
24. Arimori, S., Takeuchi, M., and Shinkai, S. (1996) Sugar-sensing by chiral orientation of dimeric boronic-acid-appended porphyrins which show selectivity for glucose and xylose, *Chem. Lett.*, 77-78.

25. Takeuchi, M., Imada, T., Shinkai, S. (1996) Highly selective and sensitive "sugar tweezer" designed from a boronic-acid-appended μ-oxobis[porphinatoiron(III)], *J. Am. Chem. Soc.*, **118**, 10658-10659; Idem (1998) Molecular design of highly selective and sensitive "sugar tweezers" from boronic-acid-appended μ-oxobis[porphinatoiron(III)]s, *Bull. Chem. Soc. Jpn.*, **71**, 1117-1123.

26. Takeuchi, M., Chin, Y., Imada, T., and Shinkai, S. (1996) Sugar-boronic acid interactions in the formation of novel chiral porphyrin dimers with various porphyrin-porphyrin angles, *J. Chem. Soc., Chem. Commun.*, 1867-1868. More recently, it was shown that the more likely structure for the D-threitol moiety in the D-threitol·$\mathbf{17}_2$ complex involves the two six-membered rings rather than the two five-membered rings: Ishi-i, T., Nakashima, K., Shinkai, S., and Araki, K. (1998) Structure determination of a 1:2 threitol-boronic acid complex, *Tetrahedron*, **54**, 8679-8686.

27. Mikami, M. and Shinkai, S. (1995) Synthesis of helical polymers by polycondensation of diboronic acid and chiral tetrols, *Chem. Lett.*, 603-604.

28. James, T. D., Murata, K., Harada, T., Ueda, K., and Shinkai, S. (1994) Chiral discrimination of monosaccharides through gel formation, *Chem. Lett.*, 273-276 and references cited therein.

29. Yoza, K., Ono, Y., Yoshihara, K., Akao, T., Shinmori, H., Shinkai, S., and Reinhoudt, D. N. (1998) Sugar-integrated gelators of organic fluids, *J. Chem. Soc., Chem. Commun.*, 907-908.

30. Zagres, W., Hall, J., and Lehn, J.-M. (1991) Helicity induction in helicate self-organisation from chiral tris(bipyridine) ligand strands, *Helv. Chim. Acta*, 1843-1852.

31. Woods, C. R., Benaglia, M., Cozzi, F., and Siegel, J. S. (1996) Enantioselective synthesis of copper (I) bipyridine based helicates by chiral templating of secondary structure: transmission of stereochemistry on the nanometer scale, *Angew. Chem. Int. Ed. Engl.*, **35**, 1830-1833.

32. Yamamoto, M., Takeuchi, M., and Shinkai, S. (1998) Chirality control of a Cu(I) complex of boronic-acid-appended phenanthrolines by sugars, *Tetrahedron Lett.*, **39**, 1189-1192.

33. Kimura, T., Takeuchi, M., Hamachi, I., and Shinkai, S. (1996) Sugar-induced chiral orientation of boronic-acid-appended amphiphile clusters formed in the DPPC membrane, *Israel J. Chem.*, **36**, 379-388; Kimura, T., Takeuchi, M., and Shinkai, S. (1998) Saccharide induction of the aggregate formed from boronic-acid-appended amphiphiles, *Bull. Chem. Soc. Jpn.*, in press.

34. Ishi-i, T., Nakashima, K., and Shinkai, S. (1998) Regioselective introduction of two boronic acid groups into [60]fullerene using saccharides as imprinting templates, *J. Chem. Soc., Chem. Commun.*, 1047-1048.

35. Inoue, K., Ono, Y., Kanekiyo, Y., Ishi-i, T., Yoshihara, K., and Shinkai, S. (1998) Chiroselective re-binding of saccharides to the fibrous aggregates prepared from organic gels of cholesterylphenylboronic acid, *Tetrahedron Lett.*, **39**, 2981-2984.

36. Mizuno, T., Takeuchi, M., Hamachi, I., Nakashima, K., and Shinkai, S. (1997) Chiroselective transcription of the sugar structure to Δ- or Λ-$[Co^{III}(bpy)_3]^{3+}$ using a boronic acid-sugar template interaction, *J. Chem. Soc., Chem. Commun.*, 1793-1794; Idem (1998) A boronic acid-diol interaction is useful for chiroselective transcription of the sugar structure to the Δ- versus Λ-$[Co^{III}(bpy)_3]^{3+}$ ratio, *J. Chem. Soc., Perkin Trans. 2*, in press.

SYNTHESIS AND CHARACTERIZATION OF HYDROGEN-BONDED ASSEMBLIES: TOWARD THE GENERATION OF BINDING SITE DIVERSITY

D. N. REINHOUDT,* P. TIMMERMAN, F. CARDULLO, M. CREGO-CALAMA
Laboratory of Supramolecular Chemistry and Technology, University of Twente, P.O. Box 217, 7500 AE Enschede, The Netherlands

ABSTRACT:

Discrete assemblies of nano-size dimensions were synthesized by non-covalent chemistry. Calix[4]arenes, diametrically substituted at the upper rim with two melamine units, form stable box-like assemblies in solution in the presence of two equivalents of 5,5-diethylbarbituric acid (**DEB**). Various techniques, such as NMR spectroscopy, MALDI-TOF mass spectrometry, and X-rays crystallography have been used for the characterization of these assemblies held together by 36 hydrogen bonds. Furthermore, assemblies with different types of functionalities at the upper rim of the calix[4]arene and at the melamine units were prepared. The effect of these functional groups on the stability of the assembly has been investigated. Small libraries of functionalized assemblies were formed by mixing the individual components under thermodynamically controlled conditions. This strategy provides access to functional group diversity.

1. Introduction

Organic synthesis [1,2] has, until the present time, been concerned with the construction of assemblies of atoms by the controlled, selective, and sequential formation of covalent bonds. This has led to synthesis of relatively large molecules such as vitamin B_{12} [3,4] and palytoxin [5]. The constructing molecules of nanometer-size dimensions using covalent synthesis will become increasingly difficult [6]. Therefore reversible non-covalent assembly is increasingly being considered as a valuable alternative [7]. Non-convalent assembly offers several attractive features. First of all it involves the use of relatively simple building blocks. Secondly, the assembly of the aggregates can be controlled by varing the reaction conditions and finally the reversibility of the process offers the possibility of error checking.

In 1992 we started a program for the preparation of artificial receptors [8,9] based on the spontaneous assembly *via* hydrogen-bond formation [10]. The central idea involves the assembly of discrete aggregates that are stable in solution with a functionalized cavity for specific guest binding. The modules are designed to provide ample possibilities for

R. Ungaro and E. Dalcanale (eds.), Supramolecular Science: Where It Is and Where It Is Going, 181–195.

structural variations in terms of size and shape of the nano-sized cavity, and functional groups on the concave inside, thus enabling the fine tuning of the binding properties of the receptors. Moreover, in an ambitious effort to mimic the properties of natural antibodies, binding site diversity is targeted *via* the generation of non-covalent libraries of the functionalized assemblies.

In this paper we report our recent results in this area. The following points are discussed: (*i*) the hydrogen-bonding motif and molecular modules, (*ii*) the assembly process, (*iii*) the development of suitable characterization methods for self-assembled systems, (*iv*) the generation of functional group diversity, and (*v*) the progress towards the creation of a cavity within the assembly.

2. Assembly and Characterization of Hydrogen-Bonded Molecular Boxes

2.1. SELF-ASSEMBLY OF HYDROGEN-BONDED MOLECULAR BOXES

Complementary hydrogen bonding between cyanuric acid and melamine has received considerable attention as a structural motif for self-assembly. Work from the groups of Whitesides [11] and Lehn [12] has shown that the recognition between these components can be used to form different types of planar (2D) aggregates, either linear or crinkled tapes, or cyclic rosettes. Which type of aggregate is formed critically depends on the extent of preorganization of the complementary binding units and on steric interactions between their substituents. Due to the facile synthesis and the high degree of controllability we have decided to use this recognition motif for the self-assembly of hydrogen bonded box-like assemblies [8,13].

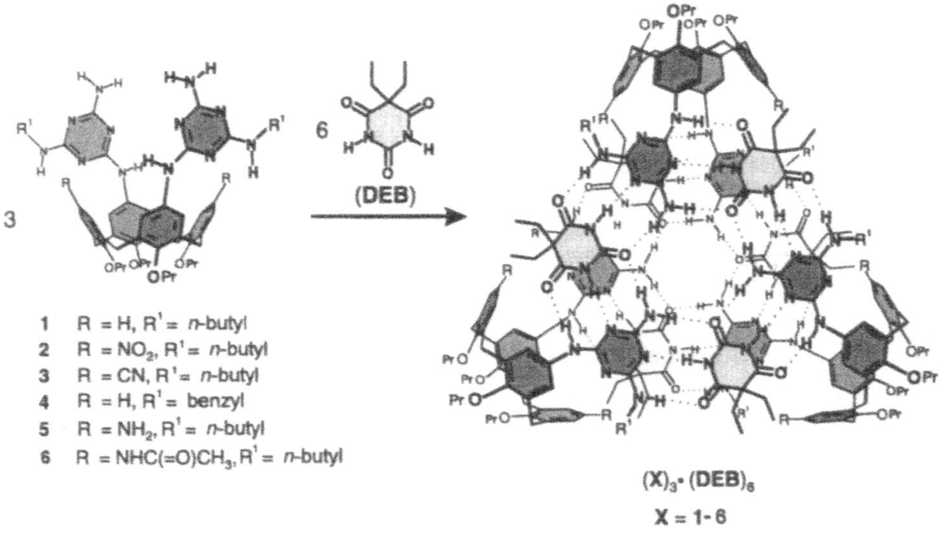

Figure 1. Calix[4]arene-based bis-melamines form well-defined assemblies with **DEB** when mixed in $CHCl_3$ (**1-4**) or toluene (**1-5**)

Calix[4]arene **1**, diametrically substituted at the upper rim with two melamine units is a perfect U-turn which has the correct structure to spontaneously form a well-defined 3D box-like assembly $(\mathbf{X})_3 \cdot (\mathbf{DEB})_6$ in the presence of two equivalents of 5,5-diethylbarbituric acid (**DEB**) (Figure 1). The assembly comprises three bismelamine and six barbiturate units and is held together by a total number of 36 hydrogen-bonds. The top and bottom of this box-like assembly consist of the cyclic hydrogen-bonded platform whereas the calix[4]arene units act as side walls. The assembly is stable in apolar solvents like $CHCl_3$ or toluene even at 0.1 mM concentration, but the stability significantly decreases when competitive solvents like Me_2SO or MeOH (10-20%) are added [8].

The characterization of large multicomponent assemblies held together by weak forces is far from straightforward and is one of the major challenges in supramolecular chemistry [14]. Structural proof is in general provided by a combination of several different techniques. Here we will mainly discuss the use of ^{1}H NMR spectroscopy and matrix assisted laser desorption ionization time of flight (MALDI-TOF) mass spectrometry for the structure assignment of the assemblies. Single crystal X-ray diffraction studies have also been performed to characterize one of the assemblies, and they will be only briefly reported in the following sections.

2.2. CHARACTERIZATION OF HYDROGEN-BONDED ASSEMBLIES BY NMR SPECTROSCOPY

Titration of bis-melamine **7** with **DEB** in $CDCl_3$ shows several characteristic features (see Figure 2) [8]. Already at low concentrations of **DEB** two signals become evident at very low field. These resonances, which remain at $\delta = 14.10$ and 13.32 ppm regardless of the **7**:**DEB** ratio, are assigned to the hydrogen-bonded NH protons of **DEB** in the complex. These protons are observed at different chemical shifts as a result of the unsymmetrical substitution of the melamine units, which gives both protons a chemically different environment in the complex. When the amount of **DEB** is increased, the signal corresponding to the NH_2 protons of free **7** decreases in intensity but remains at the same position. At the same time, two signals are observed around $\delta = 6.9$ (c) and 6.7 (d) ppm for both the NH_2 protons of **7** in the hydrogen-bonded complex. Two additional signals appear at $\delta = 8.37$ (e) and 7.43 (f) ppm, which correspond to the two secondary amine protons. The aromatic protons *ortho* to the melamine substituents of calix[4]arene **7** give rise to resonances at $\delta = 7.15$ (g) and 6.03 (h) ppm. In free **7** these protons display broad signals at $\delta = 6.65$-6.05 ppm. The resonance at $\delta = 6.03$ ppm is in accordance with a *pinched cone* structure in which the two melamine units approach each other. This conformation has also been observed in the solid state assembly which has been determined for double rosette $(\mathbf{2})_3 \cdot (\mathbf{DEB})_6$ (see Figure 8) [13]. The *pinched cone* conformation is the only conformation that allows simultaneous participation of the calix[4]arene units in both rosette motifs.
At a 1:2 ratio of **7**:**DEB** the spectrum is sharp, which indicates the absence of free **7**. The 1:2 ratio is consistent with the box-like assembly that is represented in Figure 1. When more than two equivalents of **DEB** are added, the signal for the NH protons of free **DEB** is observed

alongside the two signals for the hydrogen-bonded **DEB**. This indicates that exchange between hydrogen-bonded and free **DEB** is slow on the NMR time scale.

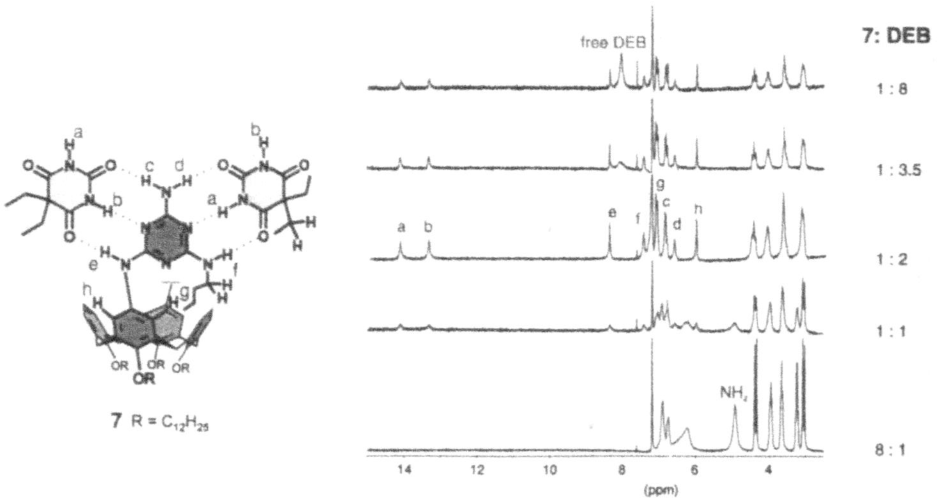

Figure 2. Titration of bismelamine derivative **7** with **DEB**

2D NOESY experiments revealed the distances between some relevant protons in the box-like assembly [8]. These studies provide sufficient evidence for the D_3-symmetric structure of assemblies $(\mathbf{X})_3\cdot(\mathbf{DEB})_6$ (**X** = **1-7**) in solution which is the result of an *anti-parallel* orientation of the two melamines at each calix[4]arene with respect to each other (Figure 1). This is in particular supported by a very strong NOE connectivity between the NCH_2 protons of **2** (R = NO_2) and one of the ethyl-CH_2 groups of **DEB**, which can only arise from the close proximity of protons which are part of the two different rosettes within the same assembly. The *anti-parallel* orientation is also observed in the solid state structure (Figure 9).

2.3. CHARACTERIZATION OF HYDROGEN-BONDED ASSEMBLIES BY MALDI-TOF MASS SPECTROMETRY

The characterization of weakly bound non-covalent assemblies by mass spectrometry, the only technique that provides quantitative data on the molecular composition, is still met with very limited success [15]. For the detection of hydrogen-bonded assemblies only few examples have been reported [16,17]. In particular ion-labeling ESMS has been applied by Lehn et al and Whitesides et al [16], but the methods they used either require covalent attachment of benzo[18]crown-6 moieties to one of the components or work only for extremely stable assemblies.

Here we describe a novel Ag^+-labeling technique for the mass spectrometric characterization of multicomponent hydrogen-bonded assemblies. The method is based on the remarkably high affinity of Ag^+ ions for cyano groups [18] and a variety of aromatic π donors [19,20] and provides for a *nondestructive* way to generate positively charged hydrogen-bonded assemblies that can be easily detected by MALDI-TOF mass spectrometry [21]. The method is applicable to very stable assemblies such as double rosettes but also to single rosettes with have a much lower thermodynamic stability.

Cyano subtituents, which are known to have an extremely high affinity for Ag^+ cation, can be easily incorporated either into the calix[4]arene or the cyanurate component of the rosette [22]. Samples prepared by stirring assemblies $(\mathbf{3})_3\cdot(\mathbf{DEB})_6$ with 1.5-2.0 equivalents of $AgCF_3COO$ in $CHCl_3$ for at least 24 hours, show an intense signal at *m/z* = 4220.0 (calcd. for $^{13}C_2{}^{12}C_{214}H_{282}N_{54}O_{30}\cdot{}^{107}Ag^+$: 4221.9) in the MALDI-TOF mass spectrum. Signals corresponding to partially formed aggregates or higher oligomers are not observed in the spectrum which revealed the mildness of this method. Formation of the silver complex of assembly $(\mathbf{3})_3\cdot(\mathbf{DEB})_6$ must result from coordination of a silver cation to one of the cyano groups, since neither $(\mathbf{1})_3\cdot(\mathbf{DEB})_6$ nor $(\mathbf{2})_3\cdot(\mathbf{DEB})_6$ shows any significant signal between *m/z* = 1500 and 8000 in the corresponding MALDI-TOF mass spectrum under the same conditions. However, samples $(\mathbf{1})_3\cdot(\mathbf{CNC})_6$ or $(\mathbf{2})_3\cdot(\mathbf{CNC})_6$ containing the *para*-cyano-substituted cyanuric acid derivative (**CNC**, Figure 3) treated as described above, give intense signals in the MALDI-TOF spectra at *m/z* = 4348.1 (calcd for $^{13}C_2{}^{12}C_{220}H_{252}N_{60}O_{30}\cdot{}^{107}Ag^+$: 4347.9) and 4618.8 (calcd for $^{13}C_2{}^{12}C_{220}H_{246}N_{66}O_{42}\cdot{}^{107}Ag^+$: 4620.4), respectively, for the monovalent silver complexes.

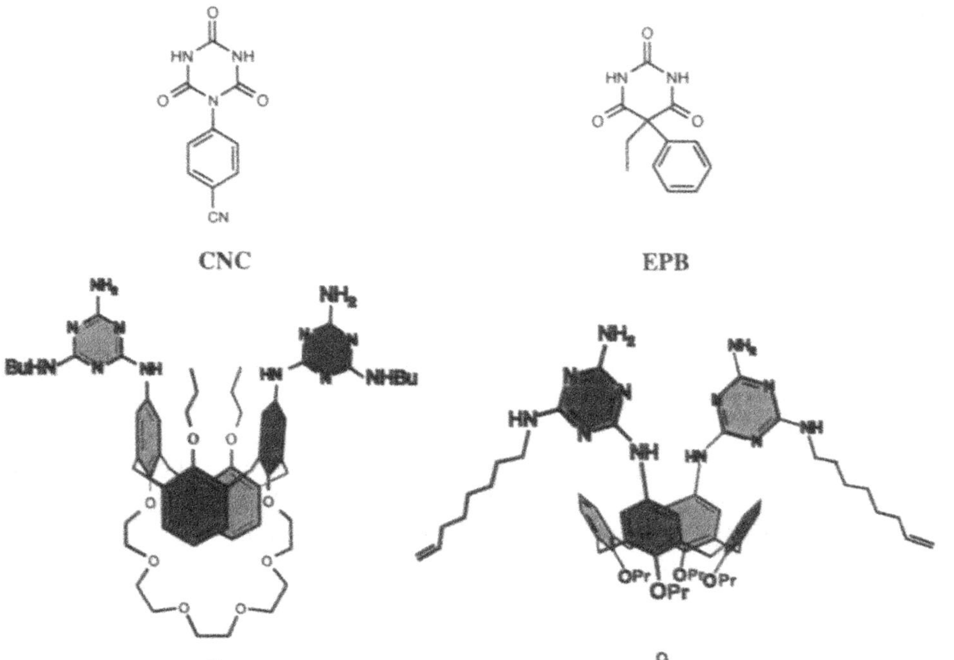

Figure 3. Examples of several bismelamines for which the corresponding assembly $(\mathbf{X})_3\cdot(\mathbf{DEB})_6$ gives a strong signal in the MALDI-TOF mass spectrum after Ag^+ labeling

Another approach for Ag^+ ion complexation takes advantage of the interaction of Ag^+ ions with π-surfaces. The MALDI-TOF mass spectrum of Ag^+-labeled benzyl-subtituted assembly $(\mathbf{4})_3 \cdot (\mathbf{DEB})_6$ shows a strong signal at *m/z* = 4278.3 (calcd for $^{13}C_2{}^{12}C_{226}H_{276}N_{48}O_{30} \cdot {}^{107}Ag^+$: 4276.1). Presumably the silver cation is complexed between a benzyl substituent and one of the phenyl rings of the calix[4]arene (see Figure 4) [22]. Assembly $(\mathbf{8})_3 \cdot (\mathbf{DEB})_6$, comprising the 1,3-*alternate* calix[4]arene bismelamine **8**, also gives a strong signal at *m/z* = 4429.3 in the MALDI-TOF mass spectrum corresponding to the Ag^+ complex (calcd for $^{13}C_2{}^{12}C_{220}H_{306}N_{48}O_{42} \cdot {}^{107}Ag^+$: 4427.0) [23]. In this case the Ag^+ ion is likely to be sandwiched between the calix[4]arene aromatic rings that are not substituted with melamines, possibly stabilized by a positive interaction with the crown-6 ether oxygen atoms. Apparently the Ag^+ complexation in $(\mathbf{8})_3 \cdot (\mathbf{DEB})_6$ occurs without significantly influencing the conformation of the calix[4]arene.
Recently we have found that components bearing ethylene groups such as melamine derivative **9** (Figure 3) are also suitable for MALDI-TOF mass spectrometry detection after Ag^+ complexation (*m/z* = 4398.3, calcd for $^{13}C_2{}^{12}C_{232}H_{324}N_{48}O_{30} \cdot {}^{107}Ag^+$: 4397.4) [24]. Double rosettes containing 5-ethyl-5-phenyl barbituric acid (**EPB**) are also detectable by this technique (Figure 3).

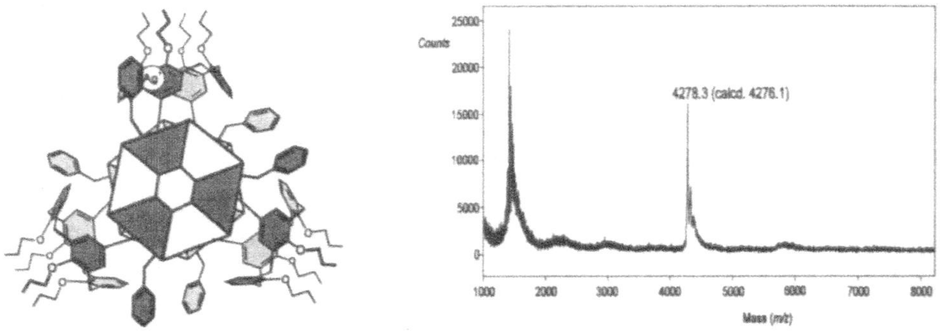

Figure 4. MALDI-TOF mass spectrum of the hydrogen-bonded assembly $(\mathbf{4})_3 \cdot (\mathbf{DEB})_6$ after complexation of Ag^+ between two aromatic π-donors (matrix: 2,5-dihydroxybenzoic acid)

We also tested the applicability of the Ag^+-labeling technique for the characterization of assemblies with much lower thermodynamic stability such as single rosettes [22]. Assemblies possessing aromatic π-donor systems in the melamine units (Figure 5) that provide a sandwich-type binding site for Ag^+, formed stable Ag^+-adducts that could be conveniently detected with MALDI-TOF MS. Even in these cases formation of fragmented assemblies was not observed.

10

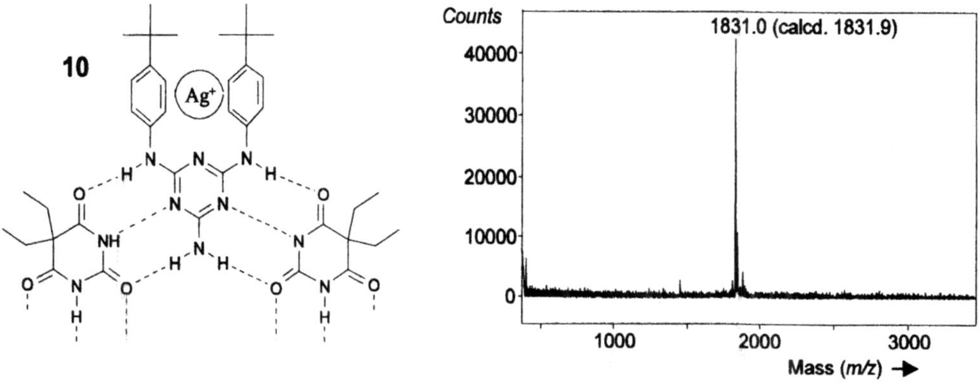

Figure 5. MALDI-TOF mass spectrum (matrix: 2,5-dihydroxybenzoic acid) of the hydrogen-bonded assembly $(\mathbf{10})_3\bullet(\mathbf{DEB})_3$ after complexation of Ag^+ between π-donors (only part of the total assembly is shown for clarity)

We conclude that MALDI-TOF mass spectrometry after Ag^+ labeling is a convenient, new tool for the mass spectrometric characterization of hydrogen-bonded assemblies. The method requires only a binding site for the soft Ag^+ ion in order to charge the noncovalent assembly in a nondestructive way. Aromatic π-donors, which can sandwich a Ag^+ ion, ethylenes or cyano groups are adequate for this purpose, but many other functionalities are known, such as acetylenes, amines and sulfur groups that interact strongly with Ag^+ ions.

3. Functionalized Hydrogen-Bonded Assemblies. Towards the Generation of Binding Site Diversity

3.1. FUNCTIONALIZATION OF HYDROGEN BONDED-ASSEMBLIES

In the hydrogen-bonded assemblies $(\mathbf{X})_3\bullet(\mathbf{DEB})_6$ *six* functionalizable sites R are gathered around the box-like cavity. In view of our objective to generate functional group diversity in or around non-covalently assembled cavities, we studied how the nature of the functionality at position R affects the stability of such calix[4]arene box-like assemblies in solution [13].
The introduction of polar substituents R, such as nitro or cyano, does not influence the stability of the cyclic hydrogen bonded aggregates in solution to any significant extent. Compounds **2** (R = NO_2) and **3** (R = CN) (Figure 1) readily assemble in $CDCl_3$ in the presence of **DEB**, the ^1H NMR spectra of which are virtually identical to that of $(\mathbf{1})_3\bullet(\mathbf{DEB})_6$ (R = H). Titration of **3** (R = CN) with **DEB** clearly proved the 1:2 stoichiometry of the box-like assembly formed.
Surprisingly, the stability of assembly $(\mathbf{5})_3\bullet(\mathbf{DEB})_6$ (R = NH_2) is highly solvent dependent. The ^1H NMR spectrum in toluene-d_8 is very well-defined and shows two singlets at $\delta = 14.55$ and 13.85 ppm for the chemically nonequivalent barbiturate NH protons. In $CDCl_3$, however, the spectrum is very broad and does not show any resonance in the region between 13 and 15 ppm, which indicates a preference for non-specific oligo- or polymer formation in this solvent.

The decreased stability of $(\mathbf{5})_3\cdot(\mathbf{DEB})_6$ in the more polar solvent $CDCl_3$ is most probably related to the hydrogen bond donating ability of the two NH_2 groups.

Compound **6** [R = $NHC(O)CH_3$] does not form the box-like assembly in either $CDCl_3$ or toluene-d_8. Both ^{1}H NMR spectra exhibit broad resonances indicating non-specific aggregation. Also in this case the formation of intramolecular hydrogen bonds, *i.e.* between the amide carbonyl and the melamine NH proton, seems to be primarily responsible for paralyzing the assembling ability of **6**. Sterically bulky groups (*e.g.* *t*-butyl) can also inhibit the formation of the double rosette depending on the exact position of the groups.

Instead of functionalization at the upper rim of the calix[4]arene, the different functional groups can also be positioned at the melamine rings (Figure 6) [25]. Compounds **11-15** all form stable assemblies $(\mathbf{X})_3\cdot(\mathbf{DEB})_6$ in $CDCl_3$ in the presence of two equivalents of **DEB**. These assemblies have been characterized by NMR spectroscopy and MALDI-TOF mass spectrometry after Ag^+ labeling. A positive response in the MALDI-TOF spectrum is due to the presence of the *m*-xylyl subtituents at the melamine rings that allows the assemblies to complex Ag^+ (see paragraph 2.3). The successful formation of the double rosette assemblies based on **11-15** clearly shows that the assembly process is fully compatible with the presence of donating and accepting hydrogen bond groups.

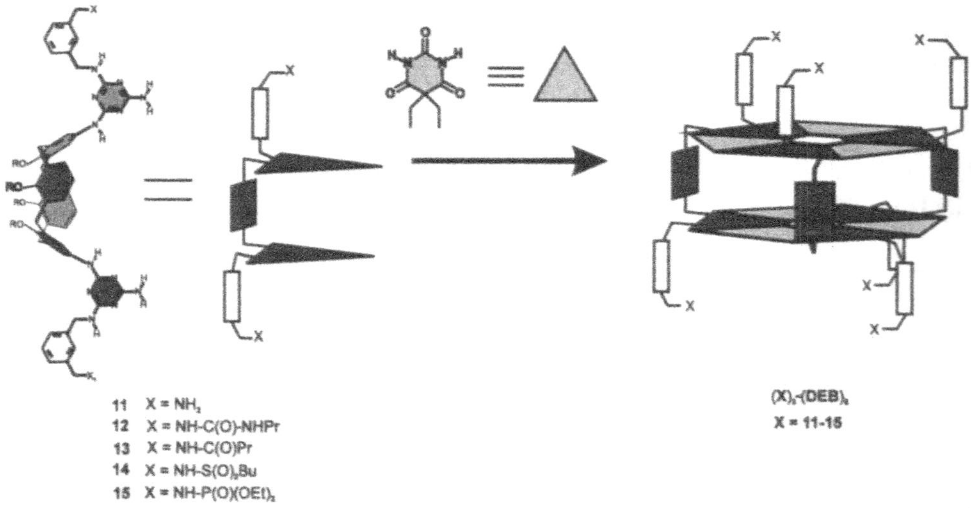

Figure 6. Hydrogen-bonded assembly with different functional groups (X) at the melamine rings

3.2. DYNAMIC LIBRARIES OF FUNCTIONALIZED ASSEMBLIES

In order to generate structural diversity inside self-assembled cavities we make use of the combinatorial strategy which has revolutionized the field of chemistry over the past decade [26,27]. Our self-assembled system allows us to develop the first example of a non-covalent synthesis of combinatorial libraries of hydrogen-bonded assemblies under thermodynamically controlled conditions [9,28].

Upon mixing in solution N different assemblies $(\mathbf{X})_3\cdot(\mathbf{DEB})_6$ ($\mathbf{X}$ = **1**, **2**, **3**,...N; Figure 1) under conditions that allow the reversible exchange of the bis-melamine components **X**, M heteromeric assemblies should theoretically form, in addition to the N homomeric assemblies (Figure 7). The total number of assemblies P (*i.e.* N + M) present in such a library rapidly increases with increasing N following equation (1):

$$P = N + N(N-1) + [N(N-1)(N-2)]/6 \tag{1}$$

For the smallest possible library (N = 2, P = 4) we have studied the combination of assemblies $(\mathbf{1})_3\cdot(\mathbf{DEB})_6$ and $(\mathbf{2})_3\cdot(\mathbf{DEB})_6$. The four possible assemblies are schematically depicted in Figure 7.

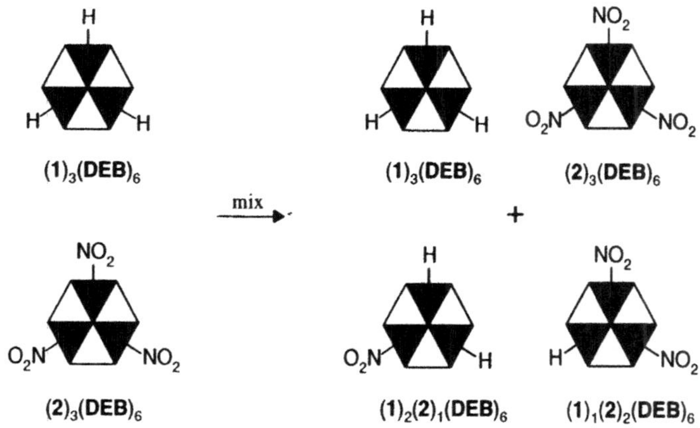

Figure 7. Four-component library obtained by mixing H-bonded assemblies $(\mathbf{1})_3\cdot(\mathbf{DEB})_6$ and $(\mathbf{2})_3\cdot(\mathbf{DEB})_6$

Mixing of 5 mM solutions of $(\mathbf{1})_3\cdot(\mathbf{DEB})_6$ and $(\mathbf{2})_3\cdot(\mathbf{DEB})_6$ at 0 °C (ratio 2:1) in toluene-d_8 gives only a mixture of the two separate homomeric assemblies. Exchange of the bismelamine calix[4]arenes is extremely slow at this temperature and, as a consequence, formation of the heteromeric assemblies is not observed. When the mixture is warmed to temperatures over 15 °C the heteromeric assemblies start to form slowly. This reflects the relatively high kinetic stability of the separate assemblies in this apolar solvent. After 2.5 hours at 25 °C the system reaches thermodynamic equilibrium (Figure 8). In the more polar solvent $CDCl_3$ the exchange of components **1** and **2** is much faster. The mixture equilibrates within seconds at 25 °C. Only below –50 °C are the homomeric assemblies kinetically inert.

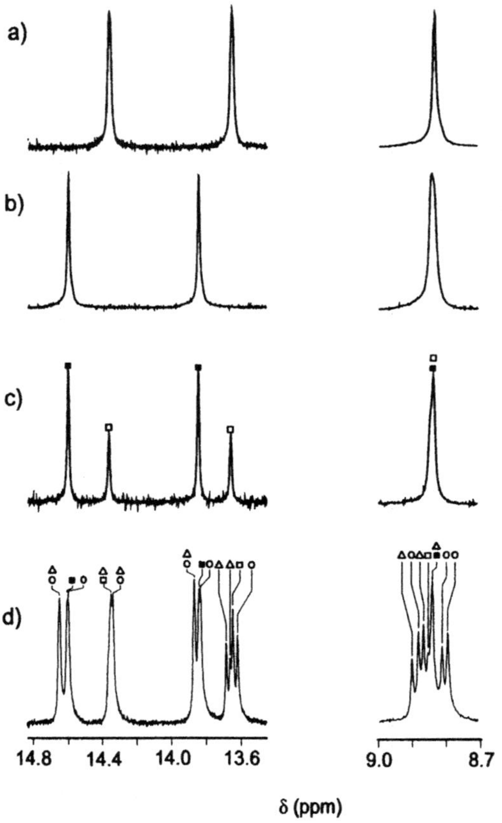

Figure 8. Temperature dependent formation of heteromeric assemblies as observed with ^{1}H NMR. Shown are the imide proton signals and one NH signal of the bismelamine of a) homomeric assembly $(\mathbf{1})_3\cdot(\mathbf{DEB})_6$, b) homomeric assembly $(\mathbf{2})_3\cdot(\mathbf{DEB})_6$, c) 2:1 mixture of $(\mathbf{1})_3\cdot(\mathbf{DEB})_6$ and $(\mathbf{2})_3\cdot(\mathbf{DEB})_6$ at 0°C and d) 2.5 hours after mixing $(\mathbf{1})_3\cdot(\mathbf{DEB})_6$ and $(\mathbf{2})_3\cdot(\mathbf{DEB})_6$ at 25°C

The unsymmetrical heteromeric rosettes $(\mathbf{1})_2(\mathbf{2})_1\cdot(\mathbf{DEB})_6$ and $(\mathbf{1})_1(\mathbf{2})_2\cdot(\mathbf{DEB})_6$ display multiple signals in the characteristic 13-15 ppm region (NH_{barb} hydrogen-bonded). Using 2D measurements such us NOESY, ROESY, COSY and TOCSY, the four component of the library were analyzed [22]. The relative concentrations of the four different members of the library were 1:3:3:1, in agreement with a statistical distribution. However, the complex ^{1}H NMR spectrum for this small library shows that NMR spectroscopy might not be the appropriate technique for analyzing larger libraries.

Characterization was also performed with MALDI-TOF mass spectrometry with Ag^+-labeling as described in section 2.3. Individual solutions of $(\mathbf{1})_3\cdot(\mathbf{CNC})_6$ or $(\mathbf{2})_3\cdot(\mathbf{CNC})_6$ comprising the isocyanurate derivative **CNC** (Figure 3) have given the expected intense signals in the MALDI-TOF mass spectra at m/z = 4348.1 and 4618 respectively, for the monovalent Ag^+ complexes. As expected, the MALDI-TOF mass spectrum of the mixture shows two additional signals for the heteromeric assemblies $(\mathbf{1})_2(\mathbf{2})_1\cdot(\mathbf{CNC})_6$ and $(\mathbf{1})_1(\mathbf{2})_2\cdot(\mathbf{CNC})_6$ at m/z = 4437.9 and 4527.7, respectively, in agreement with the calculated values.

In a similar way we have prepared libraries of other assemblies such as $(\mathbf{2})_3\cdot(\mathbf{CNC})_6$ and $(\mathbf{3})_3\cdot(\mathbf{CNC})_6$ and $(\mathbf{1})_3\cdot(\mathbf{CNC})_6$, $(\mathbf{2})_3\cdot(\mathbf{CNC})_6$ and $(\mathbf{3})_3\cdot(\mathbf{CNC})_6$. In all cases, the peaks corresponding to the different members of the libraries were detected but the relative intensity of these peaks is not in agreement with the expected concentration of the different assemblies in the mixture. This fact could pose a problem for the screening of non-covalent libraries for binding affinity to guest molecules (dynamic libraries), where the addition of an appropriate guest can template the formation of a determined rosette assuming that the assemblies are still under thermodynamic control. The problem could be overcome if the information in the different assemblies is stored by a covalent post-modification, a topic that is currently under active study in our group.

4. Hydrogen-Bonded Assemblies with Internal Cavities

In addition to the array of functionalities necessary for specific host-guest interactions, an important prerequisite in the preparation of artificial receptors is the existence of a concave cavity large enough to allow the inclusion of guest molecules. The calix[4]arene units, together with the two rosette platforms in **1-6** provide the required molecular surface for the generation of a concave cavity. X-ray crystal structure determination of assembly $(\mathbf{2})_3\cdot(\mathbf{DEB})_6$ ($R = NO_2$) revealed, however, that the two rosettes tightly stack on top of each other with an interatomic separation of 3.5 Å and 3.2 Å, at the edges and center of the box respectively. This leaves little space for inclusion of guest molecules (Figure 9) [13].

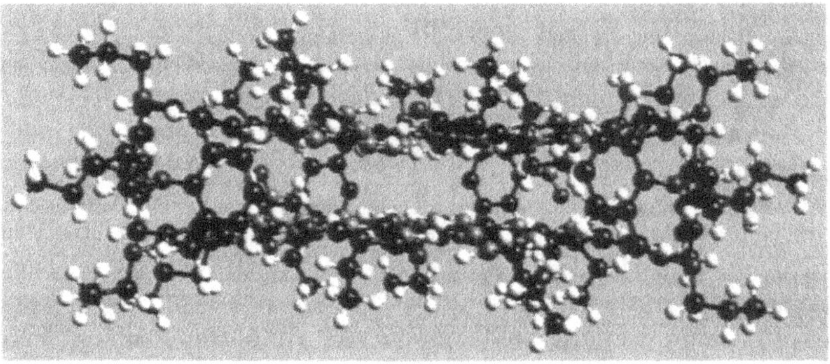

Figure 9. Side view (some atoms are omitted for clarity) of the X-ray crystal structure of assembly $(\mathbf{2})_3\cdot(\mathbf{DEB})_6$ $(\text{toluene})_{12}$ obtained by crystallization from toluene (solvent molecules are not shown)

Molecular modeling studies suggest that calix[4]arene derivatives **16-18** bearing methylene units between the calix[4]arene aromatic rings and the melamine moieties, prefer a pinched cone conformation with the melamine bearing aromatic rings twisted outward. This leads to an interatomic separation of more than 10 Å between the two rosettes. The short diethyleneoxy spacer at the lower rim in **17** and **18** was introduced to increase the preorganization of the melamine units by forcing the calix[4]arene in the favorable *pinched cone* conformation.

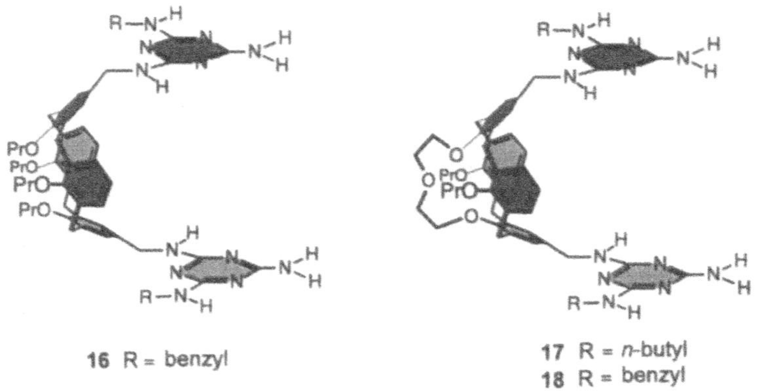

Figure 10. Bismelamine derivatives **16-18** comprising dimethyl calix[4]arene-based spacers

^{1}H NMR spectroscopic studies revealed that bismelamine derivatives **16-18** form hydrogen-bonded aggregates in the presence of two equivalents of **DEB** in $CDCl_3$; however no signals for the barbiturate NH protons in the 13-15 ppm region of the spectra were observed at 25 °C. Variable temperature ^{1}H NMR spectroscopic studies showed a resonance at low fields (δ = 14.3 ppm) below –30 °C. This supports formation of the postulated assemblies, but the stability is much lower than that of rosettes comprising **1-6** and **DEB**. The presence of only one signal, for the barbiturate NH protons in the assembly instead of the two expected for the D_3-symmetric double rosette assembly, suggests that the two rosettes act more like "single" rosettes rather than the more stable "double" rosette.

The hydrogen-bonding ability of bis(isocyanurate) derivative **19**, having less conformational freedom than bismelamines **16-18**, was investigated by its combination with melamine **10** in $CDCl_3$ (Figure 11). MALDI-TOF mass spectrometry of the 1:2 mixture of **19** and **10** showed that an aggregate of the expected mass for the double rosette is present in the gas phase (Figure 11). Combination of ^{1}H NMR studies (titration, dilution, variable temperature) supported the formation of stable aggregate(s) having 1:2 stoichiometry. The presence of at least seven signals for the isocyanurate NH protons in the 14-16 ppm region of the spectrum indicates that the cyclic assembly exists either as a mixture of diastereoisomers, or is present in a mixture with other aggregates. The characterization of these assemblies using other techniques is currently under investigation.

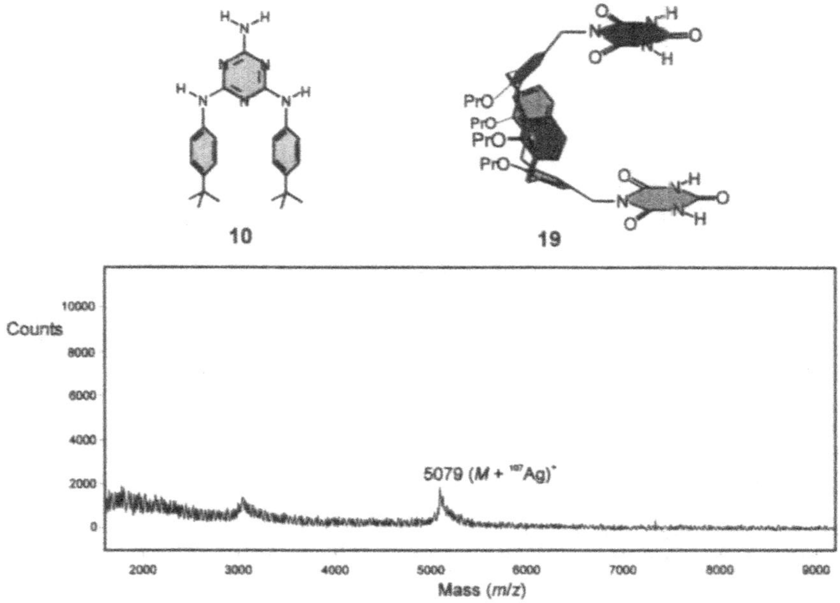

Figure 11. Molecular structures for melamine derivative **10** and bis-isocyanurate derivative **19** and MALDI-TOF mass spectrum of the 1:2 mixture of **19** and **10** after labeling with Ag^+ (matrix: 2,5-dihydroxybenzoic acid)

With the intention to increase the stability of self-assembled aggregates having an internal cavity we are currently investigating the possibility to replace the two rosettes at the top and bottom of the assembly $(\mathbf{16})_3\bullet(\mathbf{DEB})_6$, with two double rosettes, providing in this way the well-established stabilization associated with the double rosette systems.

5. Acknowledgments

We thank Dr. K. A. Jolliffe, Dr. H.-A. Klok, Dr. C. Marjo, Ir. L. J. Prins, Dr. C. Schauer, Dr. R. H. Vreekamp, Dr. W. Verboom and Dr. J.-L. Weidmann for their contribution to this work . We also thanks Mr. R. Fokkens and Prof. N. Nibbering (University of Amsterdam) for the MALDI-TOF mass spectrometric measurements. We are grateful to the European Community for the financial support with the Marie Curie Research Training Grants (M.C.C., no. ERBFMBICT 961445, F.C., no. ERBFMBICT 972555) as part of the TMR Programme.

6. References

1. a) Corey, E. J.and Cheng, X. M. (1989) The Logic of Chemical Synthesis, Wiley, New York; b) Corey, E. J., (1991) The Logic of Chemical Synthesis: Multistep Synthesis of Complex Carbogenic Molecules, *Angew. Chem., Int. Ed. Engl.*, **30**, 455-465.
2. Nicolau, K. C. and Sorensen, E. J. (1996) Classics in Total Synthesis, VCH, Weinheim

3. Woodward, R. B. (1973) The Total Synthesis of Vitamin B_{12}, *Pure Appl. Chem.*, **33**, 145-177.
4. Eschenmoser, A. E. and Winter; C. E. (1977) Natural Product Synthesis and Vitamin B_{12}, *Science*, **196**, 1410-1420.
5. Armstrong, R. W., Beau, J.-M., Cheon, S. H., Christ, W. J., Fujioka, H., Ham, W. H., Hawkins, L. D., Jin, H., Kang, S. H., Kishi, Y., Martinelli, M. J., McWhorter, W. W., Mizuno, M., Nakata, M., Stutz, A. E., Talamas, F. X., Taniguchi, M., Tino, J. A., Ueda, K., Uenishi, J., White, J. B. and Yonaga, M. (1989) Total Synthesis of a Fully Protected Palytoxin Carboxylic Acid, *J. Am. Chem. Soc.*, **111**, 7525-7530.
6. The total synthesis of palytoxin spanned almost ten years involved more than 100 researchers. This level of effort was required to produce a compound with a molecular mass of 2680 Da and dimensions an order of magnitude smaller than the 10-100 nm dimensions usually regarded as nanoscale. Clearly, extrapolation of this type of synthetic strategy to a structure with a molecular mass of ca. 10 000 Da would be prohibitive both in terms of time and manpower.
7. For reviews see: a) Philp, D. and Stoddart, J. F. (1996) Self-Assembly in Natural and Unnatural Systems, *Angew. Chem., Int. Ed. Engl.* **35**, 1154-1196; b) Lawrence, D. S., Jiang, T. and Levett, M. (1995) Self-Assembling Supramolecular Complexes, *Chem. Rev.*, **95**, 2229-2260.
8. Vreekamp, R. H., Van Duynhoven, J. P. M., Hubert, M., Verboom, W. and Reinhoudt, D. N. (1996) Molecular Boxes Based on Calix[4]arene Double Rosette, *Angew. Chem., Int. Ed. Engl.*, **35**, 1215-1218.
9. Timmerman, P. and Reinhoudt, D. N. (1998) A Combinatorial Approach to Synthetic Receptors, *Adv. Mater.*, accepted.
10. For a review see: Conn, M. M. and Rebek Jr., J. (1997) Self-Assembly Capsules, *Chem. Rev.*, **97**, 1647-1668.
11. For a review see: Whitesides, G. M., Simanek, E. E., Mathias, J. P., Seto, C. T., Chin, D. N., Mammen, M. and Gordon, D. M. (1995) Noncovalent Synthesis: Using Physical-Organic Chemistry To Make Aggregates, *Acc. Chem. Res.* **28**, 37-44.
12. Lehn, J.-M., Mascal, M., DeCian, A. and Fischer, J. (1990) Molecular Recognition directed Self-Assembled of Ordered Supramolecular Strands by Cocrystallization of Complementary Molecular Components, *J. Chem. Soc., Chem .Commun.* 479-481.
13. Timmerman, P., Vreekamp, R., Hulst, R., Verboom, W., Reinhoudt, D. N., Rissanen, K., Udachin, K. A. and Ripmeester, J. (1997) Noncovalent Assembly of Functional Groups on Calix[4]arene Molecular Boxes, *Chem. Eur. J.*, **3**, 1823-1832.
14. Huck, W. T. S., Hulst, R., Timmerman, P., Van Veggel, F. C. J. M. and Reinhoudt, D. N. (1997) Noncovalent Synthesis of Nanostructures: Combining Coordination Chemistry and Hydrogen Bonding, *Angew. Chem., Int. Ed. Engl.*, **36**, 1006-1008.
15 a) Vicenti, M. (1995) Host-Guest Chemistry in the Mass Spectrometer, *J. Mass. Spec*, **30**, 925-939; b) Smith, R., D., Bruce, J. E., Wu, Q. and Lei, Q. P. (1997) New Mass Spectrometric Methods for the Study of Noncovalent Associations of Biopolymers, *Chem. Soc. Rew.*, **26**, 191-202.
16. a) Rusell, K. C., Leize, E., Van Dorsselaer, A. and Lehn, J.-M. (1995) Investigation of Self-Assembled Supramolecular Species in Solution by IL-ESMS; a New Mass Spectrometric Technique, *Angew. Chem., Int. Ed. Engl.*, **34**, 209-213; b) Cheng, X., Gao, Q., Smith, R. D., Simanek, E. E., Mammen, M. and Whitesides, G. M. (1996) Chraracterization of Hydrogen-Bonded Aggregates in Chloroform by Electronspray Ionization Mass Spectrometry, *J. Org. Chem.*, **61**, 2204-2206.
17. a) Ma, S., Rudkevich, D. M. and Rebek Jr., J. (1998) Deep-Cavity Resorcinarenes Dimerize through Hydrogen Bonding and Self-Inclusion, *J. Am. Chem. Soc*, **120**, 4977-4981; b) Scherer, M., Sessler, J. L., Moini, M., Gebauer, A. and Lynch, V. (1998) Self-Assembly of Pyrrole-Ferrocene Hybrids, Determined Inter Alia by a New Chemically Induced Electrospray Mass Spectrometry Technique, *Chem. Eur. J.*, **4**, 152-158.
18. Gemelin (1975) Handbuch der Anorganischen Chemie, Silber, Teil B6, Springer, Berlin, pp. 346-353.
19. a) Kang, H. C., Hanson, A. W., Eaton, B. and Boekelheide, V. (1985) $[2_6]$(1,2,4,5)Cyclophane (Deltaphane) and Related Compounds. Simultaneous π-Electron Interaction among Tree Benzene Rings, *J. Am. Chem. Soc.*, **107**, 1979-1985; b) Gross. J., Harder, G., Vogtle, F., Stephan, H. and Gloe, K. (1995) $C_{60}H_{60}$ and $C_{54}H_{48}$: Silver Ion Extraction with New Concave Hydrocarbons, *Angew. Chem., Int. Ed. Engl.*, **34**, 481-484.
20. a) Ikeda, A., Tsuzuki, H. and Shinkai, S. (1994) NMR Spectroscopy and X-Ray Crystallographic Studies if Calix[4]arenes • Ag^+ Complexes. Influence of Bound Ag^+ on C_{2v}-C_{2v} Interconversion in Cone-Calix[4]arenes, *J. Chem. Soc., Perkin Trans. 2*, 2073-2080; b) Struck, O., Chrisstoffels, L. A. J., Lugtenberg, R. J. W., Verboom, W., Van Hummel, G. J., Harkema, S. and Reinhoudt, D. N. (1997) Head-to-Head Linked Double Calix[4]arenes: Convenient Synthesis and Complexation Properties, *J. Org. Chem.*, **62**, 2487-2493.

21. a) Karas, M., Bachmann, D., Bahr, U. and Hillenkamp, F. (1987) Matrix-Assisted Ultraviolet Laser Desorption of Non-Volatile Compounds, *Int. J. Mass. Spectrom. Ion Processes*, **78**, 53-68; b) Hillenkamp, F. and Karas, M. (1991) Matrix-Assited Laser Desorption/Ionization Mass Spectrometry of Biopolymers, *Anal. Chem.*, **63**, 1193A-1203A.
22. Jolliffe, K. A., Crego-Calama, M., Fokkens, R., Nibbering, N. M. M., Timmerman, P. and Reinhoudt, D. N. (1998) Characterization of Hydrogen-Bonded Suparamolecular Assemblies by MALDI-TOF Mass Spectrometry after Ag^+ Labelling, *Angew. Chem., Int. Ed. Engl.*, **37**, 1247-1251.
23. Weidmann, J.-L., Timmerman, P. and Reinhoudt, D. N., unpublished results.
24. Weidmann, J.-L., Cardullo, F., Timmerman, P. and Reinhoudt, D. N., unpublished results.
25. Crego-Calama, M., Timmerman, P. and Reinhoudt, D. N., unpublished results.
26. Hsieh-Wilson, L. C., Xiang, X.-D. and Schultz, P. G. (1996) Lessons from Immune System: From Catalysis to Material Science, *Acc. Chem. Res.*, **29**, 164-170.
27. a) De Miguel, Y. R. and Sanders, J. K. M. (1998) Generation and Screening of Synthetic Receptor Libraries, *Current Opinion in Chemical Biology*, **2**, 417-421; b) Brady, P. A. and Sanders J. K. M. (1997) Selection Approaches to Catalytic Systems, *Chem. Soc. Reviews*, **26**, 327-336.
28. Crego-Calama, M., Hulst, R., Fokkens, R., Nibbering, N. M. M., Timmerman, P. and Reinhoudt, D. N. (1998) Libraries of Non-covalent Hydrogen-Bonded Assemblies; Combinatorial Synthesis of Supramolecular Systems, *Chem. Commun.*, 1021-1022.
29. Jolliffe, K. A., Prins, L. J., Timmerman, P., and Reinhoudt, D. N., unpublished results.

PROTEIN SURFACE RECOGNITION BY SYNTHETIC RECEPTORS

QING LIN, HYUNG SOON PARK, YOSHITOMO HAMURO AND
ANDREW D. HAMILTON*
Department of Chemistry, Yale University New Haven, CT, 06520 USA

1. Introduction

Protein recognition can be divided into two categories in terms of the interaction sites: (a) interactions that occur *inside* proteins, e.g. enzyme active sites; (b) interactions that occur on the *surface* of proteins. During the past two decades large numbers of synthetic molecules targeted to disrupt interior protein interactions have been shown to have medically important biological activities.[1] However, artificially designed molecules that target the protein surface and disrupt its biological activity are rare.[2] Considering the unique composition of charged, hydrophobic and hydrophilic domains on every protein's surface, synthetic molecules that match the electrostatic features and topology of the protein targets might be expected to bind to the exterior and sterically prevent protein-ligand or protein-protein interactions. This in turn can provide a novel approach to modulating biological processes.

In certain cases, particularly where well defined clefts or cavities exist, some progress has been made designing small molecules to bind to a protein surface. For example, guanidine esters have been designed to bind to interleukin-2 and block its interaction with its heterotrimeric receptor complex.[3] Similarly, small heterocycles have been shown to bind to CD4 and disrupt its binding to MHC class II proteins on the surface of antigen presenting T-cells.[4] Others have used metal complexes to target histidine residues on a protein surface.[5] Additionally, some progress has been made in the use of synthetic peptide or peptide-like libraries to target protein surfaces.[6] One successful strategy has been to attach onto known active site inhibitors hydrophobic groups[7] or alkylating agents[8] that target residues on the nearby protein surface. Other approaches have used anionic polymers,[9] or oligomers such as aurintricarboxylic acid,[10] heparin derivatives[11] and oligophenoxyacetic acid[12] to target positively charged regions on a protein surface.

Our goal is to develop a general and modular approach to protein surface recognition that will allow the targeting of a range of different protein surfaces by modifying the recognition characteristics of synthetic receptors. A model for this strategy is provided by the immune system which generates a vast number of antibodies that show high sequence and structural selectivity in binding to a wide range of protein surfaces.[13 14] All antibody binding regions share a common structural motif of six hypervariable loops held in place by the closely packed constant and variable regions of light and heavy chains. Four of these loops (L2, L3, H2 and H3) generally take up a

R. Ungaro and E. Dalcanale (eds.), Supramolecular Science: Where It Is and Where It Is Going, 197–204.

hairpin conformation with the remaining two forming more extended loops. The majority of antibody complementary determining regions contact the antigen with four to six of the hypervariable loops.

A key feature in all antibody-antigen interactions is the extent of solvent exposed surface on the proteins that is buried upon complex formation. Analysis of crystal structures[15] has shown that >600 Å^2 of the antibody surface is buried at the interface with the antigen. Similar observations have been made for other protein-protein interactions,[16] suggesting that around 1200Å^2 of protein surface must be buried in order to form strong complexes in solution. Such large areas are required because of the highly solvated nature of the polar functional groups present on the protein periphery. As a result hydrophobic residues at the protein-protein interface play a key role in stabilizing the complex through hydrophobic forces, often comprising more than one third of the interacting sites.[17] In addition, complementarity between charged residues on the two partners (for example, between Arg residues on the antibody HyHEL-5 and carboxylate groups on the surface lysozyme[18]) provides additional binding energy and is the primary origin of the high selectivity. Despite the large surfaces areas involved, Wells has identified small regions or "hot spots" on certain protein-protein interfaces that account for the majority of the binding energy.[19] This strongly suggests that synthetic agents that are smaller than a protein or antibody FAB fragment can bind strongly to a protein surface if they include or are targeted against areas of high potential binding energy.

Our approach will be based on the attachment of several peptide loops onto a core scaffold (Figure 1). The advantages of using this design for the recognition of protein surfaces are four fold: a) large surface areas can be created (>400 Å^2); b) modular design will enable us to generate a large diversity of receptors from a small number of cyclic peptide or scaffold components ; c) various non-natural groups can be incorporated into the loop structures; d) scaffolds are relatively flexible which allow an "induced fit" mechanism of binding.

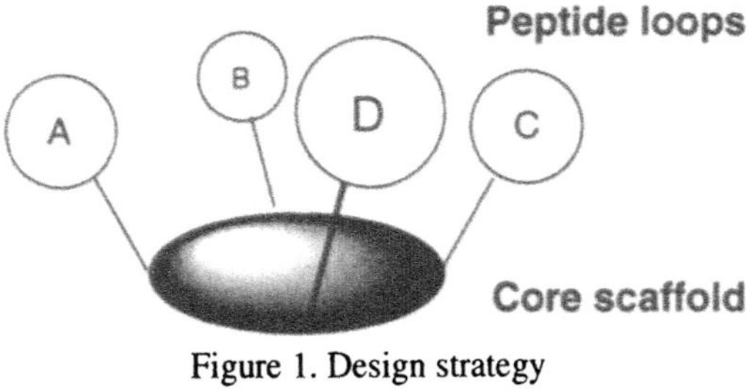

Figure 1. Design strategy

2. Protein Surface Recognition by Synthetic Receptors

We have prepared the first of a new class of protein surface receptors based on the attachment of four peptide loops to a central calix[4]arene scaffold.[20] The required tetracarboxylic acid **7** was prepared by alkylation of calix[4]arene (n-butyl bromide, NaH) followed by formylation (Cl_2CHOCH_3, $TiCl_4$) and oxidation ($NaClO_2$, H_2NSO_3H).[21] The peptide loop component was based on a cyclic hexapeptide in which two residues were replaced by a 3-aminomethylbenzoyl (3amb) dipeptide mimetic[22] containing a 5-amino substituent for linkage to the scaffold. The 5-nitro substituted dipeptide mimetic was formed by selective reduction (BH_3) of methyl 3-amidocarbonyl-5-nitrobenzoic acid followed by deesterification (LiOH, THF) and reaction with sequentially Fmoc-Asp-(tBu)-OH and H-Gly-Asp(tBu)-Gly-OH (DCC, HOSu) to yield Fmoc-Asp(tBu)-5$NO_2$3amb-Gly-Asp(tBu)-GlyOH. Cyclization was achieved in 70% yield (DMAP, TBTU) followed by reduction (H_2, Pd/C) to give the amino substituted peptide loop **2** which was coupled to the tetraacid chloride derivative of **1** (($(COCl)_2$, DMF) and deprotected (TFA) to give the tetraloop structure **3**.

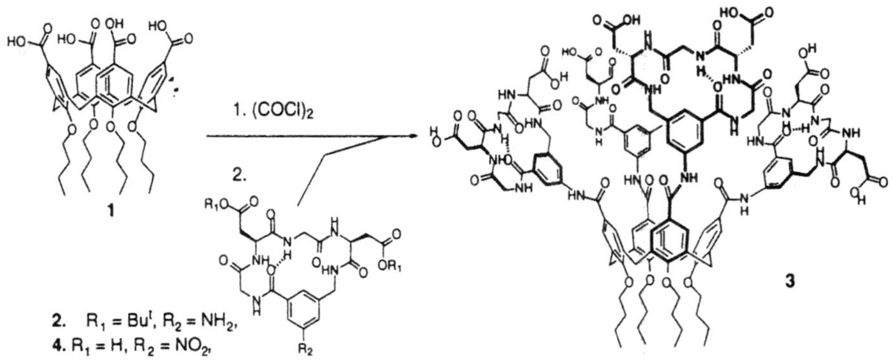

3. Recognition of the Heme Edge Surface of Cytochrome c

Our first protein target was cytochrome c since it contains a positively charged surface made up of several lysine and arginine residues. We used the sequence GlyAspGlyAsp in the loops to complement the charge distribution on the protein. Support for strong binding came from gel permeation chromatography on Sephadex G-50 in 5mM phosphate buffer at pH 7.4. A 2:1 mixture of **3**: cytochrome c (horse heart) gave a sharp band with a shorter retention time than cytochrome c (Figure 2). A corresponding 1:2 mixture showed both bands in the chromatogram, confirming the presence of both complexed and free cytochrome c in solution.

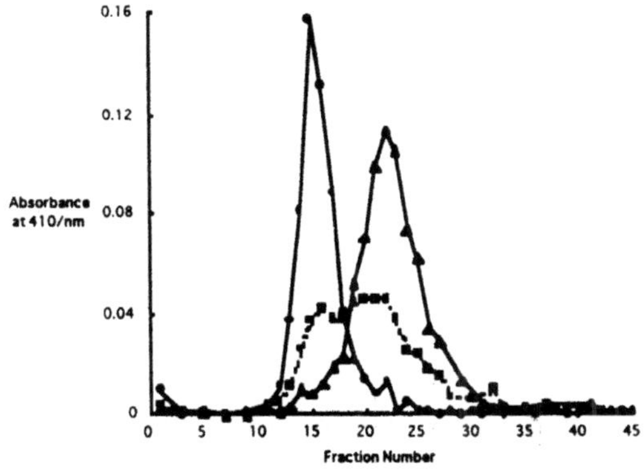

Figure 2. GPC of; ▲, cytochrome c; ●, 1:2 cytochrome c:**9**; and ■, 2.1 cytochrome c:**3**.

Evidence for the binding geometry came from ^1H NMR titration experiments in which changes in the position of the hyperfine shifted heme-CH_3 resonances were followed as a function of increasing concentration of **3** (Figure 3). The heme 8-CH_3 and

3-CH_3 resonances shifted upfield (-0.27 ppm) and downfield (+0.25 ppm), respectively, reaching saturation at one equivalent of added **3**, in a manner similar to an equivalent

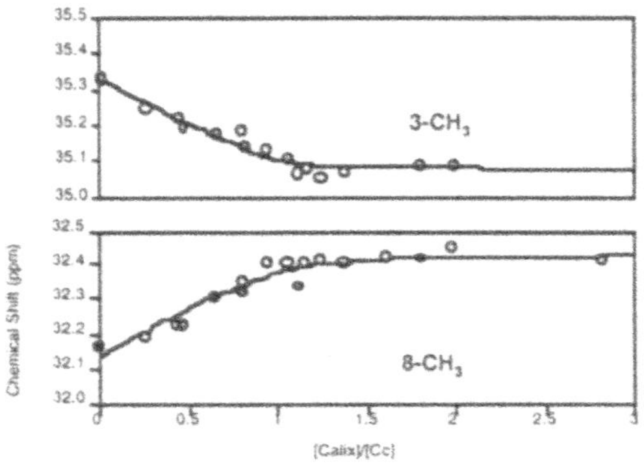

Figure 3. ^{1}H NMR changes of heme 8-CH_3 and 3-CH_3 on titration of cytochrome c with **3**.

titration with cytochrome c peroxidase.[23] This suggests that **3** is binding to a region on cytochrome c close to the heme edge containing the group of positively charged residues (Lys-17, 18, 21, 77 and 88)[24] which forms the primary interaction surface with cytochrome c peroxidase. A calculated structure for **3** docked with this region on cytochrome c (Figure 4) suggests that the four peptide loops can contact most of the lysines and cover a large area of the protein surface. In this arrangement **3** would be

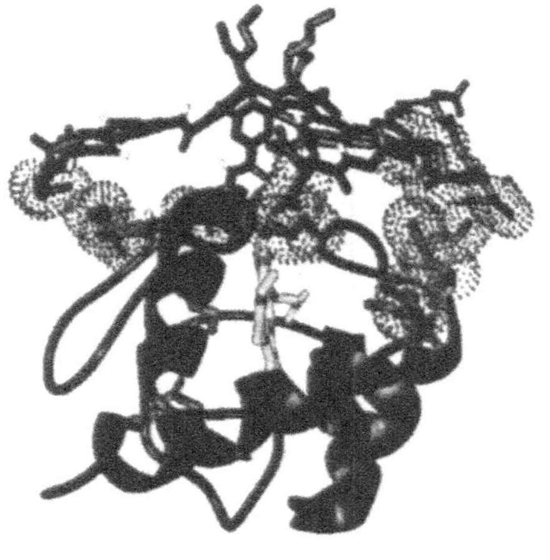

Figure 4. Proposed structure for the complex between **3** and cytochrome c.

expected to block the approach of reducing agents to the heme edge. In phosphate buffer Fe(III) cytochrome c (1.57×10^{-5} M) is rapidly reduced by excess ascorbate (2.0×10^{-3} M) with a pseudo first order rate constant of 0.109 ± 0.0001 sec^{-1} (Figure 5). In the presence of **3** (1.91×10^{-5} M) the rate of cyt c reduction is diminished ten-fold (k_{obs} = 0.010 ± 0.001 sec^{-1}) in a similar manner to that seen with the cytochrome c-cytochrome c peroxidase complex.[25] Single loop **4** (5.36×10^{-4} M) had little effect on the reduction rate, as did excess concentrations of inositol hexaphosphate (Figure 6) or polyglutamate, suggesting the interaction is not a simple polyanion effect. A plot of k_{obs} vs. [3] gave a saturation curve that indicated 1:1 binding and gave a K_a of $3.05 \pm 0.20 \times 10^6$ M^{-1}.

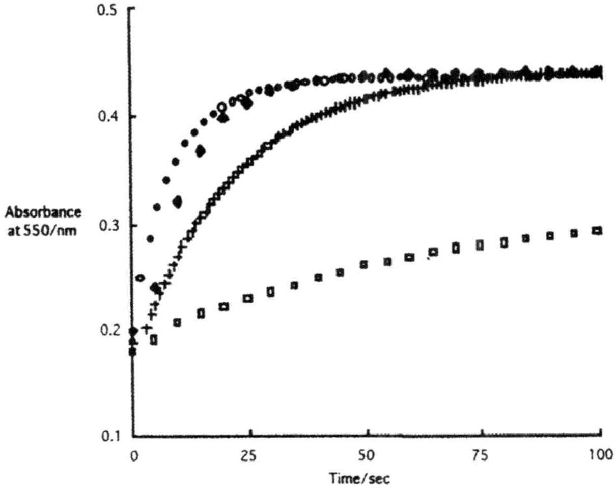

Figure 5. Kinetics of reduction of Fe(III) cyt. c by ascorbate ○, no receptor; ❑, **3**; ◆, **4** (5.36×10^{-4}M); +, inositol hexaphosphate.

We have now begun an investigation of the key features controlling the binding of these multi-loop receptors to cytochrome c. We have prepared analogs of **3** in which the sequence of the cyclic peptide is varied to include (in a *first generation* series) anionic, hydrophobic and cationic residues, GDGD (**3**), GDGY (**5**), GKGF (**6**) and GKGK (**7**). The relative effects of these four fold symmetrical receptors (at 1.91×10^{-5} M) on the rate of Fe(III) cytochrome c (at 1.57×10^{-5} M) reduction is shown in Figure 6. This clearly confirms the importance of a strongly negatively charged surface in binding to the

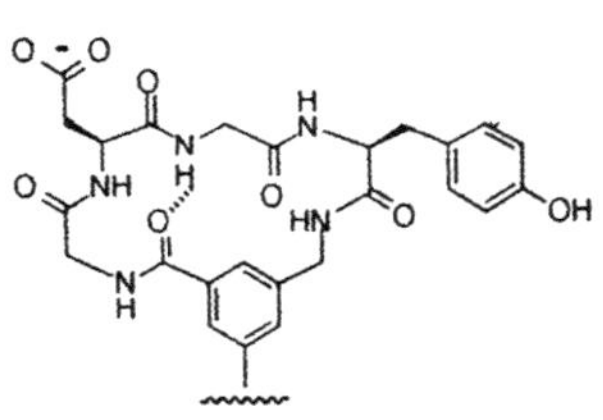

5 GDGY

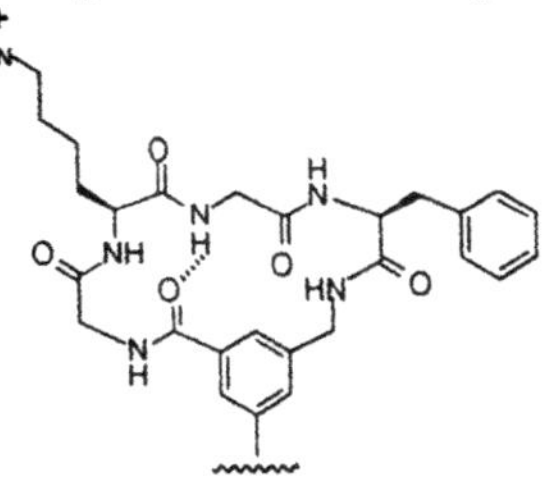

6 GKGF

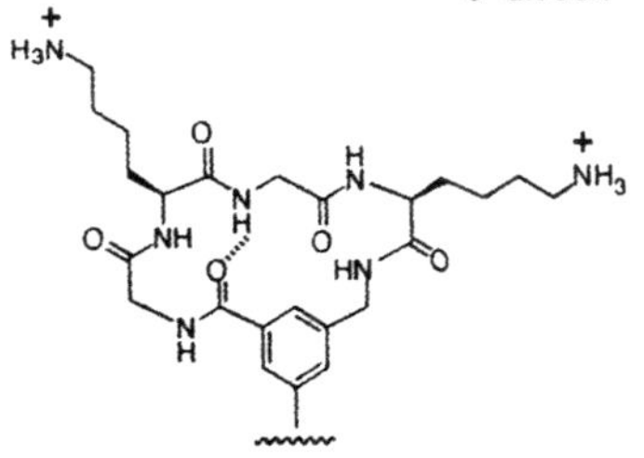

7 GKGK

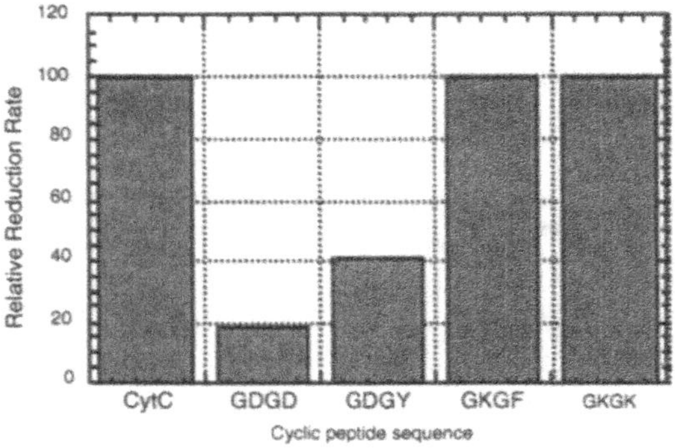

Figure 6. Relative reduction rates for Fe(III)cytochrome c in presence of 1.2 equiv. of cyclic peptides.

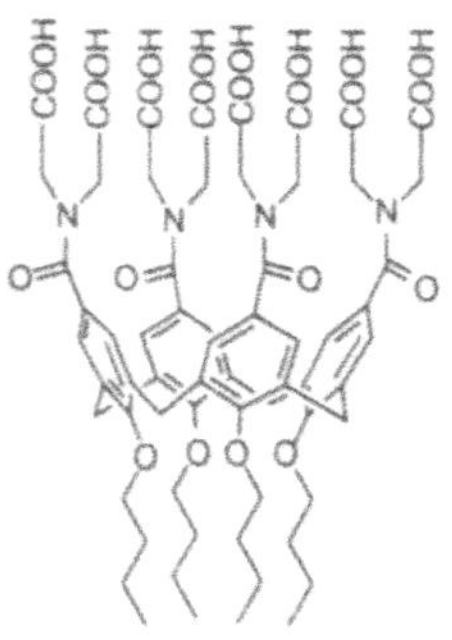

8

protein.[26] However, eight carboxylate groups are not the only contributors to recognition. We have prepared the calixarene octacarboxylate derivative **8** (OctAc, Figure 7) and shown it to have only a marginal (~75% relative rate) effect on cytochrome c reduction under equivalent conditions.

With the lead provided by the GDGD sequence (**3**) we have prepared a second generation of protein surface receptors in which minor modifications of the cyclic peptide sequence were made. In particular, we have added into the basic bis-Asp containing sequence various natural and non-natural residues in place of Gly, including Ala (A), Leu (L), Asn (N), Ser (S), D-Ala (a), D-2-naphthylalanine (2n), D-2-aminobutyric acid (abu). All compounds were four fold symmetrical as in **9**. Their relative efficiencies in blocking

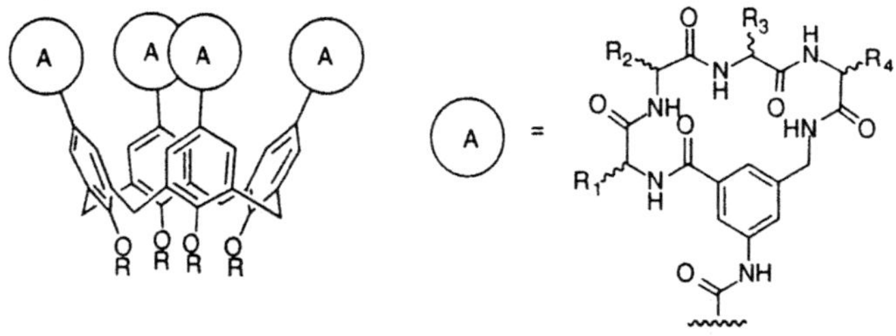

9

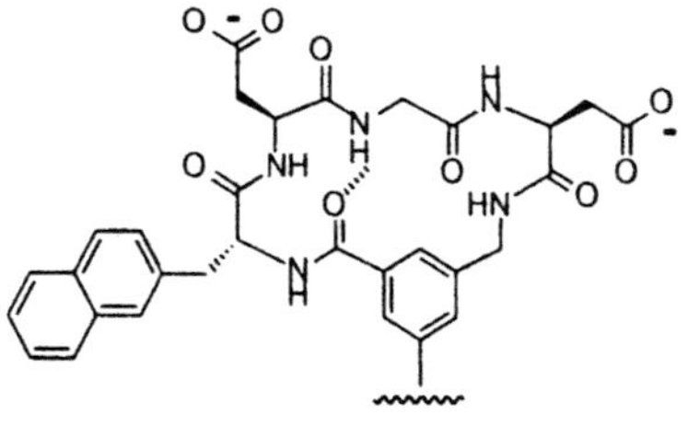

2-nal-DGD **10**

Fe(III) cytochrome c reduction are seen in Figure 7. A small improvement occurred with GDDG vs GDGD (14.0% vs. 16.4%), suggesting that the positioning of the charged groups may be more optimal in the former. A more significant improvement was seen with D-2-NalDGD (**10**) vs. GDGD (11.8% vs. 16.4%), suggesting that the naphthalene may provide a hydrophobic surface for favorable interaction with the protein exterior.

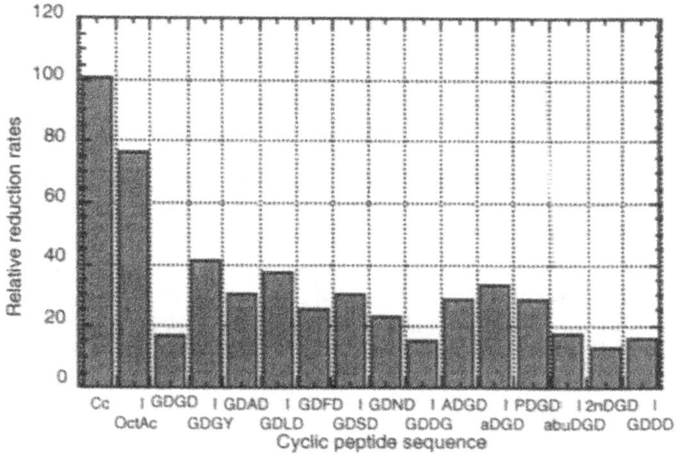

Figure 7. Relative reduction rates for second generation protein surface receptors.

In these initial experiments on protein surface recognition we have established that synthetic agents that define a large functionalized surface can be prepared and shown to bind to a protein exterior with high affinity (Kd ~ 300nM). We have shown that this binding can block the interaction of the protein with its redox partners and that this inhibition activity can be optimized through the screening of second generation libraries of receptors.

Acknowledgment

We thank the National Institutes of Health (GM53579) for support of this work.

References

1. Babine, R. E.; Bender, S. L. *Chem. Rev.* **1997**, *97*, 1359-1472.
2. Indeed, in a recent Science article it was stated "..it remains a formidable task to design small molecules that alter the function...of an enzyme with a small active site. It is even more difficult to inhibit protein-protein interactions mediated over a large surface...." Hartwell, L. H.; Szankasi, P.; Roberts, C. J.; Murray, A. W.; Friend, S. H. *Science*, **1997**, *278*, 1064-1068.
3. Tilley, J. W.; Chen, L.; Fry, D. C.; Emerson, S. D.; Powers, G. D.; Biondi, D.; Varnell, T.; Trilles, R.; Guthrie, R.; Mennona, F.; Kaplan, G.; LeMahieu, R. A.; Carson, M.; Han, R. J.; Liu, C. M.; Palermo, R.; Ju, G. *J. Am. Chem. Soc.* **1997**, *119*, 7589-7590.
4. Huang, Z.; Li, S.; Gao, J.; Satoh, T.; Friedman, T. M.; Edling, A. E.; Koch, U.; Choksi, S.; Han, X.; Korngold, R. *Proc. Natl. Acad. Sci. USA*, **1997**, *94*, 73.
5. Mallik, S., Plunkett, S. D., Dhal, P. K., Johnson, R. D., Pack, D., Shnek, D., & Arnold, F. H. *New J. Chem.* **1994***18*, 299-304.
6. Blondelle, S.; Houghten, R. A.; Peréz-Payá, E. *J. Mol. Recogn.* **1996**, *9*, 163-168. Sasaki, S.; Takagi, M.; Tanaka, Y.; Maeda, M. *Tetrahedron Letts.* **1996**, *37*, 85-88.

Boyce, W.; Li, G.; Nestler, P.; Suenaga, T.; Still, W. C. *J. Am. Chem. Soc.* **1994**, *116*, 7955-7956.
7. Jain, A.; Whitesides, G. M.; Alexander, R. S.; Christianson, D. W. *J. Med. Chem.* **1994**, *37*, 2100-2105.
8. De Voss, J. J.; Sui, Z.; DeCamp, D. L.; Salto, R.; Babé, L. M.; Craik, C. S.; Ortiz de Montellano, P. R. *J. Med. Chem.* **1994**, *37*, 665-673.
9. Benezra, M.; Ben-Sasson, S. A.; Regan, J.; Chang, M.; Bar-Shavit, R.; Vlodavsky, I. *Arteriosclerosis Thromb.* **1994**, *14*, 1992-1994.
10. Cushman, M.; Kanamathareddy, S.; De Clerq, E.; Schols, D.; Goldman, M. E.; Bowen, J. A. *J. Med. Chem.* **1991**, *34*, 337-342.
11. Redini, F.; Lafuma, C.; Hornback, W.; Choay, J.; Robert, L. *Biochem. Pharmacol.* **1988**, 37, 4257-4261.
12. Regan, J.; McGarry, D.; Bruno, J.; Green, D.; Newman, J.; Hsu, C. Y.; Kline, J.; Barton, J.; Travis, J.; Choi, Y. M.; Volz, F.; Pauls, H.; Harrison, R.; Zilberstein, A.; Ben-Sasson, S. A.; Chang, M. *J. Med. Chem.* **1997**, *40*, 3408-3422.
13. Brandon, C.; Tooze, J. *Introduction to Protein Structure*; Garland: New York, 1991.
14. Wilson, I. A.; Stanfield, R. L.; Rini, J. M.; Arevalo, J. H.; Schulze-Gahmen, U.; Fremont, D. H.; Stura, E. A. *Ciba Foundation Symp.* **1991**, *159*, 13-39.
15. Davies, D. R.; Padlan, E. A.; Sheriff, S. *Ann. Rev. Biochem.* **1990**, *59*, 439-473.
16. Janin, J.; Chothia, C. *J. Biol. Chem.* **1990**, *265*, 16027-16030.
17. Stites, W. E. *Chem. Rev.* **1997**, *97*, 1233-1250.
18. Sheriff, S.; Silverton, E. W.; Padlan, E. A.; Davies, D. R. *Proc. Natl. Acad. Sci. USA*, **1987**, *84*, 8075-8079.
19. Clarkson, T.; Wells, J. A. *Science*, **1995**, *267*, 383-386.
20. Hamuro, Y.; Calama, M. C.; Park, H. S.; Hamilton, A. D. *Angew. Chemie Int. Ed. Engl.*, **1997**, *36*, 2680-2683.
21. Conner, M.; Janout, V.; Regen, S. L. *J. Org. Chem.* **1992**, *57*, 3744-3746.
22. Bach, A. C.; Eyerman, C. J.; Gross, J. D.; Bower, M. J.; Harlow, R. L.; Weber, P. C.; DeGrado, W. F. *J. Am. Chem. Soc.* **1994**, *116*, 3207-3219.
23. Satterlee, J. D.; Moench, S. J; Erman, J. E. *Biochim. Biophys. Acta* **1987**, *912*, 87-97S.
24. Mclendon, G. In *Control of Biological Electron Transport via Molecular Recognition and Binding: The "Velcro" Model*; Clark, M. J. et al. Eds., Spring-Verlag, Berlin Heidelberg, **1991**, p 165.
25. Mochan, E.; Nicholls, P. *Biochem. Biophys. Acta*, **1972**, *267*, 309-319.
26. Nicholls, P. *Biochem. Biophys. Acta*, **1974**, *346*, 261-310.

A NEW AUTOMATED METHOD FOR MODELING NON-COVALENT MACRO-MOLECULAR INTERACTIONS

P.N. PALMA[1,2], L. KRIPPAHL[1], J.E. WAMPLER[3] and J.J.G. MOURA[1]

[1] Departamento de Química, Centro Química Fina e Biotecnologia, Faculdade de Ciências e Tecnologia, Universidade Nova de Lisboa, 2825 Monte de Caparica, Portugal.

[2] Instituto Superior de Ciências da Saúde - SUL, 2825 Monte de Caparica, Portugal

[3] Department Biochemistry, University of Georgia, Athens, GA 30602, USA

ABSTRACT. Molecular recognition and non-covalent macromolecular associations are important events in Chemistry and Biology. The prediction of the structure of complexes formed within macromolecular systems is a valuable step in the understanding of chemical pathways and of mechanisms operating in living organisms.

A new effective and automated method is reported for predicting the most likely mode of binding of two proteins, using the three-dimensional structures of the interacting molecules.

The program parameterized and tested on a series of protein complexes, well documented by crystallographic data. For thirteen out of fifteen tested cases, it is shown to successfully predict the correct or an approximate binding geometry. The value and limitations of the method for the study of strong and weak macromolecular interactions, not yet revealed by experimental techniques, are discussed.

Most of the time consuming tasks rely on very fast Boolean type (binary, such as OR, AND, etc.) operations, which makes it feasible to run on inexpensive and widely accessible personal computers.

1. INTRODUCTION

Protein-protein interactions often result from a delicate balance between chemical forces of different kinds, including long range electrostatic interactions, hydrogen bonding, van der Waals interactions and non-specific "forces", like conformational change of entropy and hydrophobic effects resulting from differences in solvation energies between the separated and bound molecules. Nevertheless, molecular recognition and specific macromolecular interactions play a critical role in many biological processes. The

R. Ungaro and E. Dalcanale (eds.), Supramolecular Science: Where It Is and Where It Is Going, 205–223.

ability to predict the structure of protein-protein complexes by computer assisted modeling will be of great value in the study of many biochemical problems.

We presently report a computer-automated method capable of such predictions, by taking into account a combination of several interactions terms between the molecules at interaction (such as electrostatics, solvent exclusion, etc.). Moreover, its predictive capabilities can be used in common situations where the structures of the unbound proteins are available but the crystallization of the complex is difficult or impossible. In these cases, docking models may be used to envisage and propose action mechanisms associated with functional complexes, which can then be validated or refuted by complementary experiments.

From a practical point-of-view, the problem of protein docking addresses two distinct issues; in a first step, a series of plausible candidate structures must be, in some way, generated. The range of the search, in this step, must be wide enough to include all significant binding geometries, including the "correct" one, but the number of solutions thus retained must be reasonably small to allow interaction energies to be computed and compared among them. In a second step, the putative docked structures must be ranked according to some set of properties relevant to the association process. A valid ranking model should be able to distinguish the near-correct docked geometries from the other - false - solutions or, at least, to present a narrow set of probable solutions, which can then be further inspected.

The analysis, at the atomic level, of X-ray structures of protein complexes reveals, in most cases, a close geometric match between the molecular surfaces at the binding interface [1]. Although geometric complementarity, by itself, does not constitute any type of direct driving force for molecular association, the more extensive the surface contacts, the greater the probability of other interaction terms, such as hydrogen bonds, electrostatic interactions or solvation effects, having an affect on the overall stabilization of the complex.

Many different algorithms for geometric docking have been reported [2-14], which are generally successful in excluding a large proportion of incorrect geometries, while retaining, in most situations, a near-correct solution. However, the proposed methods are invariably demanding in computational resources when docking two macromolecules, making their use limited to high-end, high-price workstations.

As part of the present work, we propose an efficient new algorithm for generating plausible docked geometries and for evaluating them according to the extent of surface shape complementarity. This algorithm, called BoGIE (Boolean Geometric Interaction Evaluation), is part of a functional program package written for readily available PC computers, aimed at the prediction of the structure of protein-protein complexes.

A distinct and more difficult problem to deal with, however, is to define a general form of scoring criteria that enables one to distinguish the "correct" or "true" geometric solution from the remaining "false" ones, generated in the first step. Several such criteria have been implemented with different levels of success, ranging from purely geometric scoring [3], or reduced forms of electrostatic interaction energy [4-6], to more complete evaluations of relative free energy of binding [9]. Regardless of the method used for predicting the most favorable solution, it seems that none of the individual energetic contributions that have been accessed has been sufficient, to distinguish the true solution in all tested complexes. In some cases one single energetic contribution (e.g. interaction electrostatics) is able to select the correct geometry of the complex, but in other cases it leads to false solutions, presumably due to the predominance of other factors relevant to the formation of the complex.

We propose a new scoring method for ranking a large number of putative binding geometries, which relies on a set of empirically adjusted interaction terms, pertinent to the process of molecular association. These include contributions from explicit electrostatics and solvation effects, as well as molecular surface complementarity and effective interaction terms related to the relative propensity of amino acid pairwise contacts.

The implementation of each of these individual interaction terms is discussed in the next section. The proper scaling and the combination of the individual terms in an effective scoring function of molecular interaction are optimized during a learning stage, where the docked geometries of twelve different protein-protein complexes, with known crystallographic structure, are predicted.

Finally, the validation and limitations of the proposed method are discussed, as a tool to predict the structure of protein-protein complexes, using the known X-ray structures of the individual molecules.

2. METHODS

2.1. TOPOLOGICAL COMPLEMENTARITY OF MOLECULAR SURFACES

Shape complementarity between interacting surfaces is the only criterion that is evaluated for the complete set of billions of docked structures generated, and is thus used as a first filter to reduce the large set of possible solutions to a manageable size.

This evaluation is performed on matrix representation of the molecular shapes. Each molecule is first digitized into a cubic grid (typically 1 Å cube size), by assigning a value “1” to every matrix location within the van der Waals radii (plus 1 Å) of any protein atom and a value “0” elsewhere. This initial matrix defining the molecular volume and shape is subsequently split into two new matrices, one containing the locations corresponding to the hollow surface shell of the original representation and the other, the remaining inner core region of the represented molecule (Figure 1).

Secondly, a large set of alternative docking geometries is generated by systematically rotating and shifting the matrix representation of one molecule (the probe) relative to the other (the target). Both rotations and translations are performed in discrete steps, typically 15° for rotation and 1Å for translation.

In this process, every solution where there is overlapping of core locations from both molecules is immediately discarded and, for the remaining ones, the extent of geometric complementarity is evaluated by summing the number of overlaps between grid positions corresponding to the surface shells (Figure 2).

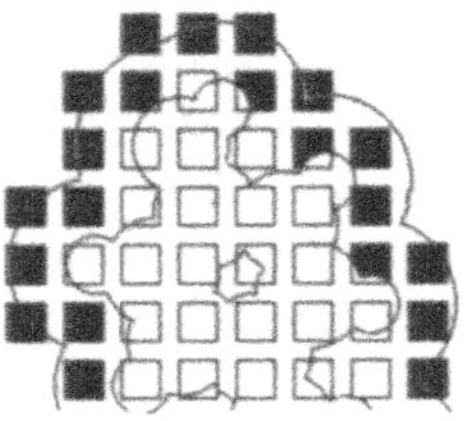

FIGURE 1. Schematic representation of a digitized molecular fragment. Black "cubes" define the molecular surface layer and white "cubes" the molecular core. In this example, the matrix resolution is 1 Å and an additional 1Å was added to the van der Waals radii during the digitalization.

Secondly, a large set of alternative docking geometries is generated by systematically rotating and shifting the matrix representation of one molecule (the probe) relative to the other (the target). Both rotations and translations are performed in discrete steps, typically 15° for rotation and 1Å for translation.

In this process, every solution where there is overlapping of core locations from both molecules is immediately discarded and, for the remaining ones, the extent of geometric complementarity is evaluated by summing the number of overlaps between grid positions corresponding to the surface shells (Figure 2).

The complete set of docked geometries so generated is sorted according to surface complementarity and a pre-defined subset (e.g. 1000) containing the best docking geometries is kept on disk for further evaluation according to the interaction evaluation described below.

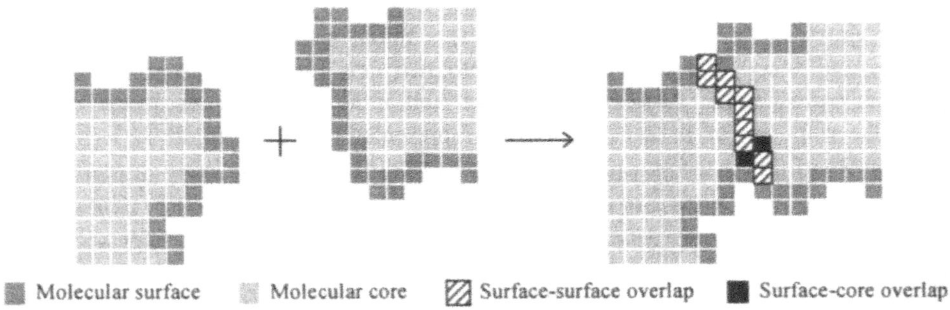

FIGURE 2. Evaluating surface contact (simplified two-dimensional representation).

By assigning some finite thickness to the surface layer and allowing the overlapping of surface cubes belonging to the two molecules, the method accounts in an implicit way, for some extent of flexibility of the surface amino acids. A soft-docking approach such as this is an important feature since the proteins are treated as rigid bodies, i.e., the explicit atomic coordinates are not allowed to change during the association process. However, in real situations, the association process induces some conformational changes, in particular, at the level of the amino acid side chains located at the molecular interface.

2.2. ELECTROSTATICS

Although charge-charge interactions play an obviously important role in macromolecular interactions, its accurate evaluation represents a serious challenge. Several electrostatic models could, in principle, be adopted, ranging from a simple Coulombic model to more elaborated multipole interaction terms, from microscopic models with explicit treatment of solvent dipoles and atomic polarizabilities [15-17] to macroscopic continuum models to evaluate the solvent reaction field [18-21]. For a more general review, see also references [22-24].

However, if electrostatics are to be evaluated for a large number of different docked geometries, such as the ones previously selected for their high surface complementarity, it is imperative to keep the model as simple as possible.

In the present implementation, the program uses a modified Coulombic potential to evaluate the electrostatic interactions between the two proteins, in each alternative association mode. The atomic point charges used are taken from the molecular mechanics force field Amber 4.1 [25-27] and a distance-dependent dielectric function is used.

There is, however, one difficulty associated with the straight use of a simple charge-charge Coulombic potential. Upon binding, the two molecules may undergo conformational changes to better accommodate side chain atoms at the interface. In the present model, however, the proteins are treated as rigid bodies and such conformational changes are modeled implicitly, by allowing a limited inter-penetration of the respective grid positions (see above). This may cause some atoms at the molecular interface to become unrealistically too close to each other. Since the Coulombic energy is inversely proportional to the square of the inter-atomic distances, this may give rise to unrealistic high values for the interaction between those atoms.

Thus, when evaluating the effective electrostatic term between a pair of atoms across the docking interface, a constant value of 1.5 Å is added to the distance separating both nuclei. This modified Coulombic term is evaluated for every pair of atom-centered charges (a 10 Å cutoff is used through this work) across the molecular interface.

2.3. SOLVATION TERM

The molecular association process may be viewed (in a similar way as for the protein folding) as a competition between protein-protein and protein-solvent interactions. As the association mode changes, the contacts between atoms of the two proteins are eventually replaced by contacts with solvent molecules and *vice versa.* A key quantity involved in this process is the balance between the free energy of solvation of each docked geometry and that of the unbound proteins.

In this work, an effective solvation potential is modeled using a method developed by Wang and coworkers [28, 29], which was originally optimized to discriminate between native and non-native structures of globular proteins.

The difference between the solvation energies of the complex (in each particular docked geometry) and that of the free proteins is estimated by the loss of solvated surface of predefined molecular fragments, due to association.

The same set of original molecular fragments and solvation parameters [29] were used, except that, for reasons of performance, a simplified and faster algorithm was implemented to compute the loss of molecular surface areas. Thus, if the distance between the surface of an atom *a* (belonging to molecular fragment *i* of one of the proteins) and any atom belonging to the other protein is equal or greater than 2.8 Å (taken as the average diameter of a water molecule), then atom *A* may still remain solvated and its solvent excluded area is set to zero. At the other extreme, the solvent excluded area of atom *a* from one molecule, which is in contact (actually touching) with atom *b* from the other molecule, is estimated as the cross section area of *b* (given by πr_b^2, r_b being the radius of atom *b*). For intermediate situations, the value is estimated by a linear function of the distances between the surfaces of *a* and *b*.

2.4. SIDE CHAINS SPECIFIC CONTACTS

Different molecular groups have different preferences for interacting with solvent or other chemical groups. The relative propensity for contact of each pair of amino acids is the result of a complex set of enthalpic and entropic effects. However, they can also be evaluated empirically, by observing the statistical frequency of naturally occurring interactions (by proximity) between pairs of amino acids in a population of well resolved X-ray protein structures.

This information is collected in the *Atlas of Protein Side-Chain Interactions* [30] and is used as the basis of an additional interaction scoring function for evaluating the quality of each docked geometry.

The underlying assumption is that the molecular packing and the nature of the intermolecular chemical interactions at the interface of two complexed proteins should be equivalent to those observed inside globular protein structures.

The program uses a simple and fast algorithm for evaluating contacts between the side chains of amino acid across the interaction surface. Although the two molecules are docked as rigid bodies, the definition of side chain contacts between proteins takes into account their conformation flexibility in an approximate way. The volume accessible to a particular side chain is taken as a sphere containing all side chain atoms and centered at the geometric center of that group of atoms. Thus, two amino acids belonging to different proteins are considered in contact if the spheres defining their accessible volumes touch or intercept.

2.5. GLOBAL SCORING FUNCTION

Up to this point we described a set of four scoring terms, that can be potentially used to rank the likeliness of any alternative docked geometry, according to different properties. As we discussed, the evaluation of the surface geometric matching, takes a precedent over the other three terms, because it is the only one that can be calculated for the entire set of docked geometries (tens of millions) in a reasonable amount of time. However, once a sizable set of geometries (e.g. 1000) with high extent of surface complementarity is defined, the other interaction terms can then be easily evaluated.

The question now is how should these four terms be properly scaled to each other and combined into an effective interaction potential capable to discriminate near-correct docked geometries out of hundreds of other false solutions. The properties evaluated, are not independent of each other and so cannot be directly compared. For example, the side chains specific contacts, which evaluates the pairwise propensities of amino acid side chains to contact or to avoid each other, already includes terms such as electrostatic interaction energies or hydrophobic effects. In addition, these four properties are obviously expressed are in distinct units. In other words, a proper way of weighting the different interaction terms is essential to calculate a global scoring term. A simple way of combining the different terms (P_i) into one unique scoring index is through a linear combination, provided that the set of weighting coefficients (c_i) is adjusted to optimally scale the different interaction terms:

$$\textit{Global scoring function} = c_1P_1 + c_2P_2 + c_3P_3 + c_4P_4 \quad (1)$$

An optimization of the adjustable parameters c_i was obtained using the information on a set of 12 protein-protein complexes with known crystallographic structures. For this purpose, the structures of bound molecules in the crystallographic complexes where taken apart, rotated at random, and re-docked as explained. In cases where the X-ray structures of the unbound proteins are available as well, the structures of the complexes where also predicted using the atomic coordinates of the unbound molecules, completing a total of 15 test cases (see Table 1). In order to assure wild applicability of the method, a selection different protein complexes was made, which are supposed to be dominated by chemical interactions of different natures. These include several stable dimers formed between identical subunits possessing extended and specific contact surfaces, stable and specific complexes between different enzymes (endonucleases and proteases) and different inhibitors, as well as a complex between two electron transfer proteins (Table 1).

The values of the four interaction terms (surface contact, side chain inter-molecular contacts, relative solvation energy and Coulombic interaction potential) were computed for the 200 best matching geometries generated by the BoGIE module, for each of the 15 docking cases.

TABLE 1. Selected protein complexes used as test cases.

Dock Designation	Comment	*Pdb Files*
bound structures		
1. hem-dimr	hemoglobin homo-dimer (Ark clam)	4sdh
2. trp-dimr	troponin-*c* homo-dimer (*Gallus gallus*)	1cta
3. txn-dimr	β-erabutoxin homo-dimer (sea snake venom)	6ebx
4. brn-bst	bacterial barnase / barstar (mutant C40A, C82A)	1bgs
5. tryp-inhib	trypsin / trypsin inhibitor (bovine pancreas).	2ptc
6. chym-ovo	α-chymotrypsin / ovomucoid third domain	1cho
7. subt-inhw	subtilisin BPN' / subt. inhibitor (wild type *streptomyces*)	2sic
8. subt-inhm	subtilisin BPN' / subt. inhibitor (mutant M70G, M73R)	5sic
9. subt-eglin	subtilisin Novo BPN' / Eglin c (mutant L45R)	1sbn
10. ccp-ccy	cytochrome *c* peroxidase / cytochrome *c* (yeast)	2pcc
pseudo-unbound		
11. hiv-dimr	HIV-2 protease homo-dimer (human)	2mip
12. dx-dimr	desulforedoxin homo-dimer (*Desulfovibrio gigas*)	1dxg
unbound		
13. subt-inhwf	subtilisin BPN' / subt. inhibitor (wild type *streptomyces*)	2stl 2ssi
14. brn-bstf	bacterial barnase / barstar	1bgs 1bta
15. ccp-ccyf	cytochrome *c* peroxidase / cytochrome *c* (yeast)	1ccp 1ycc

Finally, weighting coefficients were recursively optimized with a genetic algorithm [31] using the full population of 15x200 evaluated docked geometries. For this propose, the algorithm seeks to maximize the following two conditions: *i*) the binding mode possessing the highest docking score for a given complex, as computed by Equation 1, should correspond to a near-native structure of that complex (with less than 2 angstrom α-carbons r.m.s. deviation from the crystallographic coordinates) and *ii*) the difference between the values of the best docking score (associated to a near-correct geometry) and that of the best score associated with non-native structure (with more than 2 Å r.m.s.d. from X-ray), should be maximal. This is intended to sharpen the distinction between correct and incorrect solutions.

It should be pointed out that this method is tuned to assign the highest docking score to a near-native binding mode, but it does not attempt to establish any correlation between this global score and the r.m.s. deviation between the total set of solutions and the X-ray coordinates. In fact these two properties do not necessarily correlate. Although one can argue that the native binding mode is the one possessing the most favorable interaction, it is possible to find many incorrect solutions possessing large but similar values of r.m.s. deviation from the X-ray structure, which may, however, be quite distinct from each other and possess different interaction energies.

3. RESULTS AND DISCUSSION

Filtering the total population of binding modes on the basis of surface complementarity was able to retain at least one near-correct model structure of the complex within the 200 top ranked solutions, for 13 out of 15 docking cases assayed (Table 2). These approximate solutions present r.m.s. deviations from the actual X-ray structures smaller than 2 Å for all cases except the complex between cytochrome *c* peroxidase and cytochrome *c*, for which, the closest solutions found were 3.6 and 3.8 Å, when starting from the bound or the unbound structures of the two proteins, respectively.

This result clearly shows an important role of surface shape complementarity in protein associations and that this may be efficiently used as a first filtering criterion to reduce the universe of possible binding modes to a manageable set of plausible ones, with a good probability of containing a near-correct solution. There were, however, two cases where this procedure was totally unable to retain a near-native structure of the complexes. This occurred when trying to predict the structure of the erabutoxin dimer (PDB: 6ebx) and that of the subtilisin-subtilisin inhibitor complex (PDB: 2sic) and the reasons will be discussed later on.

Twelve protein complexes (all except cases 3, 13 and 15) were used as a learning set for the optimization of docking parameters. The resulting scoring function is shown to successfully score a near-correct binding mode within the five top-ranked solutions, for every test case considered (Table 2).

In order to clarify the complete procedure, it will be exemplified step by step, using the case of the Desulforedoxin homo-dimer (case 12). Finally the results of all other test cases will be summarized in Table 2. Desulforedoxin, isolated from the sulfate reducing bacteria *Desulfovibrio gigas*, is a small dimeric protein formed by two identical subunits of 4 kDa each, containing an iron atom coordinated by four cysteinyl residues, which is responsible for the oxido-reduction properties of the protein.

The process and conditions used for generating a population of alternative docked were those given in methods. The 200 docking solutions presenting the highest surface contact scores were further evaluated according to the other 3 interaction terms (electrostatics, solvation and side chain intermolecular contacts).

The plots in Figure 3 summarize the relevant aspects of these results. In the top left panel, the first 50 individual solutions (each corresponding to a particular binding mode) from BoGIE are sorted by their surface contact scores. For each one, the corresponding score value (note that this plots present the crude interaction indexes, not yet scaled with the weighting coefficients) is represented as an open circle, while its r.m.s. deviation from the actual X-ray structure of the complex is represented as a vertical bar (scale shown in inset).

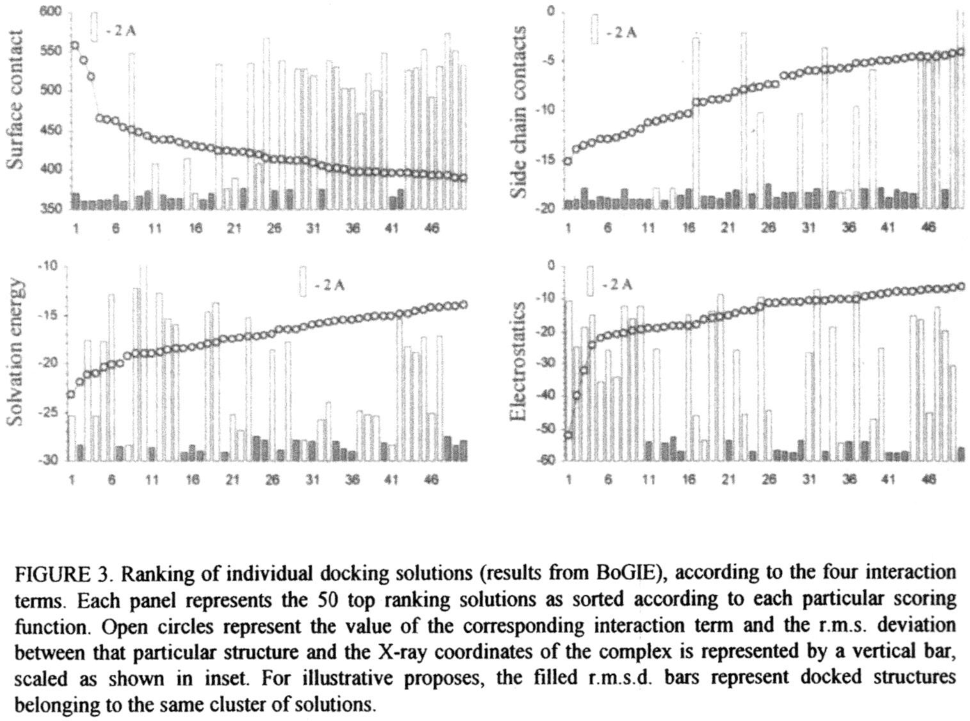

FIGURE 3. Ranking of individual docking solutions (results from BoGIE), according to the four interaction terms. Each panel represents the 50 top ranking solutions as sorted according to each particular scoring function. Open circles represent the value of the corresponding interaction term and the r.m.s. deviation between that particular structure and the X-ray coordinates of the complex is represented by a vertical bar, scaled as shown in inset. For illustrative proposes, the filled r.m.s.d. bars represent docked structures belonging to the same cluster of solutions.

As can be seen, the geometries with higher degree of surface contact are very similar to the X-ray structure of the dimer, presenting r.m.s.d. values of less than 1.2 Å. A geometric fitting algorithm would be, in this case, sufficient to find the correct structure of the complex. However, it must be pointed out that this is not so for every complex (as will be shown), so other type of interaction terms must be weighted as well.

Additionally, this panel also shows that there are several solutions (e.g. solutions 22, 26, 41) similar to the X-ray structure (and also similar to each other) which may however, possess a surface contact score much lower than other false solutions. In fact, due to the fact that the protein are treated as non-flexible structures, a slight shift or rotation of one protein relative to the other can result in a significant loss of surface contact, causing similar solutions to appear quite separated in the geometric scoring table.

The other three panels in Figure 3 show the 50 top-ranked solutions, as sorted by the other three evaluation criteria: side chain contacts, solvation energy and Coulombic interaction potentials. The structural solutions presented in one panel are not necessarily the same shown in the other panels, since, for instance, the solutions with higher surface contact do not necessarily present the strongest Coulombic interaction potentials and *vice versa*. Thus, a global scoring function with proper weighting of the different interaction terms is essential to reliably distinguish the valid approximate structures from the incorrect ones, in situations where the actual X-ray structure of the complex would be unknown.

The four different interaction terms presented in Figure 3 are shown combined in Figure 4 (left Panel), using Equation 1 and the set of optimized weighting coefficients. As shown in this figure, the linear combination of the four interaction terms (global scoring function), still retains the ability to distinguish the correct solutions. The left chart shows the 50 top-ranked binding modes, according to the global scoring function. Due to the discrete nature of the molecular representation and of the orientational searching, several distinct but similar docked structures may be generated. They may, however, be considered a family of variations of the same significant binding mode and can be easily clustered together to simplify the analysis of the results. The clusters may then be represented either by the ensemble of structures superimposed on each other, as shown in the right Panel of Figure 4, or by a single structure, representative of that family.

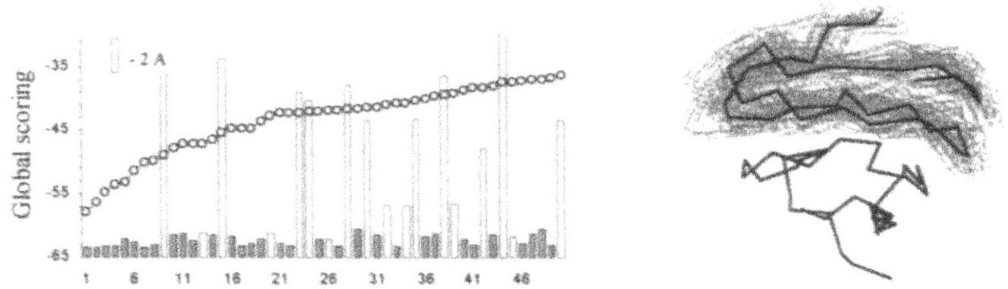

FIGURE 4. *Left Panel*: Ranking of the 50 top ranking solutions as sorted according the global scoring function. Symbols as in Figure 3. Filled bars represent docked structures belonging to the best cluster of solutions. *Right Panel*: Cluster of 19 similar binding modes (thin lines) represented by filled bars in left chart, superimposed on the X-ray structure (thick lines) of the dimeric form of Desulforedoxin.

Table 2 summarizes the docking results for the complete set of protein complexes tested, using a similar procedure. This table presents, for each complex, the ranking position of the predicted near-native docked geometry, according to the combined docking index, as well as to the individual interaction terms.

It should be pointed out that the solutions presented are not necessarily the ones possessing the greatest similarity to the crystallographic complexes. The weighting coefficients were tuned with the criterion of seeking approximate solution structures, within 2 Å of r.m.s. deviation from the X-ray coordinates (α-carbons only), and not necessarily the one with lower r.m.s.d. However, every docking geometries with r.m.s. deviation from X-ray of less than 2 Å should be considered indistinctly good approximate solutions, since this is the order of magnitude of the imprecision introduced by the digitizing method itself.

TABLE 2. Protein docking results of test cases. The table presents, for each complex, the ranking position of a near-correct docked geometry, as predicted according to the combined docking score, as well as to the individual interaction terms.

	Ranking position of near-correct binding modes [a]					
Protein complex	Global score	surface contacts	side chains spec. contacts	solvation energy	Coulombic interaction	αC rmsd from X-ray
bound structures						
1. hem-dimr	**1**	2	22	25	2	0.53
2. trp-dimr	**1**	48	1	1	200	0.83
3. txn-dimr						
4. brn-bst	**1**	4	1	28	15	0.64
5. tryp-inhib	**1**	5	47	3	6	0.96
6. chym-ovo	**1**	1	51	7	25	1.49
7. subt-inhw	**1**	6	5	1	6	0.92
8. subt-inhm	**1**	90	83	1	60	0.91
9. subt-eglin	**1**	10	32	8	3	1.31
10. ccp-ccy	**3**	167	10	61	4	3.67
pseudo-unbound						
11. hiv-dimr	**1**	8	2	2	144	1.19
12. dx-dimr	**1**	3	4	15	41	0.63
unbound						
12. brn-bstf	**5**	133	2	58	35	1.16
13. ccp-ccyf	**2**	167	3	39	8	5.36

[a] Solutions with less than 2Å of r.m.s. deviation from the native structure are considered indistinctly approximate. In the case of the cytochrome *c* peroxidase - cytochrome *c* complex, no binding modes could be found with that level of precision and the solutions indicated represent the best approximation found.

These results clearly show that none of the individual interaction terms, alone, may constitute a reliable criterion to distinguish the native-like structure from the false ones. This indicates, not only that a complex balance between different interaction forces are involved in driving the formation of these complexes, but also that the nature of the predominant forces, are not the same for all of them.

For example, the dimeric form of troponin-*c* (case 2) does not show a great extent of intermolecular surface complementarity, ranking in the 48th position when compared to other non-native binding modes tested by the program. In addition, the native-like complex proposed presents an unfavorable Coulombic interaction potential. However, the strong and favorable intermolecular amino acid contacts and solvation terms, may overcome the unfavorable components of the interaction thus explaining the actual geometry found in the crystallographic structure. In some other cases, none of the computed individual terms seems particularly strong for the native-like geometries, but their combination is, however, stronger than that of any other incorrect docked structures.

Hence, a proper combination of the four interaction terms presented in this work, proves to successfully narrow to a great extent, the set of alternative binding modes generated, in such a way that a reasonably approximate association mode is likely to be found within the few top ranking solutions.

A rough, but global visualization of the docking solutions is exemplified in Figure 5, for the case of desulforedoxin dimer. The backbone of the target protein is represented surrounded by small spheres, each representing an alternative docking position of the probe molecule. The spheres are placed at the center of mass of the probe molecule and colored from white to black according to the corresponding global interaction score.

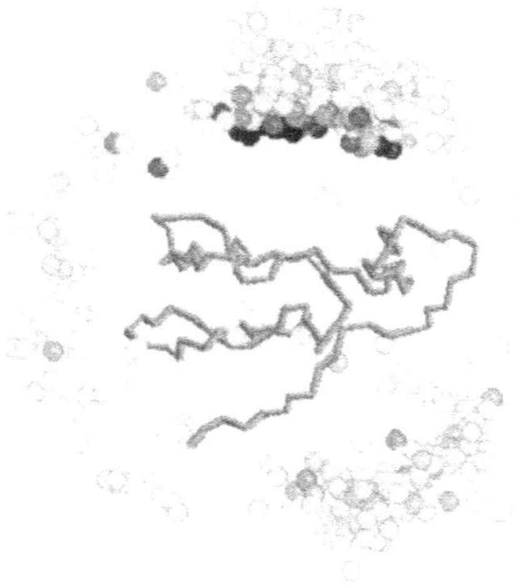

FIGURE 5. Global visualization of the docking solutions obtained for the case of the desulforedoxin dimer. The backbone of the target protein is shown surrounded by small spheres, each representing an alternative docking position of the probe molecule. The spheres are placed at the center of mass of the probe molecule and colored from white to black according to the corresponding global interaction score (darker color represent higher scoring).

The figure shows that most of the binding modes initially selected on the criterion of maximal surface contact are actually unfavorable solutions, represented as light shaded spheres. The docking sites assigned with higher interaction scores (darker spheres), however, are mostly localized at a well defined region on the protein surface and this corresponds to the actual docking site observed in the X-ray structure of the dimer. Furthermore, these solutions are part of a numerous and tight group (cluster) of similar binding modes, already shown in Figure 4. A similar type of result is observed for most of the protein docking experiments assessed, for which the program can identify the interaction surface with few ambiguities and almost no need for subjective analysis.

The complex between yeast cytochrome *c* peroxidase and cytochrome *c*, however, seams to present a different type of behavior with respect to the association process. As for most known complexes formed between electron transfer proteins, this is a week complex with a binding constant of the order of 10^3. In contrast to the other cases, these two proteins do not show a great extent of surface matching at the docking interface, nor do they show very specific amino acid interactions across the intermolecular surface. As a result, the predicted model of the structure of this complex is less well defined.

It has been suggested [32] that electron transfer between redox proteins may occur through the dynamic formation of an ensemble of alternative weak complexes, rather than through a unique and very specific complex. Our results on the cytochrome *c* peroxidase - cytochrome *c* complex as well as for other complexes between electron transfer proteins (to be published) also support this argument. In fact, for this class of complexes, a single, well-defined, docking site is not found, but rather a set of alternative favorable solutions, which are generally constrained to the region surrounding the redox cofactors, but which allow for some rotational and translational freedom of the proteins.

The computed individual interaction terms do not account for all the chemical forces that might contribute for the stabilization of such complexes. However, they can give some insights into the nature of the predominant interactions determining the specificity and stability of different types of complexes. From our results, every interaction term except the electrostatics, seam to be more important for the strong complexes than for the weaker electron transfer complex. In particular, the extent of surface complementarity seams to be one of the predominant issues in the case of protein dimers or specific enzyme-inhibitor complexes, while the transient complexes formed between the electron transfer couple appears to be much less shape specific. The same is not true, however, for the electrostatic stabilization forces. Electrostatics appears, indeed, to constitute a major motif for stabilizing the interaction between the peroxidase and the cytochrome *c*, overcoming, in this case, other chemical contributions. A strong electrostatic component is also observed for other complexes between electron transfer proteins (our non-published data).

The distinct roles of electrostatics in guiding different types of molecular interactions, can also be inferred from Figure 6, which shows iso-potential contour maps for the yeast cytochrome *c* and cytochrome *c* peroxidase and for the two identical monomers of bacterial desulforedoxin. Each pair of interacting molecules is shown oriented as in the complex and the iso-potential cross-sections pass through the centers of mass of the molecules.

The acidic Desulforedoxin monomers are surrounded by a grossly negative electrostatic field, which does not show any apparent long range complementarity. Although the complex between the two monomers possesses a favorable negative electrostatic energy, due to the formation of short-range ionic pairs, their approach is probably counteracted by the interaction of their negative electrostatic fields. This type of behavior is also observed for many of the proteins present in Table 2, that form strong complexes. On the other hand, cytochrome *c* peroxidase induces a highly polar electrostatic field in its environment, being particularly negative in the region surrounding the heme crevice. This extended polar field can work as a guiding force, attracting and pre-orienting the positively charged portion of cytochrome *c*, containing the heme group.

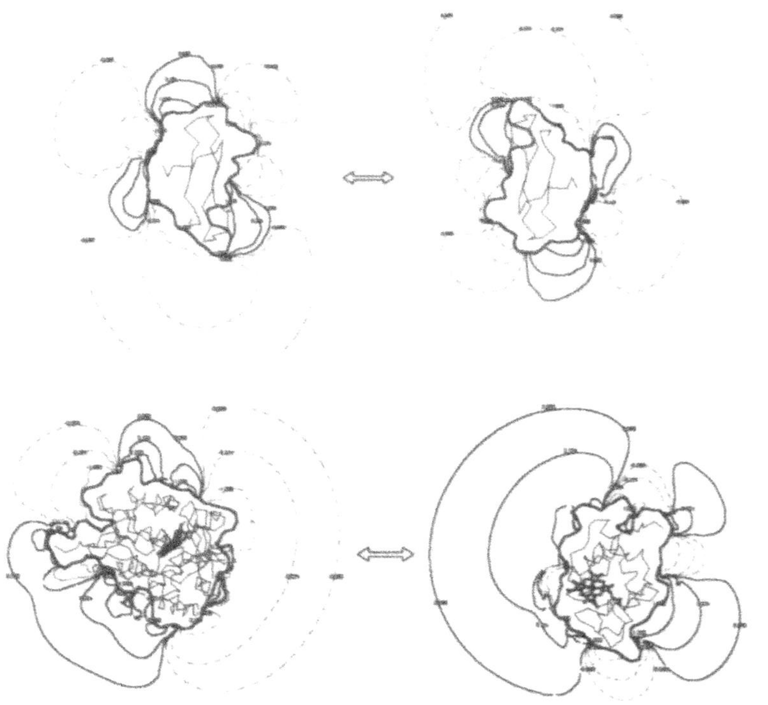

FIGURE 6. Iso-potential contour maps for the two identical monomers of bacterial Desulforedoxin (top) and for the yeast cytochrome *c* and cytochrome *c* peroxidase (bottom). Each pair of interacting molecules is shown oriented as in the complex and the iso-potential cross-sections pass through the center of mass of the molecules. Electrostatic potentials were computed through the numerical solution of the finite-difference Poisson-Boltzmann equation, set up on a cubic grid and using the dielectric constants of 2 and 80 for the inner and outer regions of the protein and a solution ionic strength of 100 mM. Plain and dashed lines represent, respectively, positive and negative iso-potentials at 1.00, 0.22 and 0.05 kcal/mol.

These results generally agree with the fact that the electron transfer rate between many of these types of proteins is ionic strength dependent. In protein interactions where the kinetics of the molecular association/dissociation is physiologically important, the existence of a strong and polarized electrostatic field around the molecules, may work to increase the number of effective encounters over that of the (mostly non-productive) diffusion-controlled collisions. In fact, the electrostatic field generated by the protein in its surroundings may provide a molecular approaching pathway where the unfavorable loss of entropy associated to the decrease of translational and rotational freedom of the molecules may be smoothly split along that pathway and compensated by a electrostatic interactions.

Finally, a few but well characterized limitations of the method proposed ought to be discussed at this point. BoGIE searches the complete rotational and translational space (with the limitations of a discrete step approximation) of one molecule relative to the other and stores the thousand solutions presenting a higher effective surface contact. However, in cases where the correct structure of a complex presents a small contact region between the two molecules while, at the same time, the program can find many incorrect binding modes with more extensive surface contacts, then the correct solution will be pushed down the ranking list and may, eventually, be excluded from the set of retained solutions. This occurs in the case of the β-erabutoxin dimer (case 3 in Table 2), where the two monomers interact through an edge-shaped contacting surface, forming Hydrogen bonded intermolecular beta sheet. The existence of a large number of alternative binding modes between the two molecules, with much higher levels of surface contact than that of the native-like solution, makes the program to miss the correct binding mode.

The conjunction of these two factors can cause the BoGIE algorithm to fail, but either one, alone, may be overcome. Two examples of this are the complexes between the two monomers of the muscle protein troponin *c* and between cytochrome *c* peroxidase and cytochrome *c*. In the first case, there are many possible, but incorrect, interaction geometries, presenting higher geometric complementarity than the native-like solution (see Table 2). However, the correct solution also possesses a large surface contact between the monomers, enabling it to be retained (in the 48th position) for further analysis. On the other hand, the crystallographic complex between cytochrome *c* peroxidase and cytochrome *c* does not present a large surface complementarity between the two molecules, but so doesn't any other possible docking geometries, so again, a valid solution, although roughly approximated, could be found. The αCarbon r.m.s. deviation between this docked structure and the crystallographic structure of the complex is 3.67 Å, due to a slight rotation around the axis connecting the two molecules. However, the interaction surfaces were correctly identified and the intermolecular distance between the heme groups was approximately predicted (14.8 Å vs. 15.5 Å in the X-ray and predicted structures, respectively).

In our opinion, the greatest potential limitation of the BoGIE algorithm arises from the treatment of the proteins as non-flexible chemical entities. This might not be very important when trying to predict protein complexes starting from structures of the bound molecules, which are already self accommodated, but it can represent a serious limitation if docking unbound molecules. In truth, the method actually assigns some "softness" to the molecular shapes by enabling a certain amount of overlap between the surface representations of both proteins. This softness is apparently sufficient to account for, in an implicit way, most of the conformational changes relevant in the unbound (and pseudo-unbound) docking experiments.

In the cases of Desulforedoxin and HIV Protease, the prediction of the interaction geometries was fully successful, although two copies of the same monomer were used instead of one of each. Note that the atomic coordinates of the two monomers differ by 0.99 and 1.20 Å r.m.s.d. (all heavy atoms), for desulforedoxin and HIV protease, respectively. Also, the r.m.s.d. between the atomic coordinates of the isolated (NMR average structure) and complexed barstar molecule is 1.85Å.

However, if the unbound and complexed structures differ too much, the binding mode may be incorrectly predicted. The difficulty in predicting the correct interaction geometry between cytochrome *c* peroxidase and cytochrome *c*, due to the moderate extent of surface complementarity is, in fact, more evident when starting from the structures of the unbound proteins, which show, respectively, 0.86 and 1.06Å of r.m.s. deviation (all heavy atoms) from the corresponding conformations, in the complex. In this case, the interaction surfaces were grossly identified, but the position of cytochrome *c* is somewhat rotated and translated, relative to that adopted in the crystallographic complex.

The worst result, however, was obtained when docking (case 13 in Table 2) the unbound forms of subtilisin and its inhibitor (respectively, 0.72Å and 1.72Å r.m.s. deviation from complexed structures). The interactions between proteases, such as subtilisin, chymotrypsin or trypsin, and their inhibitors, are commonly made through an exposed loop of the inhibitor polypeptide, which fits into a tight grove that constitutes the active center of the protease. Thus, the specificity of the interaction between subtilisin and its inhibitor, is assured by the perfect matching between the side chains of only five residues of the inhibitor's interacting loop and the active site of subtilisin. However, three of those five residues (D68, M70 and M73) show quite different side chain conformations, when comparing the unbound and complexed structures of the inhibitor. Without allowing these side chains to re-accommodate, they cause too strong steric conflicts inside the active site of the protease and cause the correct solution to be missed.

4. CONCLUDING REMARKS

Although the structure of protein-protein complexes is not fully explained by geometric complementarity, this does appear to be a fairly common and relevant feature of macromolecular associations. The results presented support the assumption that surface shape complementarity, as evaluated by surface contact, can be a useful filtering criterion to reduce the number of possible alternative geometries to a manageable size, which may then be further evaluated.

It was also shown that the four interaction indexes introduced (surface contact, relative solvation, electrostatics and the relative propensity of amino acid pairs to interact) can be conveniently combined into an effective interaction potential function, able to distinguish native from non-native binding modes and to propose a limited set of plausible candidate models of protein complexes. For every complex tested (except for the two where the correct solutions were excluded from the ones retained by BoGIE), a near-correct docked structure was found within the five top-ranked solutions indicated by the global interaction function.

Together with the innovative character of the fast Boolean-based algorithm and its modest computational requirements, this protein docking approach achieves many of the goals of the ideal computational method described in the introduction. As with many other tools of computational chemistry, the availability of faster machines and faster algorithms running on less expensive machines should allow broader exploration of the issues of macromolecular interaction.

Some limitations are known though, which are presently the focus of our current investigation: first, the treatment of the proteins as non-flexible molecules may fail to capture subtle but important features of some protein complexes. Although this limited approach is rooted in computational feasibility reasons, we are improving the algorithm to account for different amino acid flexibility. Secondly, our approach assumes that the topological surface complementarity take a precedent over any other interaction term. This again is set for reasons of computational performance, since it is the only feature that could be calculated for the complete universe of alternative binding modes. Fast evaluation of other properties is being tested in order to improve the liability of the method. Finally, a new method based on neural networks is being implemented to introduce a new form of effective interaction potential function. These mew approaches will be the focus of future publications.

ACKNOWLEDGMENTS: PRAXIS, NATO, COST, HCM(JJGM), NIH(JEW)

5. REFERENCES

1. Hubbard, S.J., Campbell, S.F. and Thornton, J.M. (1991) Molecular recognition. Conformational analysis of limited proteolytic sites and serine proteinase protein inhibitors, *J Mol Biol* **220**, 507-30.
2. Kuntz, I.D., Meng, E.C. and Shoichet, B.K. (1994) Structure based molecular design., *Acc Chem Res* **27**, 117-123.
3. Katchalski-Katzir, E., Shariv, I., Eisenstein, M., Friesem, A.A., Aflalo, C. and Vakser, I.A. (1992) Molecular surface recognition: determination of geometric fit between proteins and their ligands by correlation techniques, *Proc Natl Acad Sci U S A* **89**, 2195-9.
4. Jiang, F. and Kim, S.H. (1991) "Soft docking": matching of molecular surface cubes, *J Mol Biol* **219**, 79-102.
5. Bacon, D.J. and Moult, J. (1992) Docking by least-squares fitting of molecular surface patterns, *J Mol Biol* **225**, 849-58.
6. Walls, P.H. and Sternberg, M.J. (1992) New algorithm to model protein-protein recognition based on surface complementarity. Applications to antibody-antigen docking, *J Mol Biol* **228**, 277-97.

7. Jones, G., Willett, P. and Glen, R.C. (1995) Molecular recognition of receptor sites using a genetic algorithm with a description of desolvation, *J Mol Biol* **245**, 43-53.
8. Norel, R., Lin, S.L., Wolfson, H.J. and Nussinov, R. (1994) Shape complementarity at protein-protein interfaces, *Biopolymers* **34**, 933-40.
9. Jackson, R.M. and Sternberg, M.J. (1995) A continuum model for protein-protein interactions: application to the docking problem, *J Mol Biol* **250**, 258-75.
10. Lin, S.L., Nussinov, R., Fischer, D. and Wolfson, H.J. (1994) Molecular surface representations by sparse critical points, *Proteins* **18**, 94-101.
11. Lawrence, M.C. and Davis, P.C. (1992) CLIX: a search algorithm for finding novel ligands capable of binding proteins of known three-dimensional structure, *Proteins* **12**, 31-41.
12. Sezerman, U., Vajda, S., Cornette, J. and DeLisi, C. (1993) Toward computational determination of peptide-receptor structure, *Protein Sci* **2**, 1827-43.
13. Shoichet, B.K. and Kuntz, I.D. (1991) Protein docking and complementarity, *J Mol Biol* **221**, 327-46.
14. Connolly, M.L. (1992) Shape distributions of protein topography, *Biopolymers* **32**, 1215-36.
15. Russell, S.T. and Warshel, A. (1985) Calculations of electrostatic energies in proteins. The energetics of ionized groups in bovine pancreatic trypsin inhibitor, *J Mol Biol* **185**, 389-404.
16. Warshel, A. and Aqvist, J. (1991) Electrostatic energy and macromolecular function, *Annu Rev Biophys Biophys Chem* **20**, 267-98.
17. Lee, F.S., Chu, Z.T. and Warshel, A. (1993) Microscopic and semimicroscopic calculations of electrostatic energies in proteins by POLARIS and ENZYMIX programs., *J Comp Chem* **14**, 161-85.
18. Gilson, M.K. and Honig, B. (1988) Energetics of charge-charge interactions in proteins., *Proteins* **3**, 32-52.
19. Gilson, M.K. and Honig, B. (1988) Calculation of the total electrostatic energy of a macromolecular system: Slvation energies, binding energies and conformational analysis., *Proteins* **4**, 7-18.
20. Honig, B., Sharp, K. and Yang, A.-S. (1993) Macroscopic models of aqueous solutions: Biological and chemical applications., *J Phys Chem* **97**, 1101-1109.
21. Honig, B. and Nicholls, A. (1995) Classical electrostatics in biology and chemistry, *Science* **268**, 1144-9.
22. Davis, M. and McCammon, J.A. (1090) Electrostatics in biomolecular structure and dynamics., *Chem Rev* **90**, 509-521.
23. Harvey, S.C. (1989) Treatment of electrostatic effects in macromolecular modeling, *Proteins* **5**, 78-92.
24. Burley, S.K. and Petsko, G.A. (1988) Weakly polar interactions in proteins, *Adv Protein Chem* **39**, 125-89.
25. Cornell, W.D., Cieplak, P., Bayly, C.I., Gould, I.R., Merz, K.M.J., Ferguson, D.M., Spellmeyer, D.C., Fox, T., Caldwell, J.W. and Kollman, P.A. (1985) A second generation force field for the simulation of proteins and nucleic acids., *J Am Chem Soc* **117**, 5179-5197.
26. Weiner, S.J., Kollman, P.A., Case, D.A., Singh, U.C., C., G., Alagona, G., Profeta, S. and Weiner, P. (1984) A New Force Field for Molecular Mechanical Simulation of Nucleic Acids and Proteins., *J Am Chem Soc* **106**, 765-784.
27. Weiner, S.J., Kollman, P.A., Nguyen, D.T. and Case, D.A. (1986) An All Atom Force Field for Simulations of Proteins and Nucleic Acids, *J Comp Chem* **7**, 230-252.
28. Wang, Y., Zhang, H., Li, W. and Scott, R.A. (1995) Discriminating compact nonnative structures from the native structure of globular proteins, *Proc Natl Acad Sci U S A* **92**, 709-13.
29. Wang, Y., Zhang, H. and Scott, R.A. (1995) A new computational model for protein folding based on atomic solvation, *Protein Sci* **4**, 1402-11.
30. Singh, J. and Thornton, J.M. (1992) *Atlas of Protein Side-Chain Interactions.*, Vol. I & II, IRL press, Oxford.
31. Michalevicz, Z. (1992) *Genetic algorithms + Data structures = Evolution programs.*, Spring Verlag, Berlin.
32. Northrup, S.H. (1996) in *Protein electron transfer.* (Bendall, D.S., Ed.) pp 69-97, BIOS Scientific Publishers Ltd., Oxford.

RIGID ÅNGSTRÖM CLEFTS IN LIPID MEMBRANES ON SOLID SURFACES

L. RUHLMANN, J. ZIMMERMANN, C. MESSERSCHMIDT,
W. FUDICKAR, J.-H. FUHRHOP

1 *Institut für Organische Chemie*
der Freien Universität Berlin
Takustr 3, D-14195 Berlin, Germany

1. Clefts on Protein Surfaces

Enzymatic catalysis and reversible binding of external compounds by globular proteins takes either place on the surface of a single or between several rigid β-pleated sheets [1] or, more common, in surface clefts between helices with a width of a few Ångströms. The best known examples are lysozyme, which catalyzes the hydrolysis of poly(acetals) and myoglobin, which binds oxygen [2]. In both cases the clefts are open to the bulk water phase, but nevertheless contain quite a number of hydrophobic amino acids. Furthermore the gap is rigid with well-defined widths and contractions at binding sites. This rigidity is obtained by helical substructures of the protein, to which the active side-chains are bound. As an example, myoglobin [2] and a model peptide containing heme [3] are reproduced in Figure 1. The porphyrin is fixated by two histidine units which bind to the central iron atom and the hydrophobicity of the myoglobin pocket allows for reversible oxygen addition. From the model structure the alternation of rigid binding sites and empty space for the addition of substrates becomes obvious (Figure 1).

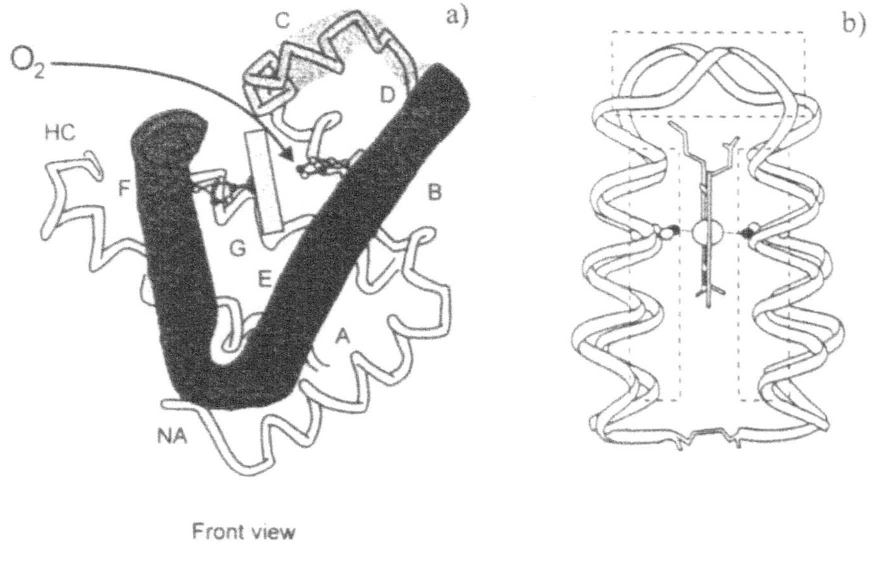

Figure 1: Model of myoglobin and an (S-S)-connected model peptide

R. Ungaro and E. Dalcanale (eds.), Supramolecular Science: Where It Is and Where It Is Going, 225–232.

It must be emphasized here that the protein surface clefts have no particular common shape. They are neither concave nor do they look like an imprint of the substrate or co-enzyme. "Fitting" just means the size of the gap is large enough for the substrate and the orientation of polar and apolar regions correspond positively to each other.

2. Macrocyclic and Concave Binding-Sites

A large variety of receptor molecules and catalytic sites is based on the idea that the inner cavity of macrocycles entraps metal ions or guest molecules if several binding groups are present. Protonation or thermal movement of these binding sites may then cause release of the metal ion or guest molecule [4]. Cyclic molecules usually only bind ions in water, e.g. metal ions, amines or phosphates, tubular molecules, in particular cyclodextrin, also recognize hydrophobic surfaces. Kemp acid derivatives finally provide a concave cavity with sizes up to about 10 Å, in which dipole-dipole interactions are sufficient to bind molecules [5].

Some of the artificial receptors provide a rigid cavity, e.g. the ring of pyranose units in the cyclodextrins, the resorcinol units in calixarenes and Kemp's acid with a *trans*-decalin basis. The poly(ethylene glycols) of the crown ethers and the corresponding nitrogen-containing macrocycles, on the other hand, have flexible conformations. In the latter case, it is the binding moiety, which defines the shape of the complex and it is in general quite easy to destroy it by change of external conditions. In case of the rigid hosts, care must be taken to allow for rotational freedom of the binding functional groups or the uptake and release of guests will be extremely slow [6,7]

3. Hydrophobic Ångström Gaps in Flexible Lipid Monolayers

LB-Monolayers of fatty acid salts, e.g. cadmium arachidate, approach the rigidity of crystalline monlayers at both the water/air and gold/air interfaces. Infrared spectra and various other physical data indicate a perfect ordering of the C_{20}-chains. The situation changes, however, at the edges of the monolayer. The alkyl chains will be bound to their neighbors as well as to the neighboring subphase. Statistical coil like conformations will prevail in the border regions. Such a monolayer can hardly be used as a basis for the formation of defined Ångström clefts. The danger that the molecular gaps in the membrane becomes glued up by the alkyl chains, which form the walls of the gaps, is always present.

Experiments with a steroid bound flatly to a gold subphase and surrounded by octadecanethiol fences were, however, quite successful. If a gold electrode covered about 50 % by the steroid and 50 % by the C_{18}-thiol was plunged into aqueous ferrocyanide solution, the cyclovoltammogram of the iron complex showed about 30 % of the peak current of the naked gold electrode. Upon addition of 1,2-*trans*-cyclohexanediol the current was practically abolished. *cis*-Cyclohexanediol had no such effect. Washing with distilled water did not remove the water-soluble *trans*-1,2-cyclohexanediol from the cavity, but HCl (pH $\leq$ 3) did so. The explanation given for this behavior was derived from model calculations of the water structure on hydrophobic surfaces. These calculations showed that 3-4 water layers become immobilized on the surface [8]. In 7 Å-wide hydrophobic gap filled with water cell water-molecules would therefore be immobilized. An ice-like water volume would form, which may be strongly stabilized by stereochemically fitting cyclohexane units only bearing equatorial hydroxyl groups. Axial hydroxyl groups, on the other hnad, would not fit the hexagonal ice structure and protons would destroy the net of hydrogen bonds (Figure 2) [9].

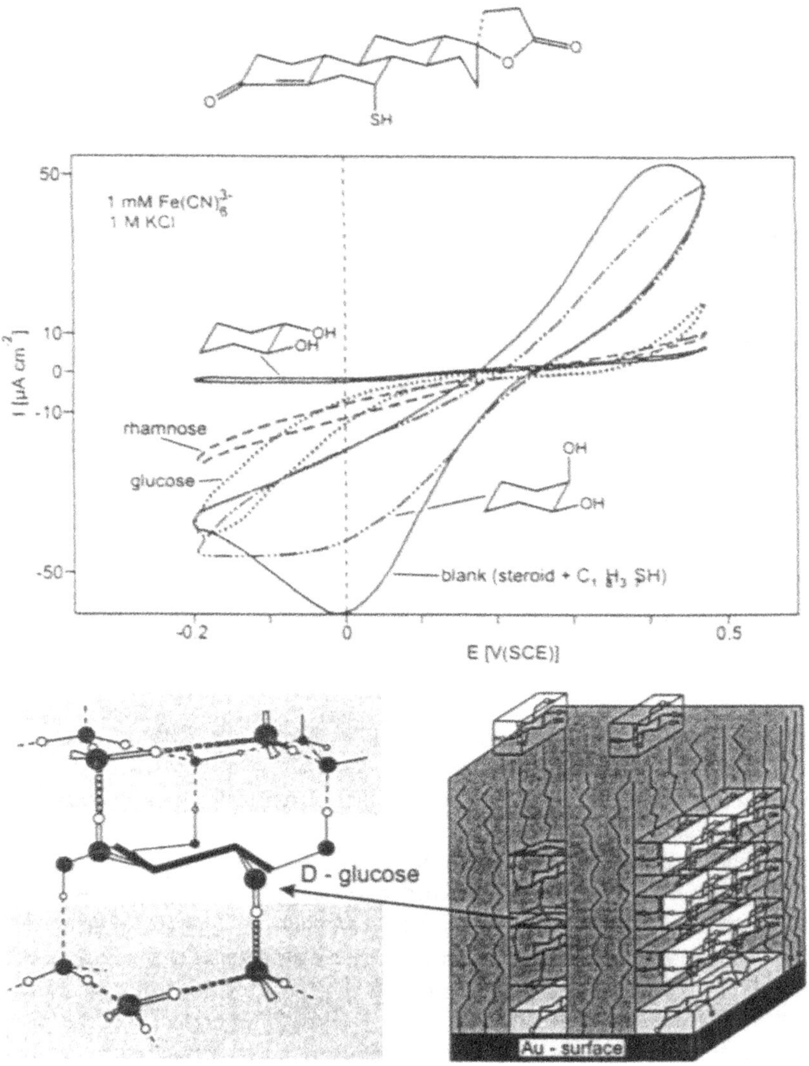

Figure 2: Cyclic voltammograms comparing the blocking behavior of *cis*- and *trans*-1,2-cyclohexanediol. Solid water model containing *trans*-1,2-cyclohexanediol.Hydrophobic gaps in surface monolayers. Entrapped 1,2-*trans*-cyclohexanediol in the steroid gap and model of ice-like water entrapping D-glucose.

The membrane structure depicted in Figure 2 consists of very small C_{18}-SH domains. Almost each single molecule is an edge molecule or on two or three alkyl chains away from the edge. Nevertheless a rigid water structure is built up after addition of stabilizing diol molecules. D-Glucose and other monosaccharides have similar, although less pronounced, effects for the transfer of ferrocyanide ions from bulk water to the gold surface.

4. Rigid Monolayers

We have realized two different ways to form stable lipid monolayers:

(i) A polyene bolaamphiphile made of bixin and *meso*-tetra-(o-phenol)-porphyrin produces stiff vesicle membranes, which remain stable after lyophilization. The dry vesicles have been characterized by AFM on solid surfaces. Upon irradiation an equally stiff polymer sphere is formed, which survives organic solvents [10]. Attempts to produce rigid surface monolayers of such a polymer failed so far. Only swollen bilayers were observed.

(ii) Bolaamphiphiles with reactive benzoylazide head groups and two secondary amide groups in the hydrophobic core were self-assembled on a polacrylnitrile surface. Upon addition of methylamine in the gap phase only the outer azide was amidated within minutes. The inner head group remained untouched for several hours. Two amide hydrogen bond chains obviously render the monolayer totally impermeable to amines which dissolve readily in normal lipid membranes [11]. Similar crystallization effects have been observed in Au-S band bolaamphiphiles with an even-numbered connecting oligomethylene chain. Odd-numbered homologues are fluid [12].

Only the diamide monomers seem to form well-organized, rigid monolayers. Amphiphilic polymers tend toward swelling and irregular structures.

5. Rigid Monolayers with Ångström Gaps

In order to characterize the shape and accessibility of Ångström gaps we have introduced porphyrins instead of steroids lying flat on metal or semiconductor electrode surfaces. It turns out that such porphyrins on gold still show strong rest fluorescence, so that their accessibility can be tested by fluorescence quenching with porphyrins with oppositely charged porphyrins. Most suitable for such tests are pairs of β-tetrapyridinium- and *meso*-tetra(m-phthalate)-porphyrins. The octacarboxylic acid porphyrin binds tightly to gold surfaces at pH 12 and forms non-fluorescing heterodimers with the β-tetrapyridinium porphyrin with a binding constant > 10^7 M [13]. If the *meso*-octacarboxylate porphyrin is fenced in with a secondary diamide bolaamphiphile, it is still accessible by the pyridinium porphyrin as shown by quantitative fluorescence quenching [14] (Figure 3, on the next page).

6. Heterodimers at a 10-15 Å Distance

If the rigid fence containing diamide hydrogen bond chains contains double bonds in the central hydrophobic core one can functionalize them there by oxidation with water-soluble oxidants. OH- or NH_2-Groups may be introduced and bind to correspondingly substituted porphyrins. Although the amphiphiles thus become much more hydrophilic and water-soluble, they should still stand upright at the gap's edge because of the amide hydrogen bonds. The olefinic double bonds inside the domains should not be attached by ionic oxidants in the aqueous phase.

The porphyrin pairs can then be used for light-induced charge separation, the aqueous volume in between can be filled up with phenol relay systems. Such systems will be especially useful if they are transferred from electrodes to colloidal particles, such as gold or silicate (Figure 4).

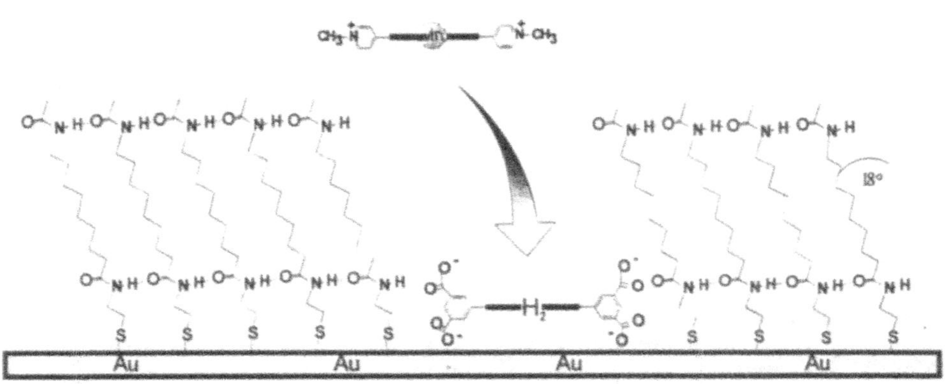

Figure 3: Rigid gaps made of diamide thiols on gold. Fitting porphyrins quantitatively occupy the trap within about 30 min.

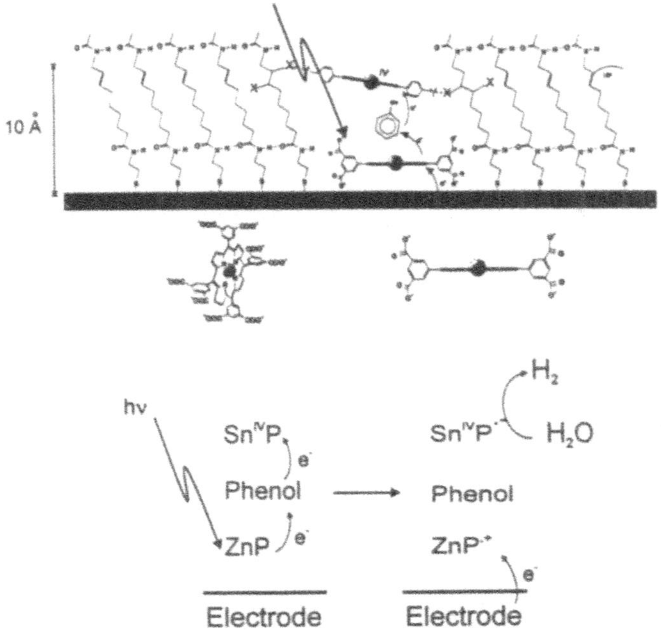

Figure 4: Polar substituents at the stiff gap's periphery will allow the fixation of a second porphyrin at a fixed distance.

7. Immobile Water

The water entrapped in hydrophobic membrane gaps does not release entrapped 1,2-*trans*-cyclohexanediol or D-glucopyranose for days into the bulk water. This probably means that the ice-like water volume also does not exchange water molecules with the bulk. In this sense the gap-water has a hydrophobic character, which needs, however, to be proven. Furthermore the structure of the entrapped water may or may not depend on the structure and rigidity of the enveloping walls. It is therefore mandatory to transfer the porphyrin-based or steroid-based gaps from gold electrodes to smooth colloidal particles in order to provide enough entrapped immobile water for solid state NMR studies.

It is hoped that the quasi-crystalline water framework between electron donors and electron acceptors will also fixate electron transfer agents, e.g. tyrosine or other phenol derivatives. Alignment may be possible. The distant pairs of electron donors and acceptors proposed in section 6 can thus be connected and hopefully produce useful, non-covalent charge separation systems.

8. Charge Separation

If light-induced charge separation is successful in rigid membranes, the question of energy harvesting arises. This is a difficult problem and has been solved by Bolton many years ago with a fluid mercury electrode. A highly speculative system based on Bolton's electrochemical arrangement [15]coupled with our synkinetic membrane gap is sketched in Figure 6. Its fundamental advantage over other more realistic photovoltaic systems is the fact that it can be obtained on a very large scale by a succession of self-assembly processes of appropriate lipids and chlorophyll derivatives. Photodegraded components could be replaced by flushing with acids or solvents and reassembly (Figure 5).

9. Differences between Fixed Monolayer, Protein Clefts and Artificial Receptors

The magic of the protein clefts (Figure 1) is the possibility to adjust their size and shape by movement of the backbone connecting the helices of β-sheets. A monolayer cannot achieve this in a comparable degree. Other properties, such as tight fitting, hydrophobicity, reversible uptake and release of up to three substrates in a given order can, however, be mimicked. This is made possible by the immobility of the entrapped water volume containing small solutes which fit the "ice" structure. An advantage of the monolayer systems is, that they are of macroscopic size in two dimensions and can therefore be integrated into photoelectric assemblies.

In comparison to artificial receptors with concave binding sites (see section 2) the membrane gaps provide the advantage of a much larger size. Depending on the height of the membrane at least three to five different molecules can be placed in variable order. Furthermore analysis by cyclic voltammetry is easy and the self-assembled systems can be prepared easily on a large scale if colloidal particles are used as subphase.

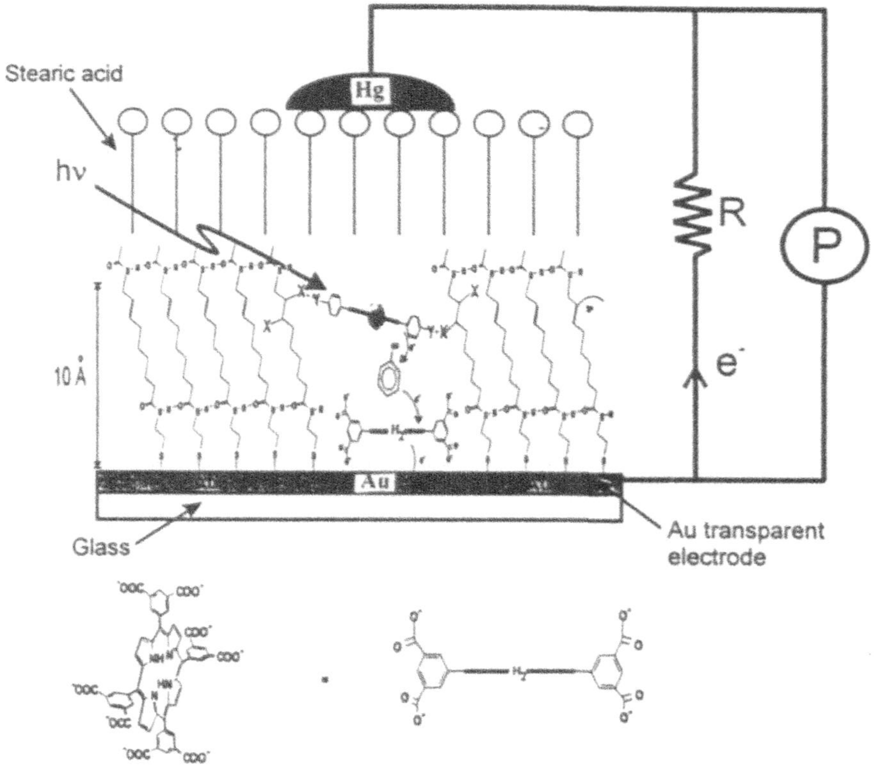

Figure 5: Model of a photovoltaic element based on Bolton's design. The top electrolyte layer is symbolized here by a porous amphiphile layer.

Acknowledgement: This work was supported by the European Commission within the TMR network "Artificial Photosynthesis", the Deutsche Forschungsgemeinschaft (SFB 312 "Vectorial Membrane Processes" and SFB 348 "Mesoscopic Systems") and the FNK of the Free University Berlin.

10. References

1. Fuhrhop, J.-H. and Arlt, M. (1990) Stereoselective Recognition of Pyranosides and their Tetraacetates by Concanavalin A, *Angew. Chem.* **102**, 699-701; *Angew. Chem. Int. Ed. Engl.* **29**, 672-674.
2. Dickersen, R.E. and Geis, I. (1969) The Structure and Action of Proteins, Harper & Row, Publishers, New York, Evanston, London.
3. Choma, D.T., Lear, J.D., Nelson, M.J., Dutton, P.L., Robertson, D.E. and DeGrado, W.F. (1994) Design of a Heme-Binding Four-Helix Bundle,*J. Am. Chem. Soc.* **116**, 856.
4. a) Cram, D.J. and Cram, J.M. (1978) Design of Complexes between Synthetic Hosts and Organic Guests, *Acc. Chem. Res.* **11**, 8.
   b) Timko, J.M., Helgeson, R.C. and Cram, D.J. (1978) Host-Guest Complexation: Recognition in Solution between Carboxyl-Containing Macrocyclic Polyethers and an α-Amino Acid, *J. Am. Chem. Soc.* **100**, 2828.
5. Tjivikua, T., Ballester, P. and Rebek, Jr., J. (1990) A Self-Replicating System, *J. Am. Chem. Soc.* **112**, 1249 and references therein.

6. Dietrich, B., Viout, P. and Lehn, J.-M. (1993), Macrocyclic Chemistry, VCH, Weinheim
7. Lippard, S.J. and Berg, I.M. (1994), Principles of Bioinorganic Chemistry, University Science Books, Mill Valley, Calif.
8. Vossen, M. and Forstmann, F. (1994) The structure of water at a planar wall: An integral equation approach with the central force model, *J. Chem. Phys.* **101** (3), 2379.
9. Fuhrhop, J.-H., Bedurke, T., Gnade, M., Schneider, J. and Doblhofer, K. (1997), Hydrophobic Gaps of Steroid size in a Surface Monolayer Collect 1,2-trans-Cyclohexanediol and Glucose from Bulk Water, *Langmuir* **13**, 455.
10. Komatsu, T., Tsuchida, E., Böttcher, C., Donner, D., Messerschmidt, C., Siggel, U., Stocker, W, Rabe, J.P. and Fuhrhop, J.-H. (1997) Solid Vesicle Membrane Made of *meso*-Tetrakis[(bixinylamino)-*o*-phenyl]porphyrins, *J. Am. Chem. Soc.* **119**, 11660.
11. Böhme, P., Hicke, H.-G., Böttcher, C., and Fuhrhop, J.-H. (1995) Reactive and Rigid Molecular Monolayers of Bisaroylazide diamide Bolaamphiphiles on Polyacrylonitrile Surfaces, *J. Am. Chem. Soc.* **117**, 5824-5828.
12. Schneider, J., Gnade, M., Messerschmidt, C. and Fuhrhop, J.-H., Tilt Dependent Odd-Even Effects in Monolayers of Gold Adsorbed Bolaamphiphilic Diamides, submitted.
13. Endisch, C., Fuhrhop, J.-H., Buschmann, J., Luger, P., and Siggel, U. (1996) β-Tetraethyl-β-tetrapyridin-5-yl-porphyrins, their N-methylated Tetracations and Heterodimers with *ms*-Tetraphenylsulfonato Porphyrins, *J. Am. Chem. Soc.* **118**, 6671-6680.
14. Ruhlmann, L., Zimmermann, J., Fudicker, w., Roeder, B. and Fuhrhop, J.-H., Hydrophobic Membrane 'Gaps on Gold with a Fluorescing Porphyrin Base, *J. Am. Chem. Soc.*, submitted for publication.
15. Janzen, A.F. and Bolton, J.R. (1979) Photochemical Electron Transfer in Monolayer Assemblies. 2. Photoelectric Behavior in chlorophyll a/Acceptor Systems, *J. Am. Chem. Soc.* **101**, 6342.

LOW MOLECULAR WEIGHT GELATORS FOR ORGANIC SOLVENTS

From serendipity towards design

JAN VAN ESCH*, FRANCK SCHOONBEEK,
MAAIKE DE LOOS, E. MARC VEEN,
RICHARD M. KELLOGG, AND BEN L. FERINGA*≠

Laboratory for Organic Chemistry, University of Groningen
Nijenborgh 4, 9747 AG Groningen, The Netherlands

1 Introduction

Everyone knows what a gel is, but from a scientific point of view the term gel encompasses chemically very diverse systems. Well known gel systems include, for instance, dilute solutions of polymers, proteins, and surfactants in water and organic solvents. These gel systems are important in medicine, biology, chemistry, and physics, and find many applications in the photographic, cosmetics, food, and petroleum industries [1]. However, as D. Jordan Lloyd already wrote in 1926: 'The colloidal condition, the "gel", is one which is easier to recognise than to define,...'. And although an exact definition of a gel is still a problem, from a topological point of view gels can be defined as dilute mixtures of at least two components, in which both components form a separate continuous phase throughout the system [2]. This definition includes not only gels composed of a solid-like and a liquid phase, but also those composed of a solid and a gas phase (so called aerogels). For most gels a solid-like phase is the minor component which forms a three dimensional network structure within the fluid or gas phase. For solid-fluid gels it can be said that the network structure prevents the fluid from flowing, whereas the liquid phase prevents the network from collapsing [3]. The coexistence of a solid network structure together with a liquid phase distinguishes gels from pure solid, liquid crystalline, or fluid materials and gives gels their unique elastic properties.

Often gels are divided into two groups depending on the type of interactions which hold the network structure together. In chemical gels, both the individual filaments of which the network consists, as well as the connections (junction zones) between the filaments to form the network, are created through the formation of covalent bonds. Cross-linked polymer gels belong to this class of gels, but also many gels composed of inorganic oxides, like vanadium pentoxide and silica. The properties of

≠ To whom correspondence should be addressed.

R. Ungaro and E. Dalcanale (eds.), Supramolecular Science: Where It Is and Where It Is Going, 233–259.

chemical gels can be as different as their constituents, but they have in common that their formation is irreversible. Opposed to chemical gels are the physical gels. In this type of gels the network structure is build up from small molecular subunits, which are held together by non-covalent interactions. The attractive forces between the molecular subunits can be very specific, like hydrogen bonding or π-π stacking, but also solvophobic and entropy effects play an important role. Because the non-covalent interactions stabilizing the network are orders of magnitude weaker than covalent bonds and comparable to the thermal energy ($RT \approx 0.6$ kcal/mol at 300 K), physical gels exhibit a characteristic reversible transition from the gel phase to a solution at moderate temperatures. Many gels containing mineral clays, polymers and proteins belong to this class. Properties of physical gels like thermal stability and viscoelastic behavior are therefore the result of a delicate balance between the properties of their constituents and the interactions between them.

Also certain low molecular weight organic compounds are capable of forming gels. Whereas with macromolecular gels the smallest molecular entity has at least a size of several kDa, the size of most of these compounds, often called organogelators, ranges from 300-1000 Da. Gelation of solvents by organogelators is the subject of increasing attention, not only because of the numerous applications of gels, but in particular because these compounds represent a class of gelators that exhibit striking properties with respect to self-assembly phenomena [4]. Although many aspects of the mechanism of gelation are unclear and there is a great variety in the structure of low molecular weight gelators it appears that these compounds have certain features in common. Gelation of a solvent by organogelators occurs through self-assembly of the gelator molecules into elongated fiber like structures, which then form an entangled network in the solvent [4]. In these networks, the fibers consist of infinite arrays of small molecules, solely held together by non-covalent interactions. Organogels thus belong to the physical gels. Despite impressive achievements of supramolecular chemistry in the controlled self-assembly of small molecules [5,6,7], most low-molecular weight gelators so far have, however, been found by serendipity rather than design. The control of gelation phenomena induced by small organic molecules and the design of new gelators are challenging goals, and it has only been recently that a number of successes have been reported.

Because the literature on organogelators till 1996 has been covered in an excellent review by Terech and Weiss[4], we will give in the following section only a brief overview with some representative examples of the older literature. In the next sections we discuss some of the background and principles used for the design of organogelators, and give an overview of our work on the design of novel organogelators. Finally we will briefly discuss the development of functional organogelators, which is an area of large potential and whose development only has become possible through a better understanding of organogels. It should be stressed, however, that the field of organogelators, although still in its infancy, only has developed to what it is nowadays thanks to the work of a limited number of research groups, including Weiss, Rabolt, Terech, Hanabusa, Shinkai, and ours, who recognized in an early stage the large potential of organogels.

2 Examples of Organogelators

2.1 INTRODUCTION

The group of low molecular weight organogelators consists of a great number of very different compounds, which have in common that they self-assemble into fiber-like structures, which in turn form a three dimensional network in the solvent. The non-covalent interactions which provide the driving force for the self-assembly process include ion-ion and dipole-dipole electrostatic interactions, hydrogen bonding, π-stacking interactions, and van der Waals interactions. It is therefore tempting to classify organogelators according to the nature of interaction responsible for self-assembly. However, in most cases it is not well understood what kind of structure a specific organogelator forms nor what the contribution is of each kind of interaction. It is even more likely that the often very efficient aggregation of organogelators is the result of several forces acting in a cooperative way. Therefore we classified the organogelators according to their chemical constitution, in similar way as has been done by Terech and Weiss [4].

2.2 FATTY ACID AND SURFACTANT TYPE GELATORS

Most of the earlier reports on organogelators deal with compounds that are derived form fatty acids or surfactants. A few examples are given in Figure 1. Compound **1** is an example of a fatty acid type of organogelator [8]. It forms thixotropic gels with organic solvents like benzene or chloroform. According to FT-IR measurements the gels are stabilized by intermolecular hydrogen bonds. The gels form a lyotropic mesophase and exhibit a supramolecular helical structure. Fluorocarbon-hydrocarbon diblock compounds **2** form gels with saturated and aromatic hydrocarbon solvents, whereas inverted micelles were formed in fluorinated solvents. The gelating capability and surfactant like behavior is contributed to the mutual solvophicity of the hydrocarbon and fluorocarbon moiety [9,10]. The organogelator **3** is clear example of gelation of organic solvents by surfactant type of molecules [11,12]. A wide range of solvents can be gelated by this type of compounds and some of these compounds show polymorphic behavior.

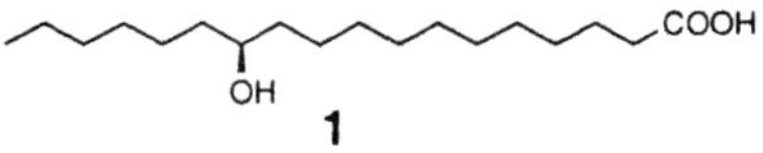

1

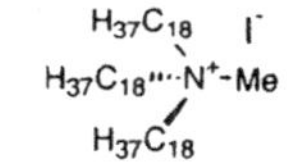

2 $CH_3(CH_2)_n\text{-}(CF_2)_mCF_3$ **3**

Figure1 *Fatty acid and surfactant type organogelators.*

2.3 ANTHRACENE, ANTHRAQUINONE, TETRALINE AND STEROID DERIVATIVES

A well-studied group of organogelators are derived from anthracene and anthraquinone (**4,5**) [13], or steroids (**6**) [14], and include also compounds which contain both groups linked via a spacer (**7,8**) [15,16], Other organogelators of this type are an azobenzene-steroid derivatives (**9**) [17] and a chiral tetraline derivative (**10**) [18]. Aggregation of these compounds is based upon π-π stacking and solvophobic effects. Figure 2 presents some examples.

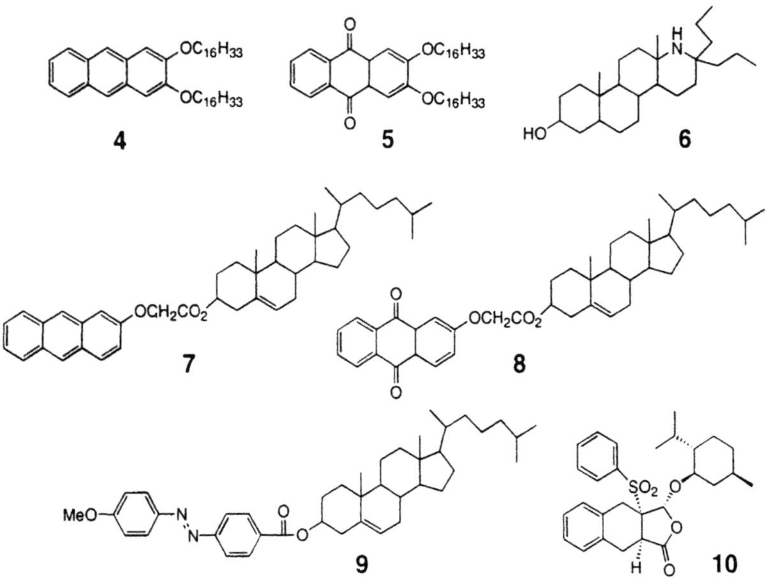

Figure 2 *Anthracene, anthraquinone, tetraline, and steroid based organogelators.*

Compound **5** forms thermoreversible gels in various aliphatic alcohols and amines at low concentrations. A three dimensional network is formed, which consists of assembled head-to-tail aggregates of the aromatic component with the solvent. The organization mainly depends on dipolar forces and π-π stacking. This compound is of special interest because of its photochromic properties.

The steroid derivative **6** was found to gelate only a few hydrocarbons. It aggregates into helical fibers, which intertwine at the junction zones thus forming a three dimensional network.

Compounds **7** and **8** are members of a family of gelators (ALS) in which an anthryl or anthraquinonyl group (A) is connected to a steroid (S) via a linker (L). Both form thermoreversible gels with a wide range of organic solvents. The stability of these gels depends on the concentration and structure of the gelator, as well as the properties of the solvent. The gel network consists of long intertwined fibers, which are built from several stacks of molecules. In these molecular stacks the anthryl or anthraquinonyl groups overlap. Some of their derivatives form luminescent gels.

Another steroid derivative is the azobenzene **9**, which is able to gelate a large variety of organic solvents. The thermoreversible gels are built from helical intertwining fibers. Furthermore, it was possible to switch reversibly from the gel to the sol phase by means of light. This transition could be read-out by means of CD-spectroscopy.

The chiral compound **10** was found by accident in our laboratory. It is able to gelate a series of alkanes and alcohols via π-π stacking. X-ray analysis indicated the formation of helical fibers, which formed a three dimensional network through the fluid as was shown by electron microscopy (EM).

2.4 METAL-BASED ORGANOGELATORS

This class contains compounds for which gelation is induced by the formation of coordination complexes. Two examples are shown in Figure 3, however more compounds are known [19].

Compound **11** forms a gel with cyclohexane at room temperature. From small-angle neutron scattering (SANS) measurements it was found that the molecules are stacked on top of each other linked by axial copper-oxygen coordination bonds. The stacks consist of a polar core (copper and oxygen) surrounded by a hydrophobic shell (aliphatic tails). These stacks in turn are part of the three dimensional network which forms the gel.

Compounds **12**a and **12**b were found to gelate solvents like tetrahydrofuran, toluene and methanol at low concentrations [20]. The ligand itself does not gelate any solvent. Transmission electron microscopy proved that the gels consist of (helical) fibers. These fibers in turn can self-assemble into super helices.

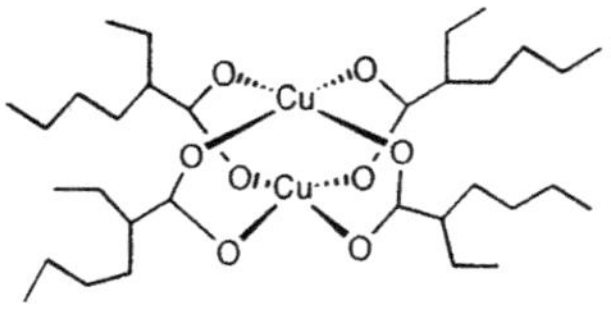

11

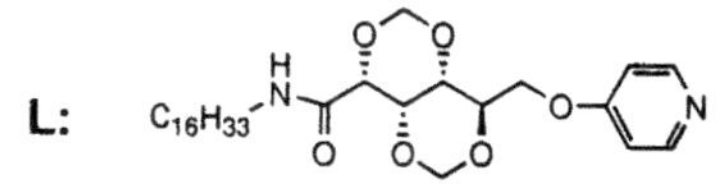

12

a: L_2PdCl_2

b: L_2PtCl_2

Figure 3 *Metal-based gelators:* ***11*** *is a binuclear copper(II)-tetra((2-ethyl) hexanoate) complex;* ***12a*** *and* ***12b*** *are Pd(II) or Pt(II) bis-gluconamide complexes, respectively.*

2.5 AMIDE AND CARBOHYDRATE CONTAINING GELATORS

The gelators for which aggregation is best understood are those which are able to form highly directional hydrogen bonds. Among these are amino acid, amide and urea derivatives. A few examples are shown in Figure 4.

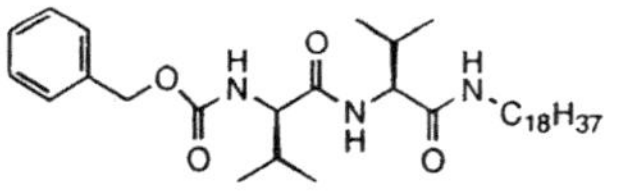

13

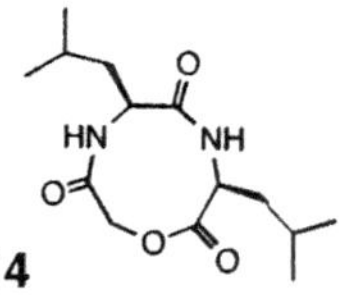

14

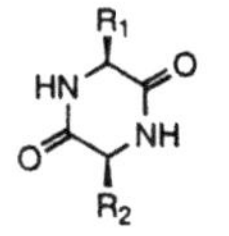

15

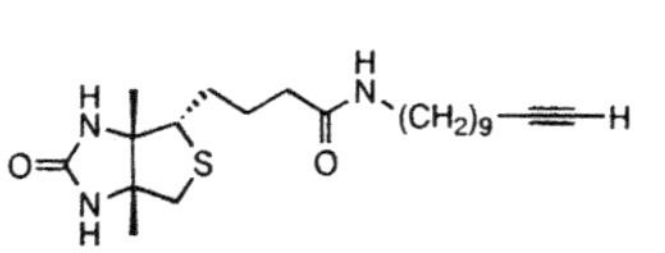

16

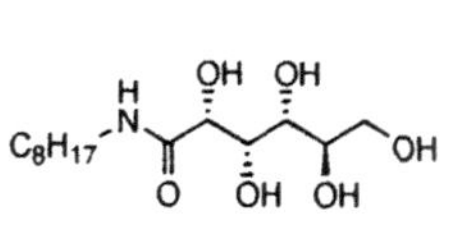

17

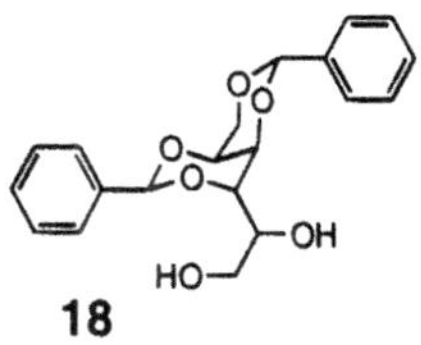

18

Figure 4 *Amide and carbohydrate containing gelators*

Compound **13** is a long-chain alkyl-amide derivative, which was found to gelate a wide range of solvents at low concentrations [21]. Infrared measurements proved that in the gel the amide groups are intermolecular hydrogen bonded. It was found that some important factors for gelation are: the structure of the amino acid residue, optical activity and the balance between the hydrophilic amino acid residue and the lipophilic tail. From transmission electron microscopy (TEM) it was found that the gel consists of a network of fibers.

The depsipeptide **14** was found to gelate a few solvents, among which polar (MeCN) and apolar (Et_2O) solvents [22]. Because **14** does not dissolve in these solvents, a pre-solubilization step with a droplet of methanol is required. Recently it was found that the gelating structure was not the cyclic compound but its ring opened form.

The cyclic dipeptide **15** [23] contains two self-complementary hydrogen bonding sites and gelate a wide range of solvents. Many different R-groups are tolerated but optimal gelating properties were observed when R_1 and R_2 are different.

Compound **16** was found by chance to gelate a few organic solvents [24]. Several other derivatives were synthesized but the range of gelated solvents remained quite small. From ^{1}H-NMR spectroscopy it was found that gelation was induced by the formation of hydrogen bonds. For surfactant **17** the only organic solvent known to be gelated is *o*-xylene [25]. In these gels long fibers were observed, which consist of multiple concentrically stacked sheets. The fibers were stabilized by amide hydrogen bonds and solvophobic effects.

The benzylidene-sorbitol derivative **18** is one of the oldest organogelators known, and is capable of gelating many different organic solvents, ranging from heptane to DMSO and glycerol [26]. Recent investigations showed that only the pure enantiomer forms gels whereas the racemate crystallizes, and that both hydrogen bonding and π-stacking interactions are important for gelation [27]. In addition, D-glucose, D-galactose and D-mannose-benzylidene derivatives have recently been reported as potent organogelators [28], as well as a number of other compounds which gelate organic solvents via intermolecular hydrogen bonds [29,30,31,32, 33, 34].

2.6 Two component systems

Some systems have been reported in which two different components are needed to obtain gelation (Figure 5).

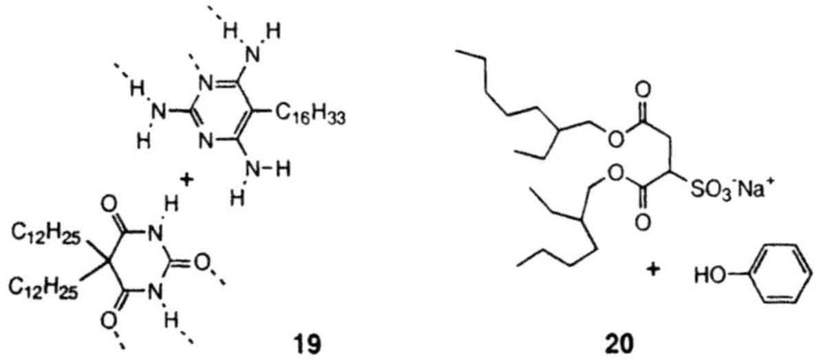

19 **20**

Figure 5 *Two types of two component systems:* ***19*** *is an equimolar mixture of 5-hexadecyl-2,4,6-triaminopyrimidine with 5,5-didodecylbarbituric acid;* ***20*** *is an equimolar mixture of sodium bis(2-ethylhexyl)sulfosuccinate (AOT) with phenol.*

System **19** forms gels with organic solvents resulting from the formation of hydrogen bonds between both molecules [35]. Long fibers are formed, which assemble into a three dimensional network through the fluid. Crystallization is avoided due to the flexibility of the long alkyl chains.

During investigations on reversed micelles of the surfactant AOT, system **20** was found to gelate organic solvents into completely clear gels [36]. Interestingly, these gels melt upon exposure to a trace of moisture. Also for this system formation of the gel results from hydrogen bond formation between the different molecules.

3 Design of Organogelators

3.1 Design Principles

For most compounds whose gelating properties have been accidentally discovered, the structure of the aggregates and mode of aggregation remains obscure. Many attempts to improve the performance of these gelators or to alter characteristics as solvent compatibility by structural modification of these gelators have failed, and have often been hampered by synthetic difficulties. In a different approach one can design new gelators for organic solvents, starting from criteria derived from some common features of known gelators [4]. In a proper design the geometry of the building blocks and the spatial arrangement and nature of the intermolecular bonds will determine the structure and properties of the supramolecular aggregate [6]. A priori knowledge of the possible modes of aggregation of the designed compounds offers a working model with which one can explain the successes and failures of gelation experiments and with which one can design new compounds.
In order to develop molecular design criteria one should consider the events and intermolecular interactions which occur during the process of gelation (Figure 6).

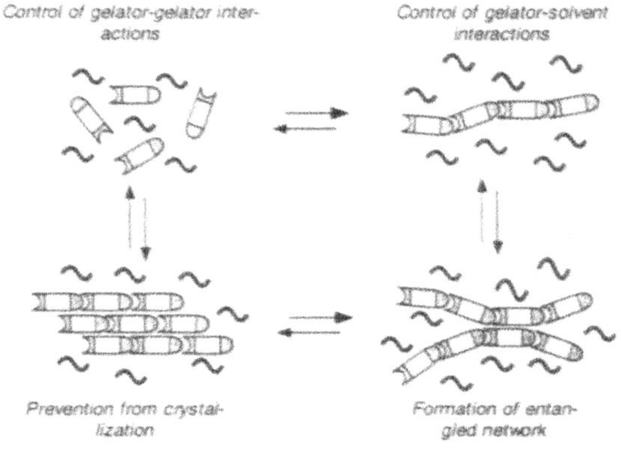

Figure 6 *Schematic view of gelation of organic solvents by small organic molecules*

The first step in the gelation process is most likely the self-assembly of the gelator molecules into extremely elongated and perhaps even one-dimensional aggregates [4]. These elongated or one-dimensional strands of gelator molecules subsequently assemble into fiber like structures, which in turn assemble into the three dimensional network in the liquid. The latter processes of fiber and network formation reflect a delicate balance between gelator-gelator interactions and gelator-solvent interactions: stronger gelator-gelator interactions induce stronger contacts between fibers which may lead to a more dense network or even crystallization, whereas stronger gelator-solvent interactions might prohibit aggregation of strands into fibers and network.

Critical factors in the design of low molecular weight gelators are therefore [see also 37,38]: (i) fiber formation by control of an anisotropic growth process. Predominant self-assembly along one direction can be controlled by employing highly directional intermolecular interactions, like hydrogen bonding, π-stacking of arene moieties, and metal-ligand interactions. Simultaneously, self-assembly in the other directions should be less favorable. This can be achieved by using solvent-like moieties, *e.g.* aliphatic chains for hydrocarbon solvents. In an optimal case, such molecules will self-assemble into aggregates in which the strongly interacting groups are exposed at one face, whereas the solvent-interacting groups are exposed at the remaining interface of the fiber. As a result the interfacial free energy of the fiber will be highly anisotropic, which under kinetically controlled growth conditions will boost the one-dimensional shape of the fibers. (ii) Intertwining of the aggregates to form a three dimensional network. A few types of linking nodes have been distinquished [39], *e.g.* (pseudo)crystalline microdomains, entanglements or spatially limited organised microdomains. The formation and stability of such junctions zones are governed by a delicate balance between fiber-fiber and fiber-solvent interactions, and its understanding requires a detailed knowledge of the structure of the junction zones. It remains therefore the most difficult factor in the design of novel gelators. As an alternative, one can consider the use of non-covalent crosslinker molecules, which would allow a more direct control of the stability and number of the junctions via the nature and amount of the cross-linker. (iii) Crystallization of the self-assembled aggregates has to be precluded and therefore a delicate balance between order and disorder has to be found. Although it is obvious that a certain degree of (one-dimensional) order is required to achieve self-assembly in one direction, the packing should be far from ideal. Prevention of crystallization remains a little obscure in the design of new gelators and approaches that are used are to frustrate crystallization by certain peripheral groups, *e.g.* by the introduction of flexible or branched alkyl chains or to introduce (deliberately) a tendency to form polymorphic structures [23].

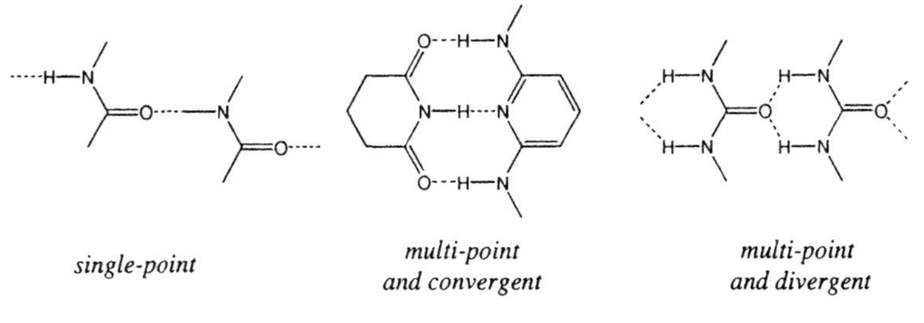

Figure 7 *Topography of hydrogen bonded assemblies formed by an amide group, the complementary couple barbituric acid-diaminopyrimidine, and urea moieties.*

There are numerous approaches to design molecules with self-complementary binding groups that can assemble into one, two or three-dimensional constructs [5,40]. The nature, orientation and the number of the interacting subunits present is essential for the size and shape that is found. Particular successful was the use of multiple hydrogen bonding in the self-assembly of macromolecule-like fibers [41,42], sheets [43,44], ribbons and rozettes [45,46,47], cages [48,49], and nanotubes [50]. Arrays of multiple hydrogen bonds where obtained by employing *e.g.* amides, urea, guanines, barbiturates, and ureidopyrimidones. Some examples of amide and urea moieties with specific single- and multi-point interactions are shown in Figure 7.

Pertinent examples related to our work of control of molecular aggregation by the design of self-complementary sub-units involving multiple hydrogen bonds are shown in Figure 8. Hamilton and co-workers obtained linear rod-shaped structures in the solid state by triple hydrogen bonded interactions of the cyclohexane-1,3,5-triamide **21** [51]. Cyclohexane bis- and tris-alkylamide derivatives **22** and **23** were found by Hanabusa to be excellent gelators [37,52]. Compounds **22** and **23** are able to gelate a wide range of organic solvents into completely clear gels. Electron microscopy revealed that the gels exist of long fibers with a width of 40-70 nm. From FT-IR and molecular modelling studies it was deduced that the molecules form intermolecular hydrogen bonds which forces them to stack right on top of each other, most likely very similar to the hydrogen-bonded arrays in crystals of **21**. In gels of **23** close packing of the alkyl chains significantly enhances the stability of the hydrogen-bonded stacks, which makes this compound also a potent gelator for polar solvents.

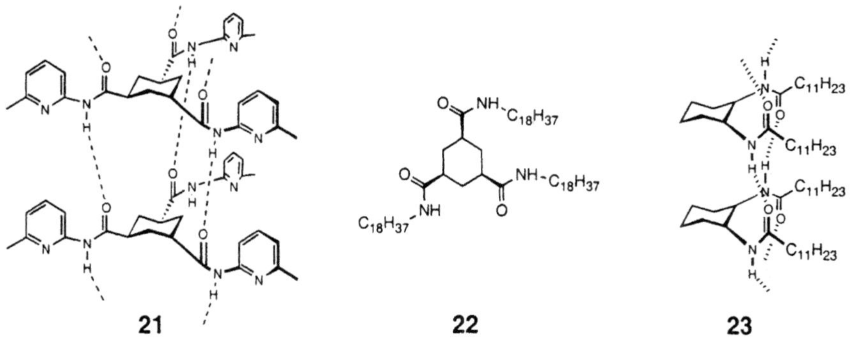

Figure 8 *Hydrogen bonded stacks formed by cyclohexyl-amide derivatives.*

3.2 LINEAR BIS-UREA GELATORS

A class of compounds particularly known to form extended one dimensional networks of hydrogen bonds are ureas. For instance, the hydrogen bond directed (co-)crystallization of diaryl-urea to form linear arrays has been demonstrated by Etter et al. [53]. A comparison of 27 crystal structures of simple urea compounds deposited in the Cambridge Structure Database revealed that urea groups preferentially form infinite hydrogen bonded arrays, the structure of which is very well maintained in the different crystals (Figure 9). The arrays have a planar structure with the carbonyl moieties all lying in one line.

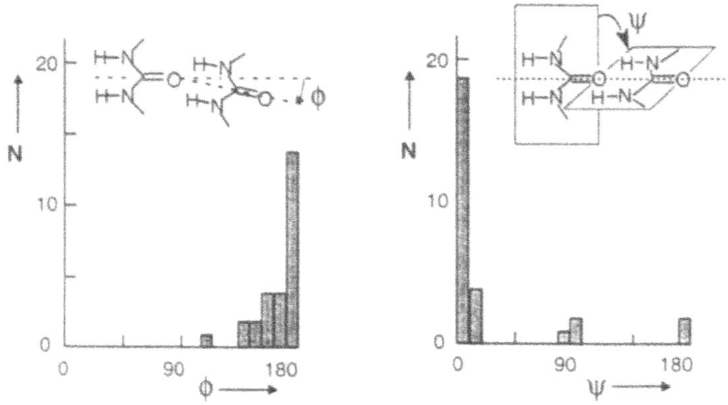

Figure 9 *Geometry of hydrogen-bonded arrays of urea groups in crystal structures.*

The well-defined geometry of aggregated ureas makes it an ideal group to be used as the main and highly directional interacting unit in the new gelators. An additional advantage is that substituted urea compounds can easily be prepared in high yields by the addition of amines to isocyanates (Figure 10) [54]. Many isocyanates are commercial available or are readily accessible from, for instance, carboxylic acids via the Curtius rearrangement of the corresponding azides.

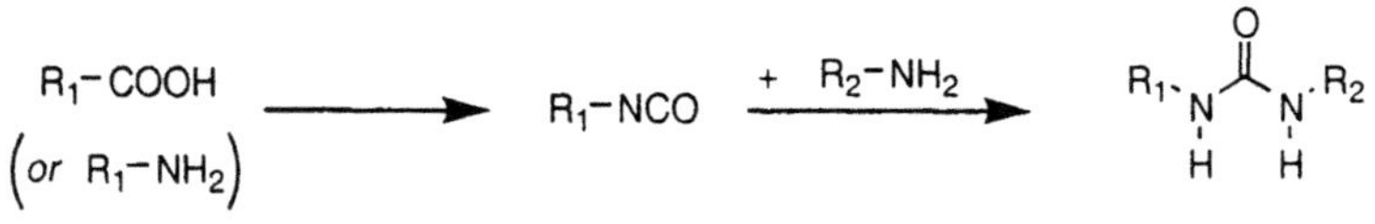

Figure 10 *Synthesis of urea compounds*

The capacity of urea groups to form extended chains of hydrogen bonds was explored by Hanabusa [55] and our group[38] in a new approach to design gelators and self-assembled nanosize structures. First we examined the gelating ability of some mono- and bis-urea compounds and quickly established that the presence of two urea groups in one molecule is a sufficient condition to enforce aggregation in dilute solutions of a range of organic solvents. When N-benzyl mono-urea **24** and N-benzyl bis-urea **25** and **26**, containing a long alkyl chain or alkyl spacer, respectively, are compared the advantage of the introduction of a second urea group becomes clear.

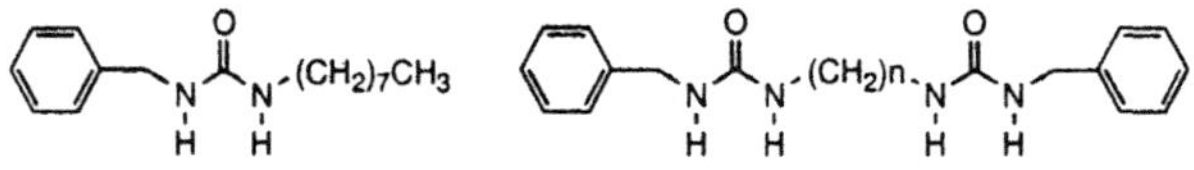

24

25 n=9
26 n=12

Figure 11 *Mono- and linear bis-urea compounds tested for gelation.*

Only in the case of **24** in hexadecane a gel was obtained. In all other solvents precipitation occurred after cooling of the solutions to room temperature. On the other hand the bis-urea compounds **25** and **26** are potent gelators for a wide range of solvents

(Table 1). Although sparingly soluble at room temperature they gradually dissolve upon heating and form gels upon cooling. Notable features are: (i) these gels are stable for at least weeks when stored at room temperature; (ii) gel formation is completely reversible; (iii) only very low amounts are needed for gelation (typical critical gelation concentrations are <1 (w/v)%); (iv) decreasing the spacer length between the urea groups to n=3, 4 or 6 makes solutions of these compounds less susceptible to gelation.

The concept of aggregation of bis-urea appears to be quite general and extremely powerful in the design of new self-assembled fibers, tapes and networks. General structural features are two urea units connected by a spacer and decorated with pending groups on each side. Upon aggregation a linear strand of bis-urea can be formed in which each molecule can form up to eight hydrogen bonds with neighboring groups in the aggregates (Figure 12).

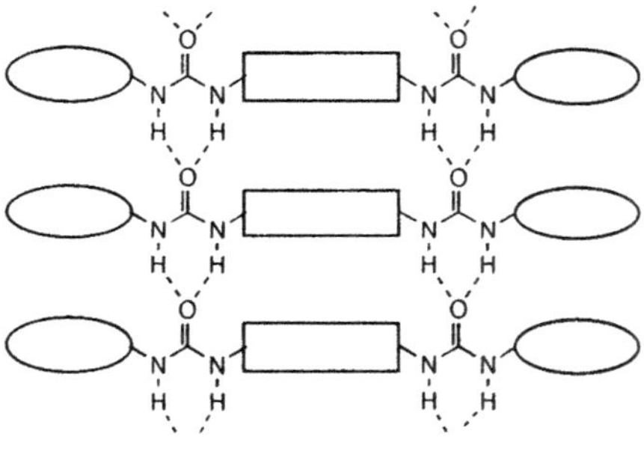

Figure 12 *Schematic representation of hydrogen bonded stack of linear bis-urea gelators.*

Figure 13 shows a number of bis-urea compounds, that were prepared from the corresponding isocyanates and bisamines, which are not only capable of aggregation but are also illustrative for the possibility to tune the assembly process via simple structural modifications [38]. The ability of the bis-urea compounds to act as gelators is shown in Table 1. The minimum gelation concentration for a typical example like **28** is remarkably low (3mg.ml^{-1}) and is independent of the solvent that is gelated. When the dodecyl side chain is replaced by a more sterically demanding group or the alkyl linker is replaced by a shorter or a more rigid linker the gelating abilities change.

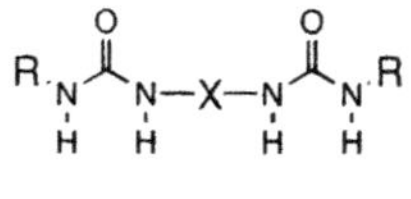

27 R=n-$C_{12}H_{25}$, X=-$(CH_2)_6$-

28 R=n-$C_{12}H_{25}$, X=-$(CH_2)_9$-

29 R=n-$C_{12}H_{25}$, X=-$(CH_2)_{12}$-

30 R=n-$C_{12}H_{25}$, X=4,4'-biphenyl

31 R=(*R*)-1-phenyl-ethyl, X=-$(CH_2)_9$-

Figure 13 *Linear bis-urea compounds*

Some characteristic features of these gels need to be emphasized. The bis-urea form transparent gels with aromatic solvents like *p*-xylene or tetralin but with other solvents turbidity is observed. Although the gels formed by these linear bis-urea compounds are stable for months they are irreversibly disrupted by mechanical agitation. The gelation process is however thermoreversible. The melting temperatures of the gels increase with increasing concentration but level off at higher concentrations. FTIR spectra of dried gels of **28** and **29** show amide bands at 1616 cm^{-1} and 1575 cm^{-1} and a NH stretch vibration at 3337 cm^{-1} which are characteristic of the presence of hydrogen bonded urea groups [56]. A broad endotherm with a maximum at 125°C for the same gel, measured by differential scanning calorimetry, points to a less cooperative phase transition.

TABLE 1 Gelating capability of linear bis-urea compounds [a]

	hexadecane	p-xylene	1-octanol	n-butyl acetate	cyclo-hexanone	tetralin
25	i	g(10)	g(10)	p	g	g(7)
26	i	p	g(10)	-	-	g(6)
27	p	p	p	p	p	g
28	g(3)	g(3)	p	g(3)	g(3)	g(3)
29	g	g	p	p	g	g
30	i	g	p	p	p	g
31	p	p	p	p	p	g

[a] The following abbreviations are used: g: gelation (minimum gelation concentration at 20°C in mg compound / ml solvent); i: insoluble at solvent reflux temperature; p: precipitate.

The structures of the aggregates depend on the type of gelator and the solvent. Figure 14 shows typical light microscopy photographs of the entangled networks of fiber-like structures present in gels of **28** and tetralin and butyl acetate, respectively. The fibers are only 2-5 μm thick but can be as long as 300-400 μm. The strong birefringence of the fibers indicate that they consist of well-ordered arrays of molecules. Furthermore the elongated shape of the fibers must be the result of a strong anisotropic growth again pointing to well ordered molecular aggregation. The single fibers show repetitions of strong birefringent regions indicating that the fibers are highly twisted. Both the width of the fibers and the degree of twisting depend on the structure of the gelator as well as the solvent. For instance fibers of **29** are somewhat broader than those formed by **28** and fibers in butyl acetate are more strongly twisted than those in tetralin.

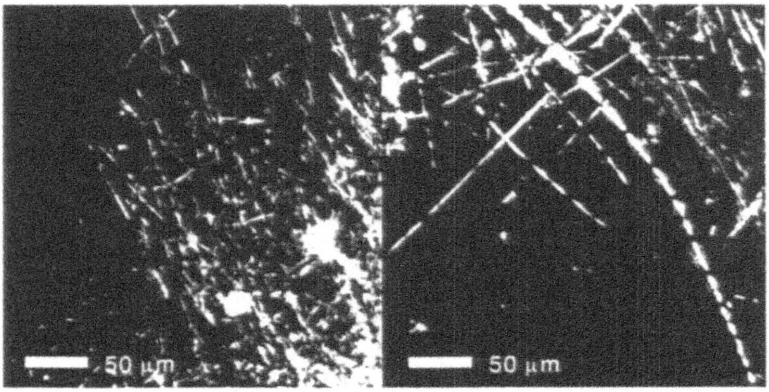

Figure 14 *Light-microscopy of a gel of **28** in tetralin (left) and butyl acetate (right).*

Further information on the organization within the fibers was obtained by electron microscopy (EM) and X-ray powder diffraction. Figure 15a shows an EM picture of a gel of **28** in tetralin clearly indicating that the fibers have a multilayered structure.

Typically 5-20 sheets are stacked in a single fiber. Based on EM data the thickness of a single sheet was estimated to be approximately 5 nm. Powder diffraction of gels of **28** and **29**, after most of the solvent was removed, confirmed a lamellar structure with spacings of 3.65-3.89 nm. Figure 15b shows a single fiber in a gel of **28** in hexadecane. It is clearly seen that the twisted structure observed in the light micrographs is a single turn of the sheets along the long axis of the fiber. Twisting has been observed in other fiber type aggregates, but the mechanism that is responsible for the formation of the twists in the fibers is not clear [25,57]. Most likely the formation of twists is driven by a reduction of the interfacial free energy, in which view a preferred screw sense can be explained by an anisotropy of the interfacial free energy in aggregates of chiral molecules. Obviously, such a preference for left- or right-handed twists was not observed for fibers of non-chiral bis-urea molecules as shown here.

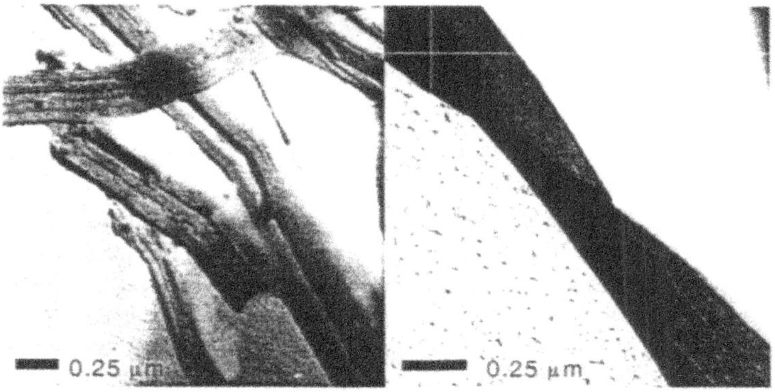

Figure 15 *Electron microscopy of a gel of* ***28*** *in tetralin (left) and n-hexadecane (right).*

A schematic view of the molecular arrangement of the bis-urea gelators, based on cumulative evidence as summarized above, is given in Figure 16. In this model the bis-urea molecules aggregate through hydrogen bond formation into strands or ribbons which assemble into sheets. Several sheets stack to form the twisted long thin fibers observed by light microscopy. It should be noted that in the sheets the ribbons are tilted.

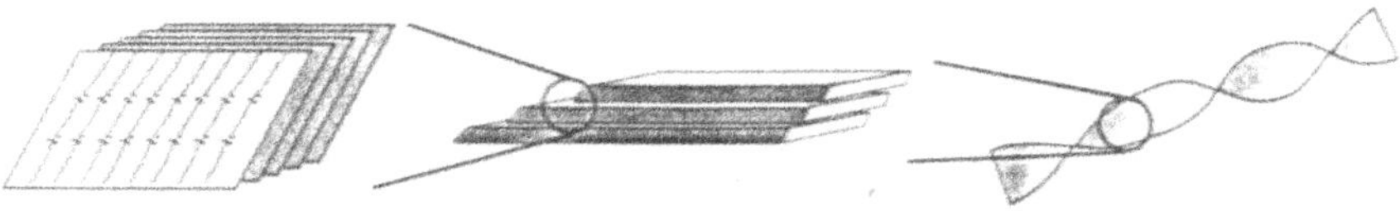

Figure 16 *Proposed arrangement of linear bis-urea compounds in sheets and fibers.*

Scanning tunneling microscopy (STM) provides a powerful tool to study the two dimensional organization of molecules physisorbed on a conductive surface [see for instance 58]. The structure of layers of the bis-urea gelators deposited on graphite was studied to gain a more detailed information on the organization on the molecular level. Monolayers of linear bis-urea physisorbed at the graphite/octanol interface imaged by

STM revealed a closely packed arrangement on the graphite surface with sub-molecular resolution [38a,59]. Figure 17 shows a typical STM image of an ordered monolayer of **27** and a two dimensional packing model with the ribbons formed by linear aggregation of bis-urea molecules.

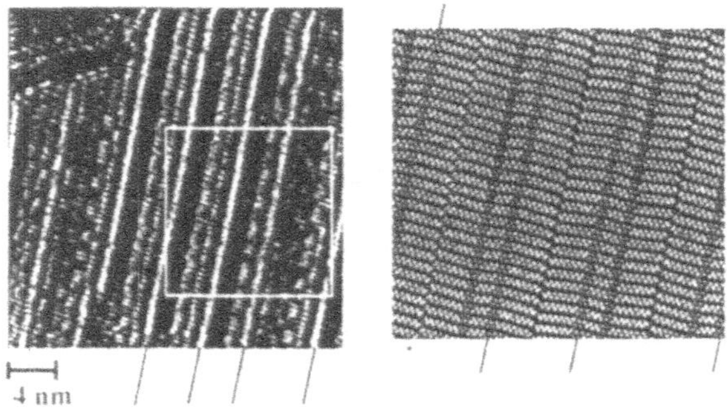

Figure 17 *STM micrograph of* **27** *physisorbed on graphite (left) and molecular packing model of the square region (right).*

A more detailed arrangement of the bis-urea molecules is seen in Figure 18. Within one ribbon the darker region corresponding with the hexyl-spacer is flanked by urea moieties but with distinct contrasts. One chain of urea appears as bright spots whereas the other chain appears as a row of relatively dark spots. Realizing that a bis-urea with a spacer containing an *even number* of methylene groups, *i.e.* **27** with a hexyl spacer, in an *all-trans* configuration adopt a kinked structure on the surface, the difference in contrast points to an antiparallel orientation of the urea groups in a single ribbon. This means that on the molecular level a distinction can be made between two similar functional groups which have a different orientation in self-assembled aggregate.

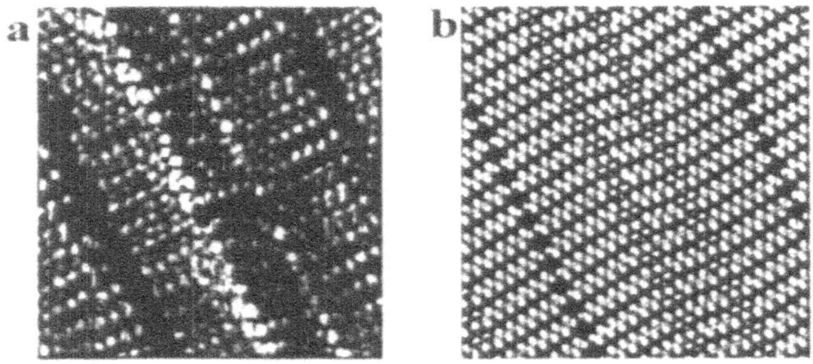

Figure 18 *STM micrograph of 27 physisorbed on graphite (a) and molecular packing model (b).*

Fascinating is also the observation at the molecular level of a change in organization upon subtle change in the structure of the bis-urea compound. In Figure 19 the STM image of bis-urea **28** with a nonane spacer is shown. Again ribbons are seen as well as defects in the two dimensional pattern. Some ribbons are characterized by bright spots of both urea moieties whereas in other ribbons the urea groups appear almost as dark as the alkyl groups. The different types of ribbons appear to coexist randomly, but within one ribbon the two chains of urea groups always have the same contrast. This is in perfect agreement with an all-trans minimum energy conformation of two urea groups separated by an odd number of methylene groups, *i.e.* as in **28** with a nonane spacer, in which both carbonyl groups point in the same direction and the molecule adopt a bend shape. A bend structure was indeed observed. Comparison of ribbons with opposite contrast with regard to the urea groups show that the ribbons are mirror imagines with the urea groups pointing in opposite directions. The distance of 0.462±0.005 nm between two neighboring molecules within a ribbon corresponds very well with the distances found for hydrogen bonded urea moieties. The width of a ribbon (5.0±0.1 nm) corresponds with the length of a single bis-urea molecule.

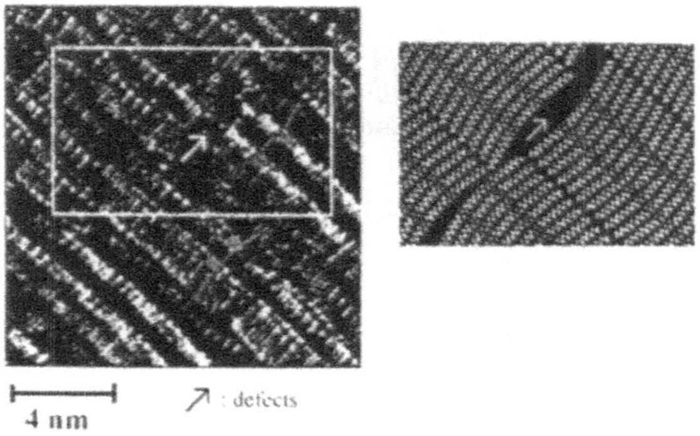

Figure 19 *STM micrograph of* ***28*** *physisorbed on graphite (left), and molecular packing model of the rectangular region (right).*

It should be emphasized that on comparison of the width of the ribbons on graphite (5.0 nm based on STM) with the thickness of the sheets observed in the gels (3.65 nm based on X-ray diffraction) of the same bis-urea compound that these are related but not identical structures. A rational is provided by a model (Figure 16) in which a sheet consist of a stack of ribbons which are tilted by approximately 45°. Further support for the arrangement shown in Figure 16 came from molecular dynamics simulations. Figure 20 shows a snapshot taken from a 200 ps molecular dynamics simulation of a rectangular box with 64 molecules of **28** arranged in 8 parallel stacked ribbons. The simulations converged to a structure in which a sheet has a thickness of 4.1 nm and the ribbons are spaced by 0.4 nm and make a tilt angle of 54° with the surface of the sheet. This arrangement is in good agreement with X-ray data.

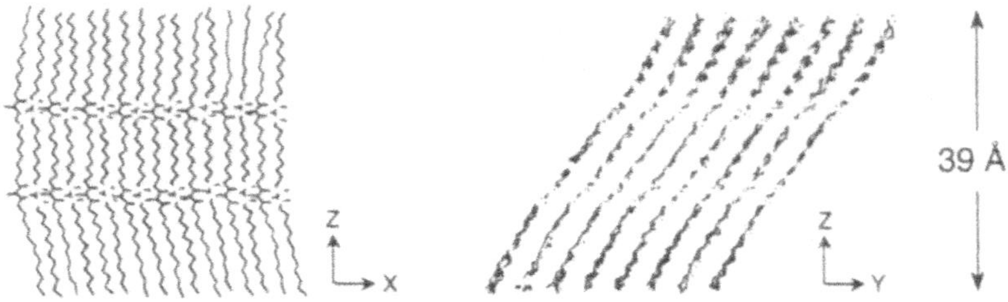

Figure 20 *Snapshot from a 100 ps MD simulation (300 K, NPT) of a box containing 64 molecules of* ***28****, showing two ribbons each consisting of 8 molecules of* ***28*** *(left, view along y-axis, the other ribbons are omitted for clarity), and a stack of 8 ribbons forming a sheet (right, view along x-axis).*

Our studies clearly show the potential of linear bis-urea in the control of molecular assemblies. For instance the intermolecular organization in a tape can be controlled by the spacer whereas the width of the tape and the interaction between different tapes might be governed by the flanking moieties, providing thus an excellent framework for the spatial organization of functional entities.

3.3 CONFORMATIONALLY CONSTRAINED BIS-UREA GELATORS

Structural studies on the micrometer long fibers of bis-urea, in which the urea groups are connected by linear alkyl spacers, indicate that in addition to hydrogen bonding the regular packing of the alkyl chains causes the formation of the well ordered structures. When the packing of the alkyl chains is distorted as in the case of non-symmetric bis-urea (Figure 21a, $R_1 \neq R_2$) less regular two dimensional structures are obtained. This behavior can be related to the conformational flexibility of the linker between two urea moieties allowing each urea group to aggregate in a particular direction.

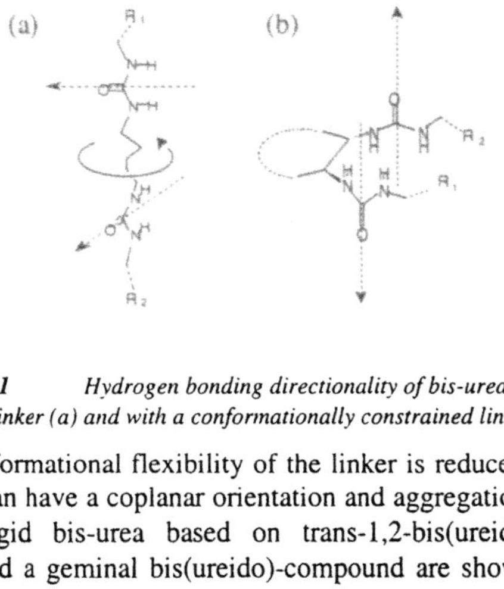

Figure 21 *Hydrogen bonding directionality of bis-urea compounds with a flexible linker (a) and with a conformationally constrained linker (b).*

When the conformational flexibility of the linker is reduced as shown in Figure 21b the urea groups can have a coplanar orientation and aggregation along one direction is enforced. Semi-rigid bis-urea based on trans-1,2-bis(ureido)cyclohexane, 1,2–bis(ureido)benzene and a geminal bis(ureido)-compound are shown in Figure 22. The antiparallel orientation of the urea groups and the different geometries due to the change in bridging unit are evident.

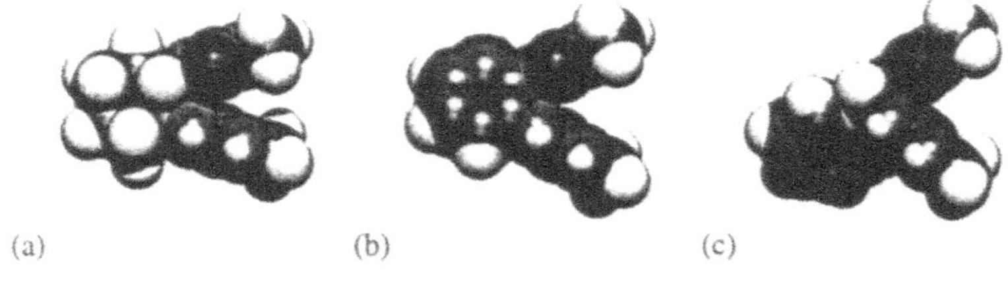

Figure 22 *CPK models of 1,2-trans-bis(3-methylureido)cycloxane (a), 1,2-bis(3-methylureido)benzene (b), and 1,1-bis(3-methylureido)-1-phenyl-methane (c).*

One of the three main conformations of (S,S) -trans-1,2- bis(methylureido)cyclohexane, obtained by molecular modeling studies, is shown in Figure 23. The urea groups have a coplanar orientation but the urea groups point in opposite directions (antiparallel conformation). Based on docking experiments it was established that the preferred sites of interaction are located above and below the urea groups (Figure 23b) [60]. Apparently, non-covalent interactions (hydrogen bond formation) between these molecules are highly anisotropic in accordance with our expectations based on the design of these systems and therefore aggregation along one direction is highly favored over other directions.

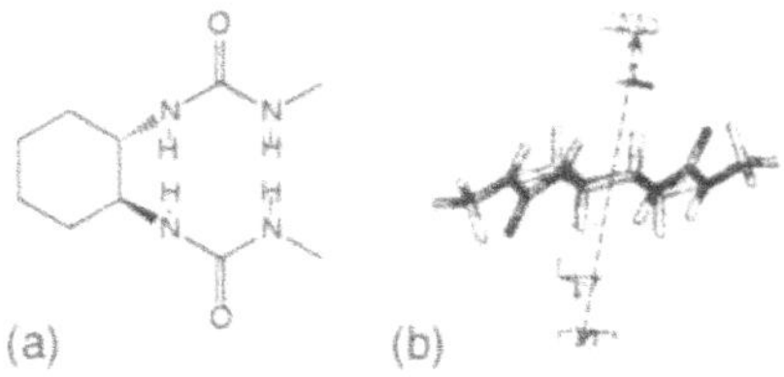

Figure 23 *Structure of the model compound trans-1,2-bis-(N-methylureido)-cyclohexane (a), and energy minimized conformation of trans-1,2-bis-(N-methylureido)-cyclohexane with a antiparallel orientation of the urea groups (b). The shaded areas in (b) are the contour levels for a interaction energy of –80 kJ/mol with a second molecule of 1,2-bis-(N-methylureido)-cyclohexane.*

The line through the most favorable sites of interaction defines the primary axis along which one dimensional aggregation most likely will take place. Molecular modeling revealed that indeed the formation of one dimensional aggregates is strongly favored by 108-122 kJ/mol relative to the most stable monomer conformation (Figure 24). In the aggregates the molecules are translated by 4.4-4.5 Å. The antiparallel conformer of the bis-urea give a translational aggregate that is stabilized by the maximum number of eight hydrogen bonds for each monomer. Surprisingly, a comparable stability was found for the screw-axis aggregate formed from the monomer with a parallel conformation of the urea groups, despite the lower stability for the parallel orientation of the urea groups in the monomer.

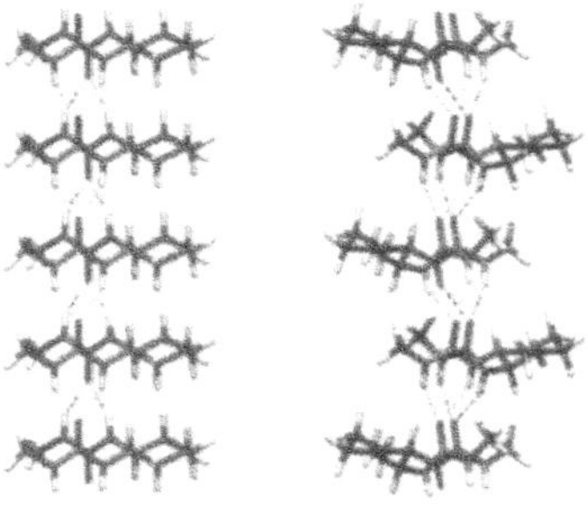

Figure 24 *Two possible hydrogen bonded aggregates of trans-1,2-bis-(N-methylureido)-cyclohexane: (a) translational aggregate with urea groups antiparallel, and (b) screw axis or glide plane aggregate with urea groups parallel.*

An important result from the modeling studies is that replacement of the methyl groups on the urea groups with longer or even branched alkyl chains does not distort the hydrogen bonding pattern which stabilizes the one-dimensional aggregates. Similar

molecular modeling experiments with 1,2-bis(methylureido)benzene gave comparable results and again one dimensional aggregates are relatively stable.
A variety of cyclohexyl- and phenyl- bis-urea compounds were indeed found to be potent gelators of organic solvents [60]. Typical examples of these gelators are shown in Figure 25 and some gelation properties and minimum gelation concentrations are given in Table 2. It is clear that a wide range of apolar and polar organic solvents can be gelated. The cyclohexyl-based compounds readily gelate aliphatic and aromatic hydrocarbons, but except for **32**, are not effective in the gelation of solvents that strongly compete for hydrogen bond formation like lower alcohols and DMSO in accordance with results from Hanabusa [55]. The gelating capability does not seem to depend much on the peripheral R- groups both for the cyclohexyl- and the phenyl- based systems.

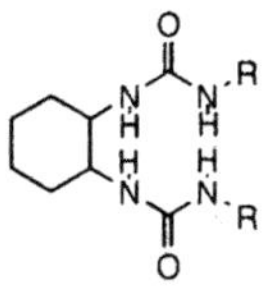

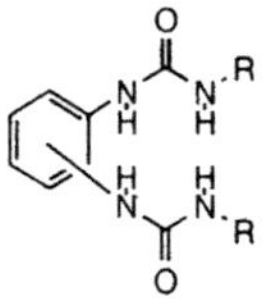

32a (*S,S*), R= $-(CH_2)_{11}CH_3$
32b (*R,R*), R= $-(CH_2)_{11}CH_3$
33 (*R,R*), R= $-CH(C_2H_5)(CH_2)_3CH_3$
34 (*R,R*), R= 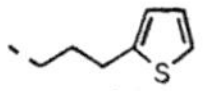
35 (*R,R*), R= 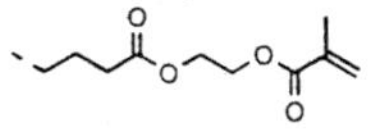
36 (*R,S*), R=R= $-(CH_2)_{11}CH_3$

37 ortho, R= $-(CH_2)_{11}CH_3$
38 ortho, R= 
39 ortho, R= 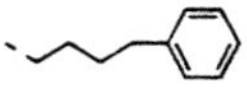
40 ortho, R= 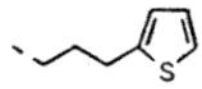
41 meta, R= $-(CH_2)_{11}CH_3$
42 para, R= $-(CH_2)_{11}CH_3$

Figure 25

Again it is seen that the geometry of the bis-urea is essential for the self- ssembly process as the cis-cyclohexyl analogue, with two adjacent urea moieties in an axial and equatorial orientation (in the chair conformation), does not lead to gelation. Similar the ortho-bis-ureido-phenyl moiety is essential for gelation whereas the meta- and para-substituted analogues fail to gelate any of the investigated solvents.

Furthermore there are some clear differences between cyclohexyl-based and phenyl-based gelators. Whereas gels of the first class are stable for months without any detoriation gels of the latter class show only limited stability. Similar trends are seen when solvent compatibility and minimum gelation concentrations are compared. The cyclohexyl-based compounds gelate a wider range of solvents and in most cases the minimum gelation concentrations required are lower. Again these gels are fully thermoreversible. A fascinating property of many gels of these bis-urea compounds is that they are thixotropic; a process that appears also to be fully reversible.

Infrared, DSC and small angle X-ray scattering (SAXS) studies on these gels reveal that aggregation is accompanied by the formation of hydrogen bonded networks, that highly cooperative and less cooperative phase transitions can occur prior to melting of the gels and that thermotropic polymorphism depends on the gelator-solvent combination.

TABLE 2A Gelation properties of cyclohexyl bis-urea derivatives [a]

solvent	32[b]	33	34	35	36
hexadecane	<10	<2	<2	p	p
cyclohexane	<2	<2	p	<2	vs
toluene	<2	<2	<10	<2	vs
p-xylene	<5	<2	<10	<2	vs
n-butyl acetate	<10	<2	p	<2	p
cyclohexanone	<2	<5	<10	<2	s
1,2-dichloroethane	<2	<2	<2	<20	p
dimethyl sulfoxide	<5	<10	s	p	p
ethanol	<2	s	s	p	s
2-propanol	<2	s	s	p	s

TABLE 2B Gelation properties of phenyl bis-urea derivatives [a]

solvent	37	38	39	40	41	42
hexadecane	<2	p	<10	p	<2	I
cyclohexane	<2	I	i	i	p	I
toluene	<2	<10	<2	<5	p	I
p-xylene	<2	p	<5	p	p	I
n-butyl acetate	p	p	p	p	p	I
cyclohexanone	s	<10	s	s	p	p
1,2-dichloroethane	p	<2	<5	<5	p	I
dimethyl sulfoxide	<5	s	s	s	p	p
ethanol	p	s	s	s	p	p
2-propanol	s	s	s	s	p	p

[a] The following abbreviations are used: gelation: g(minimum gelation concentration in mg compound /mL solvent); insoluble at solvent reflux temperature: i; precipitate: p; soluble at room temperature (solubility > 20 mg/mL): s; viscous solution: vs. [b] The same results were obtained for **32a** *and* **32b**.

Long thin fibers which form entangled networks are observed for both types of gelators by EM. The morphology show a regular shape with an extreme aspect ratio which must arise from a strong anisotropic growth process and indicating that the fibers have a well ordered molecular packing. From the pictures in Figure 26 it is clear that structural differences have a large effect on the morphology. Compound **32** forms in p-xylene untwisted thin but highly flexible straight fibers. Fibers of **33** have a less regular structure and are strongly twisted but both types of fibers are flat and consist of stacks of smaller fibers. The diameter of the smallest entity which can be distinguished is 30-50 nm and 15-25 nm, respectively, which is an order of magnitude larger than the molecular dimensions of **32** and **33**. The phenyl-bis-urea **37** and **39** also form fibers but now many thin fibers can be distinguished with diameters as small as 2-4 nm, which are comparable to the molecular dimensions. Small fibers also appear to fuse to form sheets which stack into layered structures. The calculated thickness of a single sheet was approximately 3-5 nm. There is also a clear change in morphology of the fibers depending on the solvent. Since X-ray diffraction points to the same molecular arrangement in the fibers, the different morphologies should originate from differences in interfacial energy or attachment energies in the various solvents.

From the EM micrographs in Figure 26 it is clear that both type of gelators form entangled networks and many intertwined and fused fibers are observed. Similar structural features are often observed as junction zones in gels of (bio)-polymers [39]. It has been argued that thixotropy is related to reversible disruption and formation of junction zones [61].Their presence might therefore explain the thixotropic behavior found with these gelators. In contrast, intertwined structures were not observed in gels of linear bis-urea and in accordance with the rational given above gels of linear bis-urea are not thixotropic and are easily irreversible destroyed by mechanical agitation.

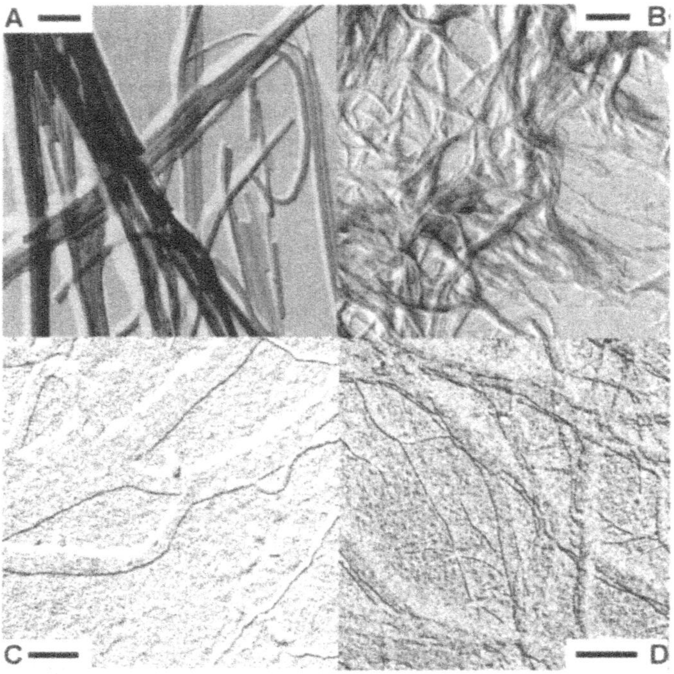

Figure 26 *Electron micrographs of* **32a** *in p-xylene (A, 3 mg/mL, Pt shadow 45°, bar = 500 nm),* **33** *in p-xylene (B, 3 mg/mL, Pt shadow 45°, bar = 500 nm),* **37** *in p-xylene (C, 3 mg/mL, Pt shadow 10°, bar = 100 nm), and* **39** *in toluene (D, 3 mg/mL, Pt shadow 10°, bar = 200 nm).*

The chirality of the gelator molecules is an intriguing aspect in view of the organization in self-assembled aggregates. The cyclohexyl-based bis-urea gelators have two stereogenic centers but rather to our surprise, the chirality is expressed at the supramolecular level only in a few cases. Only for gels of in ethanol a clear twist of the fibers is observed in the electron micrographs. Thus for (*S,S*)-**32** right handed helices are observed whereas for (*R,R*)-**32** left-handed helices are found (Figure 27). Apparently the screw-sense of the helices is related to the handedness of the gelator molecules but the pitch is not regular. This indicates that the twists do not arise from a helical arrangement at the molecular level but more likely are the result of the anisotropy of the interfacial energy.

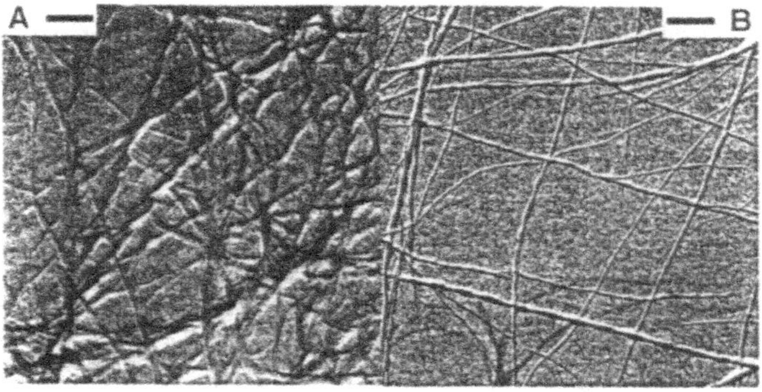

Figure 27 *Electron micrographs of* ***32a*** *in ethanol (A, 3 mg/mL, Pt shadow 45°, bar = 200 nm) and of* ***32b*** *in ethanol (B, 3 mg/mL, Pt shadow 45°, bar = 500 nm).*

X-ray diffraction measurements showed that many of the 1,2-cyclohexyl- and 1,2-phenyl-bis-urea based gels have a lamellar structure. This is in excellent agreement with the electron microscopic observations (*vide infra*), which showed that the fibers are build up of sheets with a thickness which nicely corresponds to the spacing of the lamella. As has been discussed above, the electron micrographs have revealed that the sheets consist of thin strands of only 2-4 nm thick, and the long axis of the strands runs parallel with the fiber long axis. Apparently, the fibers consist of closely packed hydrogen bonded arrays of bis-urea gelators. Within such an arrangement, however, different molecular packings are possible, as SAXS measurements of **37** and **40** provide clear evidence of polymorphism. Polymorphism can be related to different packings of strands of the gelator molecules, e.g. in a rectangular lattice or a hexagonal lattice, but it can also be the result of different arrangements of the bis-urea gelator molecules in each strand. For instance, in crystals of **40**, the two urea moieties in each molecule have a parallel orientation, and the hydrogen bonded aggregate is build up by a glide plane. Molecular modeling showed, as already has been discussed above, that other arrangements, *i.e.* translational or screw aggregate of the parallel conformation of the 1,2-bis(ureido)benzene moiety, or aggregates built up from the antiparallel conformation of 1,2-bis(ureido)benzene via translation or inversion operations, are equally stable within a window of 8 kJ/mol. For gelators **37-40**, based on the data yet available, we cannot determine which of these arrangements dominate in gels of these compounds, and probably two or more of these structures coexist in gels.

For the 1,2-bis(ureido)cyclohexane based gelators the number of possible arrangements is limited, because these molecules are *chiral* and *non-racemic*, and therefore aggregates can only be constructed by application of translation or screw axis operations. Molecular modeling studies showed that translational aggregates built up from molecules with the urea groups in an antiparallel conformation and screw axis constructed from molecules with the urea groups in a parallel conformation are equally stable (see Figure 24). Two possible lamellar arrangements of molecules of **32** are depicted in Figure 28. In the translational aggregate, hydrogen bonding between the two 1,2-bis(ureido)cyclohexane moieties allows for a close packing of the alkyl chains. For the screw axis aggregate close packing can only be achieved by intercalation The experimentally observed spacing of lamella of **32** (31.5Å) does however, neither fit with a single layer structure nor with an intercalated structure. A more likely arrangement of molecules of **32** in lamella is in a double layered structure (Figure 28) in which a tilt of the molecules can explain the discrepancy between the theoretical width of 42 Å and the observed spacing of 31.5 Å, although other arrangements can not be excluded.

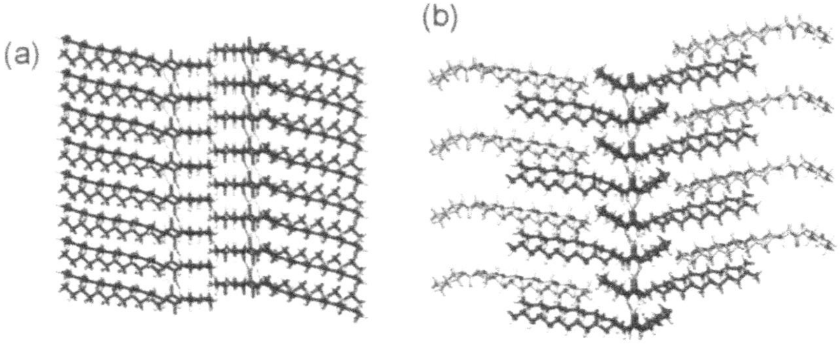

Figure 28 *Tentative arrangement of* ***32*** *in a double layer structure, constructed from translational aggregates (a), and an intercalated layer structure constructed from screw axis aggregates (b).*

It is evident that starting from molecules which preferentially self-assemble in one dimension, we have succeeded in the design of new gelators for organic solvents. The derivatives of 1,2-bis(ureido)benzene and trans-1,2-bis(ureido)cyclohexane presented here are very potent gelators for a wide range of organic solvents. Although the morphology of the fibrous network within the gels depends both on the nature of the substituents on the urea groups and on the solvent, the molecular arrangement of these bis-urea compounds is dominated by intermolecular hydrogen bond formation between the urea moieties. These one dimensional strands of hydrogen bonded bis-urea compounds assemble into sheets and lamella, which in turn stack into fiber like structures. To what extent this secondary assembly process has taken place is determined by the interfacial energy of the strands, which mainly depends on the nature of the substituents and on the solvent. As a result, these bis-urea compounds display a rich variety of morphologies.

The bis-urea compounds presented have many properties in common with other gelators. They are, however, very easy to synthesize, and many structural variations are possible without losing the gelating ability. For these reasons, the bis-urea compounds are not only excellent model compounds to study gelation phenomena in more detail, but also are excellent building blocks for the development of functional gels.

4 Towards Functional Organogelators

Many applications of gels are based on the coexistence of a three dimensional network structure and a liquid phase [1,3]. For instance, gels with a well-defined pore size are used in separation processes, and the limited diffusion in gels has been exploited for drug delivery systems. Gels with entrapped catalysts or receptors are receiving much interest in catalysis and sensor technology [62]. An exciting new development are 'smart gels', *i.e.* gels which respond to external stimuli by a change of their elasticity, size, or shape [63]. Most of these applications requires the incorporation of functional entities into the gels. In macromolecular gels this is often achieved via less selective methods, such as physical entrapment within the porous network structure or via covalent linkage of the active species to the polymer backbone. Organogels, however, have some unique features that clearly distinguish them from other gel systems: (i) the gelation process, starting with self-assembly in an isotropic solution of the gelator, up to the formation of a network structure, is completely reversible, and (ii) the self-assembly process leads to well ordered arrays of molecules, up to a length scale of micrometers, whereby the precise molecular arrangement is determined by the molecular structure of the individual molecules. These special features of organogels make them excellent systems for the development of functional gels. It is obvious that smart gels or responsive gels will greatly benefit form the dynamic and reversible formation of organogels. Also many applications will have great advantage from a well-defined spatial arrangement of the functional species within the gels. For instance, the selectivity of recognition processes will increase if receptor sites are embedded in a well-ordered matrix, and signal transduction chains will only function if the individual segments are properly aligned.

Very recently, a peptide consisting of a 24 amino acid sequence with a high propensity to form a β-sheet structure was found to form temperature and pH responsive gels in aqueous solution [64]. The first report on responsive organogels was, however, from Shinkai's group [17,65]. They prepared cholesterol-based organogelators with covalently attached azo-benzene (**9**) or crown ether moieties as light or metal responsive functions (Figure 29). The azobenzene containing organogelator **9** forms thermoreversible gels with a range of organic solvents, including aliphatic hydrocarbons, esters, alcohols, and amides, at concentrations well below 1 (w/v)%. For crown-ether containing compound **43** the gelating capability is limited to some aliphatic hydrocarbons. It was, however, found that the thermal stability of the gels strongly depend on cation complexation by the crown-ether moiety, and thus in principle represent a cation-responsive gel system. Compound **9** is not only a much better gelator, it was also found that the thermal stability of gels strongly depend on the conformation of the photo-isomerizable azo unit: gels of trans-**9** have a gel-sol transition temperature of 16°C whereas the melting temperature of gels of cis-**9** is well below 2°C. Most interestingly, by alternate irradiation at 10°C with UV and visible light a reversible switching between a gel of trans-**9** and a solution of cis-**9** has been achieved.

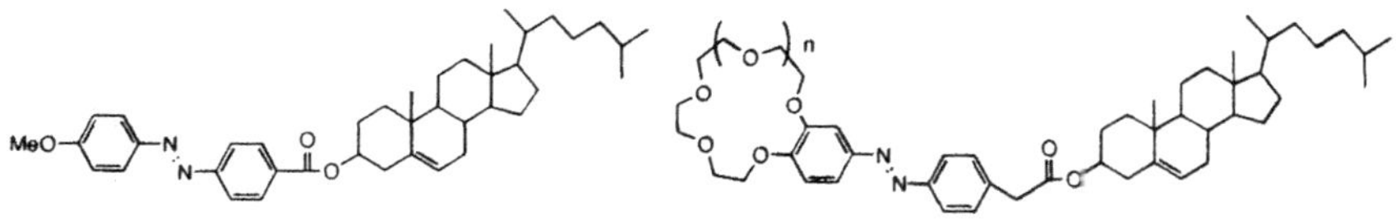

9 **43** n=1,2

Figure 29

Other examples of functional organogel systems include for instance mesogenic derivatives of cyclohexyl bis-amide **23** which form novel liquid crystaline materials with nematic liquids [66], and the tetraoctadecylammonium bromide (like **3**) and the metal-containing gluconamides **12** which were able to gelate monomers as styrene and methacrylates and were used to prepare nanoporous membranes and for polymer imprinting [67,20].

A novel approach to polymerized gel systems has been followed by our group. Based on the design principles described in the previous section, we successfully prepared a new polymerizable bis-urea gelator **35**, which is capable of gelating a many different organic solvents [68]. On the other hand, the bis-amide derivative does not gel any of the solvents investigated, indicating that the stronger hydrogen bonding urea groups are essential for gelation. Before polymerization, **35** forms thermoreversible gels which melt at temperatures below 100°C. Effective polymerization of gels of **35** was achieved by UV irradiation in the presence of a photo-initiator. Electron microscopy showed that the polymerized gels consist of a network of very thin fibers, which occasionally fuse and intertwine (Figure 30). Remarkably, after polymerization the gels are now stable up to at least 135°C, at which temperature the gels still did not show any sign of melting or shrinking. The solvent can be removed by freeze drying without causing collapse of the gel, yielding a white brittle material with a very low density, having all the characteristics of an organic aerogel.

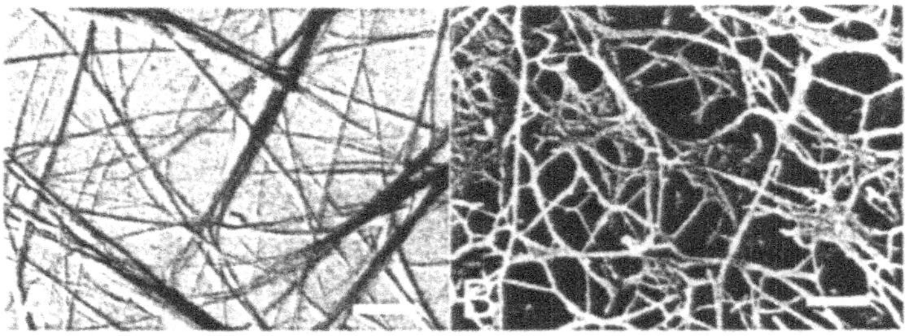

Figure 30 *Transmission electron micrograph of a butyl acetate gel of* **35** *before polymerization (A, 5 mg/ml, Pt shadowed, bar = 1 μm), and scanning electron micrograph of a benzene gel of* **35** *after irradiation for 1 h (B, 5 mg/ml, Pt coated, bar = 0.4 μm).*

5 Concluding remarks

During the last decade, the field of organogelators has rapidly developed from being guided by serendipity, towards a rich area of science, in which principles form supramolecular chemistry, macromolecular science, and colloid physics merge together. It has become clear that the gelation of organic solvents by low molecular weight organic compounds is an example of supramolecular organization par excellence. A better understanding of the structure of organogels and the mechanism of gelation will facilitate their application and will allow the development of new gelators with novel properties. Many applications of functionalized gels in catalysis, sensors, molecular electronics and material science are within reach, and especially switchable organogelators offers exciting prospects for the development of responsive (smart) gel systems.

Acknowledgement: Financial support by the Netherlands Organization for Scientific Research (NWO) and the Stichting Technische Wetenschappen (STT) is gratefully acknowledged. We are also grateful for a fellowship (to J.v.E.) of the Royal Academy of Science of the Netherlands (KNAW), and for the fruitful collaboration with Prof. Dr. F. De Schryver and Dr. S. De Feyter (KU Leuven, Belgium), and with Prof. Dr. A. Brisson and J. van Breemen (University of Groningen).

References

1. (a) *Polymer Gels: Fundamentals and Biomedical Applications*, Ed. D. Derossi, K. Kajiwara, Y. Osada, A. Yamauchi, **1991**, Plenum Press, New York; (b) J.-M. Guenet, *Thermoreversible*

Gelation of Polymers and Biopolymers, **1992**, Academic Press, London; (c) *Gels*, Ed. F. Kremer, G. Lagaly, *Prog..Coll.Polym.Sci.*, **1996**, 102.

2. P.J. Flory, *Disc. Farad. Soc.*, **1974**, *57*, 8.
3. T. Tanaka, *Scient. Amer.*, **1981**, *244*, 110-123.
4. For an excellent review see: P. Terech, R.G. Weiss, *Chem. Rev.*, **1997**, *97*, 3133-3159.
5. D. Philp, J. F. Stoddart, *Angew. Chem. Int. Ed. Eng.*, **1996**, *35*, 1154-1196.
6. J.H. Fuhrhop, J. Köning, ***Membranes and Molecular Assemblies: the Synkinetic Approach***, Royal Society of Chemistry, Cambridge, **1994**; J.M. Lehn, *Supramolecular chemistry*, VCH, Weinheim, **1995**.
7. G.M. Whitesides, J.P. Mathias, C.T. Seto, *Science*, **1991**, *254*, 1312-1319.
8. T. Tachibana, T. Mori, K. Hori, *Bull. Chem. Soc. Jpn.*, **1980**, *53*, 1714-1719.
9. R.J. Twieg, T.P. Russll, R. Siemens, J.F. Rabolt, *Macromolecules*, **1985**, *18*, 1361-1362.
10. M.P. Turberg, J.E. Brady, *J. Am. Chem. Soc.*, **1988**, *110*, 7797-7801.
11. L. Lu, , R.G. Weiss, *Chem. Comm.*, **1996**, 2029-2030.
12. L. Lu, R.G. Weiss, *Langmuir*, **1995**, *11*, 3630-3632.
13. T. Brotin, R. Utermöhlen, F. Fages, H. Bouas-Laurent, J.-P. Desvergne, *J. Chem. Soc. Chem. Comm.*, **1991**, 416-418.
14. P. Terech, R.G. Ramasseul, F.J. Volino, *J. Phys. France*, **1985**, *46*, 895.
15. Y.-C Lin, R. G.Weiss, *Macromolecules*, **1987**, *20*, 414-417.
16. a) Y.-C Lin, B. Kachar, R.G. Weiss, *J. Am. Chem. Soc.*, **1989**, *111*, 5542-5551. b) R. Mukkamala, R.G. Weiss, *J. Chem. Soc. Chem. Comm.*, **1995**, 375-376.
17. K. Murata, M. Aoki, T. Nishi, A. Ikeda, S. Shinkai, *J. Chem. Soc. Chem. Comm.*, **1991**, 1715.
18. C.S. Snijder, J.C. de Jong, A. Meetsma, F. Bolhuis, B.L. Feringa, *Chem. Eur. J.*, **1995**, *1*, 594.
19. P. Terech, C. Chachaty. et al. *J. Phys. France*, **1987**, *48*, 663.
20. R. J. H. Hafkamp,; B. P. A. Kokke, I.M. Danke, H.P.M. Geurts, A.E. Rowan, M.C. Feiters, R.J.M. Nolte, *Chem. Comm.*, **1997**, 545-546.
21. K. Hanabusa, J. Tange, Y. Taguchi, T. Koyama, H. Shirai, *J. Chem. Soc. Chem. Comm.*, **1993**, 390-392.
22. E.J. de Vries, R.M. Kellogg, *J. Chem. Soc. Chem. Comm.*, **1993**, 238-240.
23. K. Hanabusa, Y. Matsumoto, T. Miki, T. Koyama, H. Shirai, *J. Chem. Soc. Chem. Comm.*, **1994**, 1401-1402.
24. G.T. Crisp, J. Gore, *Synth. Comm.*, **1997**, *27*, 2203-2215.
25. J.-H.Fuhrhop, P. Schnieder, E. Boekema, W. Helfrich, *J. Am. Chem. Soc.*, **1988**, *110*, 2861-2867
26. S. Yamamoto, *J. Chem. Soc. Ind. Jpn.*, **1943**, *46*, 279-81 (*CA* 46, 7047h).
27. S. Yamasaki, Y. Ohashi, H. Tsutsumi, K. Tsujii, *Bull. Chem. Soc. Jpn.*, **1995**, *68*, 146-151.
28. K. Yoza, Y. Ono, K. Yoshihara, T. Akao, H. Shinmori, M. Takeuchi, S. Shinkai, D.N. Reinhoudt, *Chem. Comm.*, **1998**, *8*, 907-908.
29. (a) K. Hanabusa, Y. Naka, T. Koyama, H. Shirai, *J. Chem. Soc. Chem. Comm.*, **1994**, 2683-2684; (b) K. Hanabusa, C. Koto, M. Kimura, H. Shirai, A. Kakehi, *Chem. Lett.*, **1997**, 429-430.; (c) K. Hanabusa, K. Okui, K. Karaki, M. Kimura, H. Shirai, *J. Colloid Interface Sci.*, **1997**, *195*, 86.
30. F.M. Menger, Y. Yamasaki, K.L. Catlin, T. Nishimi, *Angew. Chem.Int. ed. Engl.*, **1995**, *34*, 585-586.
31. U. Keller, K. Müllen, S. De Feyter, F.C. De Schryver, *Adv. Mat.*, **1996**, *8*, 490-493.
32. Y. Yasusa, Y. Takebe, M. Fukumoto, H. Inada, Y. Shirota, *Adv. Mat.*, **1996**, *8*, 740-741.
33. M. Jokic, J. Makarevic, M. Zinic, *J. Chem. Soc. Chem. Comm.*, **1995**, 1732-1724.
34. H.T. Stock, N.J. Turner, R. McCague, *J. Chem. Soc. Chem. Comm.*, **1995**, 2063-2064.

35. Hanabusa, T. Miki, Y. Taguchi, T. Koyama, H. Shirai, *J. Chem. Soc. Chem. Comm.*, **1993**, 1382-1384.

36. (a) X. Xu, M. Ayyagari, M. Tata, V.T. John, G.L. McPherson, *J. Phys. Chem.*, **1993**, *97*, 11350; (b) M. Tata, V.T. John, Y.Y. Waguespack, G.L. McPherson, *J. Am. Chem. Soc.*, **1995**, *116*, 9464-9470.

37. (a) K. Hanabusa, M. Yamada, M. Kimura, H. Shirai, *Angew. Chem. Int. Ed. Engl.*, **1996**, *35*, 1949-1951; (b) Y. Yasuda, E. Iishi, H. Inada, Y. Shirota, *Chem. Lett.*, **1996**, 575-576.

38. (a) J. van Esch, R.M. Kellogg, B.L. Feringa, *Tet. Lett.*, **1996**, 281-284; (b) J. van Esch, S. de Feyter, R.M. Kellogg, F. De Schryver, B.L. Feringa, *Chem. Eur. J.*, **1997**, *3*, 1238-1243.

39. K te Nijenhuis, *Adv. Polym. Sci.*, **1997**, *130*, 1-252

40. D.S. Lawrence, T. Jiang, M. Levett, *Chem. Rev.*, **1995**, *95*, 2229-2260.

41. N. Kimizuka, S. Fujikawa, H. Kuwahara, A. Marsh, J.-M. Lehn, *J. Chem. Soc. Chem. Comm.*, **1995**, 2103-2104.

42. S. Hanessian, M. Simard, S. Roelens, *J. Am. Chem. Soc.*, **1995** , ***117***, 7630-7645.

43. J.J. Kane, R.-F. Liao, J.W. Lauher, F.W. Fowler, *J. Am. Chem. Soc.*, **1995** , ***117***, 12003-12004.

44. F.D. Lewis, J.-S. Yang, C.L. Stern, *J. Am. Chem. Soc.*, **1996** , ***118***, 2772-2773.

45. (a) J.C. MacDonald, G.M. Whitesides, *Chem. Rev.*, **1994**, *94*, 2383-2420; (b)

46. C.T. Seto, G.M. Whitesides, *J. Am. Chem. Soc.*, **1990**, *112*, 6409-6411.

47. K.C. Russell, M. Leize, A. Van Dorsselaer, J.M. Lehn, *Angew. Chem. Int. Ed. Engl.*, **1995**, *34*, 209-213.

48. P. Brunet, M. Simard, J.D. Wuest, *J. Am. Chem. Soc.*, **1997** , ***119***, 2737-2738.

49. N. Branda, R. Wyler, J. Rebek, *Science*, **1994**, *263*, 1267-1268.

50. M.R. Ghadiri, J.R. Granja, R.A. Milligan, D.E. McRee, N. Khazanovich, *Nature*, **1993**, *366*, 324-326.

51. E. Fan, J. Yang, S.J. Geib, T.C. Stoner, M. D. Hopkins, A.D. Hamilton, *J. Chem. Soc. Chem. Com.*, **1995**, 1251-1252.

52. K. Hanabusa, A. Kawakami, M. Kimura, H. Shirai, *Chem. Lett.*, **1997**, 191-192.

53. M.C. Etter, Z. Urbañczyk-Lipkowska, M. Zia-Ebrahimi, T.W. Panunto, *J. Am. Chem. Soc.*, **1990**, *112*, 8415-8426.

54. R.G. Arnold, J.A. Nelson, J.J. Verbanc, *Chem. Rev.*, **1957**, *57*, 47.

55. K. Hanabusa, K. Shimura, K. Hirose, M. Kimura, H. Shirai, *Chem. Lett.*, **1996**, 885-886.

56. Y. Mido, *Spectr. Chim. Acta*, **1972**, *28A*, 1503-1518.

57. D.A. Frankel, D.F. O'Brien, *J. Am. Chem. Soc.*, **1994**, *116*, 10057-10069.

58. S. De Feyter, P.C.M. Grim, M. Rucker, P. Vanoppen, C. Meiners, M. Sieffert, S. Valiyaveettil, K. Mullen, F.C. De Schryver, *Angew. Chem. Int. Ed. Engl.*, **1998**, *37* , 1223-1226.

59. S. De Feyter, M. Ruecker, K. Grim, P. Vanoppen, J. van Esch, B. L. Feringa, R.M. Kellogg, F. De Schrijver, *J. Phys. Chem.*, in press.

60. F. Schoonbeek, J. van Esch, M. de Loos, R. M. Kellogg, B. L. Feringa, *Chem. Eur. J.*, in press.

61. (a) A. H. Clark, S. B. Ross-Murphy, *Adv. Coll. Sci.*, **1987**, *83*, 57-192; (b) N. Willenbacher, *J. Coll. Inter. Sci.*, **1996**, *182* , 501-510; (c) G. Liu, B. White, I. Vancsó-Szmercsányi, G. J. Vancso, *J. Polym. Sci.B: Polym. Phys.*, **1996**, *34*, 277-282.

62. (a) M. T. Reetz, A. Zonta, J. Simpelkamp, *Angew. Chem. Int. Ed. Engl.*, **1995**, *34* , 301-303; (b) M.T. Reetz, *Adv. Mat.*, **1997**, *9*, 943-954.

63. (a) Y. Osada, S.B. Ross-Murphy, *Scient. Amer.*, **1993**, May, 42-47; see also R. Dagani, *Chem.Eng.News*, **1997**, June 9, 26-36.

64. A. Aggeli, M. Bell, N. Boden, J.N. Keen, P.F. Knowles, T.C.B. Leish, M. Pitkeathly, S.E. Radford, Nature , **1997**, *386*, 359-363.

65. K. Murata, M. Aoki, T. Suzuki, T. Harada, H. Kawabata, T. Komori, F. Ohseto, K. Ueda, S. Shinkai, *J. Am. Chem. Soc.*, **1994**, *116*, 6664-6676.

66. T. Kato, G. Kondo, K. Hanabusa, Chem. Lett., **1998**, 193-194.

67. W. Gu, L. Lu, G.B. Chapman, R.G. Weiss, *Chem. Comm.*, **1997**, 543-544.

68. M. de Loos, J. van Esch, I. Stokroos, R.M. Kellogg, B.L. Feringa, *J. Am. Chem. Soc.*, **1997**, *119*, 12675-12676; See also M. Jacoby, *Chem.Eng.News*, **1998**, January 26, 34-35.

SUPRAMOLECULAR DENDRIMER CHEMISTRY:
Molecular Recognition within the Dendritic Environment

DAVID K. SMITH, FRANÇOIS DIEDERICH, ADRIEN ZINGG
Laboratorium für Organische Chemie
Universitätstrasse 16
ETH Zentrum
CH 8092 Zürich
Switzerland

ABSTRACT: This paper describes the use of a dendritic shell to modify molecular recognition properties at a well-defined binding site deep within the branched architecture. Receptors based on different interactions are discussed. Dendritic Fe(II) porphyrins coordinatively bind dioxygen, and the branched architecture modulates both the strength and selectivity of binding. Dendritic cyclophanes (dendrophanes) hydrophobically snare aryl or steroid guests. The branching changes the micropolarity of the recognition unit at the core. Such dendrophanes have also been used to assemble high mass supramolecular structures, the stability and stoichiometry of which are controlled by the dendritic diameter. Dendrimers for specific applications in molecular recognition can require new, carefully designed branches, and the synthesis of such a branch is described. These new branches are then used to prepare chiral dendritic receptors based on a functionalised 9,9'-spirobi[9*H*-fluorene] core which bind glucosides via hydrogen bonding. The enantio- and diastereo-selectivities of these optically pure, dendritic clefts (dendroclefts) are altered by the presence of the dendritic branching.

1. Introduction

Molecular recognition is the foundation upon which the architectures of supramolecular chemistry have been constructed.[1] During the last thirty years, chemists have become increasingly adept at controlling interactions between molecules, and this has led to the development of an ever expanding number of elegant, substrate selective receptors.[2] Molecular recognition, however, is extremely sensitive to the surrounding environment. For example, the hydrophobic (or solvatophobic) effect is almost worthless in benzene,[3] and solvent exposed hydrogen bonds count for little in water.[4] The precise solvent environment can also change the binding selectivity.[5] Biological systems usually control the environment at enzyme recognition sites to achieve strong and selective binding. They do this by burying them within the polypeptide superstructure, and in this way, the dielectric constant, the proximate amino acid residues, and even the number of solvent molecules at the binding site are all precisely controlled.[6]

R. Ungaro and E. Dalcanale (eds.), Supramolecular Science: Where It Is and Where It Is Going, 261–272.

Dendrimers constitute a research area currently undergoing explosive growth.[7] In particular, the last five years have seen increasing interest in functional dendrimers.[8] Parallel to the first reports of the ability of dendritic branching to change the properties of an active unit located at the core,[9] we had realised the potential of a dendritic shell to mimic some aspects of an enzyme superstructure. We reported dendritic models of cytochrome c in which the branching altered the redox properties of the porphyrin core, analogously to the protein's peptide shell.[10] Naturally, our thoughts (like those of others) turned to the development of dendritic receptors.[11] We hoped that burying a binding site within a dendrimer would control its environment, modulating binding strength and selectivity. This paper describes our progress towards this goal. The results of these studies are of wide-ranging interest; providing insights into molecular recognition, the unique properties of the dendritic environment, and modelling biological systems. In this article, G-0 is used to describe an unfunctionalised core, while G-1, G-2 and G-3 refer to dendrimers containing 1, 2 and 3 generations of branching, respectively.

2. Dioxygen Binding: Heme Protein Mimics

Tetrameric hemoglobin and monomeric myoglobin are heme proteins responsible for dioxygen transport and storage. Their polypeptide shells play key roles in controlling O_2 recognition. Firstly, the steric bulk prevents the irreversible formation of μ-oxo dimers. Secondly, it is believed that amino acid residues over one face of the porphyrin destabilise the otherwise very stable carbon monoxide complex sufficiently that the body tolerates the low levels of this toxic guest which are usually present. Thirdly, allosteric effects are responsible for the cooperative O_2 binding behaviour of hemoglobin.

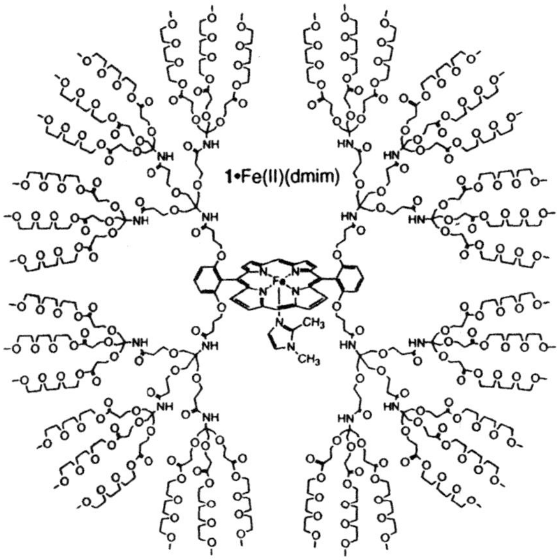

Figure 1. Second generation dendritic iron(II) porphyrin (dmim=1,2-dimethylimidazole).

The second generation (G-2) dendrimer **1** (Fig. 1) acts as a model system for globular heme proteins.[12] Similarly, Jiang and Aida proposed an iron(II) porphyrin functionalised with branched aromatic-ethers as a heme protein mimic.[13] Dendrimer **1** contains a porphyrin core, which binds iron(II), encapsulated within an aliphatic poly(ether-amide) dendritic shell.[14] The dendritic periphery is functionalised with oligo-ether solubilising groups. The affinities of first (G-1, MW= 3956 Da) and second (G-2, MW= 11296 Da) generation dendrimers for O_2 and CO were investigated in collaboration with Professor James P. Collman at Stanford University. The studies were made in toluene solution, with a large excess of 1,2-dimethylimidazole (dmim), to ensure formation of the five-coordinate high-spin iron(II) complex **1**•Fe(II)(dmim). Dendritic Fe(II) porphyrins of both first and second generation showed fast and reversible dioxygen binding (Table 1). The branching provides steric shielding, preventing irreversible μ-oxo dimer formation. Such a steric effect has, of course, also been demonstrated for 'picket-fence' porphyrins.[15] The dendrimer, however, caused a new and dramatic effect on both the strength and selectivity of binding. The O_2 affinity of the dendritic systems was vastly enhanced when compared to either T-state hemoglobin or a 'picket-fence' model system. The CO affinity, however, was comparable to that of the T-state hemoglobin, and less than that measured for the 'picket-fence' analogue. Consequently, the dendrimer strongly selects O_2 over CO. In fact, this extremely high binding strength and selectivity resembles that observed for the heme proteins of the blood worm *Ascaris*. It was proposed that the reason for this dendritic control could be a dendritic amide group which can form a hydrogen-bond with the terminal oxygen atom of O_2 bound in a bent manner. Investigations are currently underway to fully understand the role of the dendritic shell, and also to study these heme protein mimics in water.

	$P_{1/2}$ / Torr		$P_{1/2}(O_2)$/
	O_2	CO	$P_{1/2}(CO)$
Hemoglobin (T-state)[a]	40	0.30	135
Fe(II)(picket-fence)/dmim[b]	38	0.0089	4280
Ascaris[c]	0.002	0.1	0.02
G-1•Fe(II)(dmim)[b]	0.035	0.35	0.10
G-2 (**1**)•Fe(II)(dmim)[b]	0.016	0.19	0.08

Table 1. Dioxygen and carbon monoxide binding data. Conditions: a H_2O, pH 7, 25 °C; b Toluene 25 °C; c H_2O, pH 7, 20 °C. Changing solvent polarity may slightly affect the affinities.

3. Apolar Guest Binding: Dendrophanes

Many enzymes contain deeply buried apolar binding sites within their globular structures. We were interested in using dendritically shielded cyclophanes to mimic such behaviour and this led to the synthesis of a new class of molecules: the dendrophanes (*dendr*itic cycl*ophanes*).[16] The synthesis of dendrophanes such as **2** (Fig. 2) was achieved via the divergent strategy using the same branching cascade as for the dendritic iron(II) porphyrin, **1**.[14] The resultant dendrophanes are soluble in aqueous or mixed aqueous solvents at moderate to high pH. The exterior of dendrophanes such as **2** presents a large number of negatively charged carboxylate groups, which would be expected to repel one another creating a relatively open structure. Binding studies were performed with small aromatic guests in aqueous phosphate buffer containing small quantities of organic co-solvent. In all cases, recognition occurred at the dendritic core, with no binding within the branches. This validates the concept of localised molecular recognition within a large synthetic macromolecule. ^{1}H NMR analysis indicated that the host-guest exchange kinetics became a little slower as the dendrimer became larger. The strength of binding also decreased slightly with increasing dendritic generation.

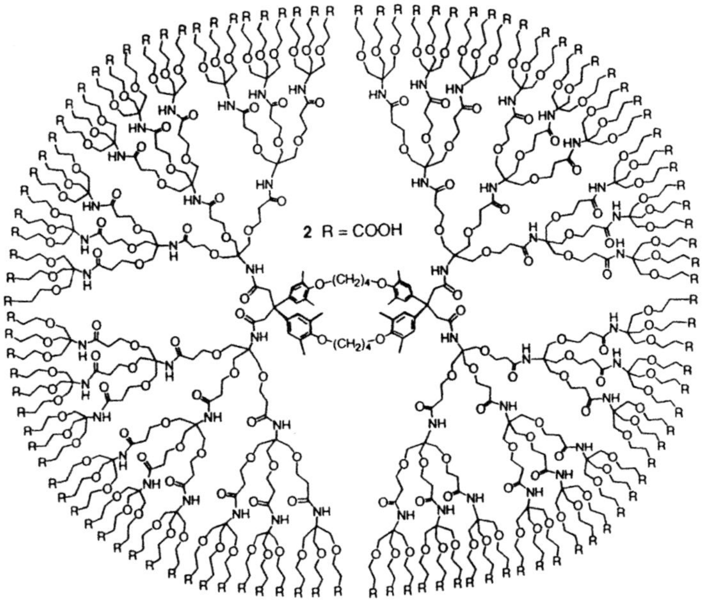

Figure 2. Water soluble, third generation dendrophane for binding small aromatic guests.

Dendritic Environment	λ_{max} [nm]	Solvent Environment	λ_{max} [nm]
G-0	450	H_2O	500
G-1	443	MeOH	443
G-2	435	EtOH	429
G-3	432		

Table 2. Emission maxima of TNS in different environments (generated either dendritically or by solvent). When bound in the dendritic environment, the surrounding solvent is aqueous phosphate buffer (pH 8).

Most interestingly, a fluorescent probe (6-(*p*-toluidino)naphthalene-2-sulfonate, TNS) was used as a guest molecule. The emission maximum of TNS is dependent on its environment, and when bound by the dendrophane it therefore reports on the environment at the core. As the dendritic shell became larger, the emission maximum shifted to shorter wavelength: indicative of decreasing polarity (Table 2). The dendritic shell therefore has a marked effect on the environment in which recognition occurs.

These results with small dendrophanes led naturally to the synthesis of dendrimers containing larger cyclophane cores, such as compound **3** (Fig. 3), which due to their increased cavity diameter are capable of binding larger, biologically relevant guests such as steroids.[17] These water soluble dendrophanes bound steroids, for example testosterone, with similar strength to the non-dendritic analogue (G-0). Amazingly, the binding kinetics were fast at all generations and this led us to investigate the use of these dendrophanes as components for the self assembly of nanometre scale architectures.

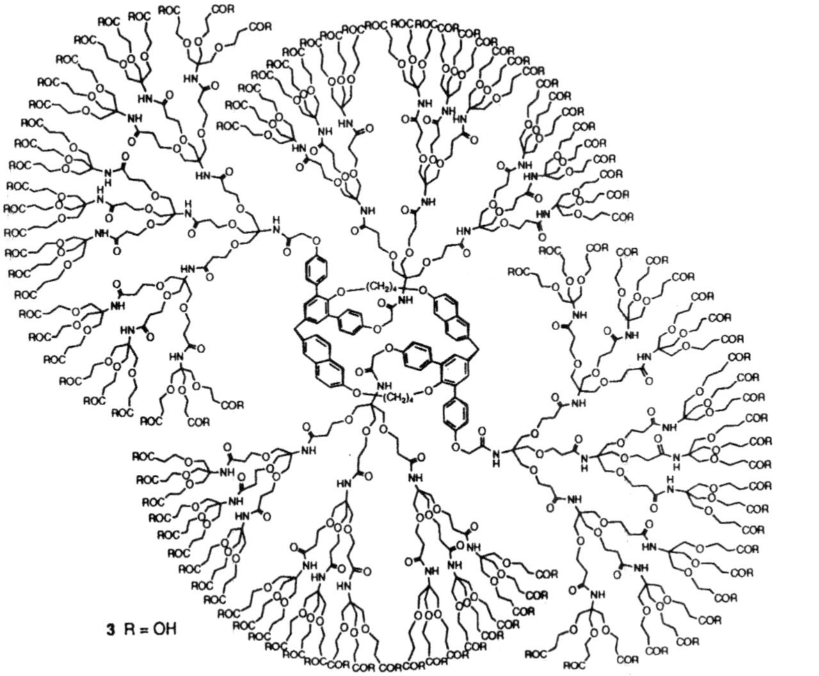

3 R = OH

Figure 3. Third generation dendrophane with an enlarged cyclophane core suitable for steroid recognition.

4. Nanometre Scale Architectures: Self assembly in Water

Self assembly processes are of great importance in the construction of large biological architectures. Zimmerman and co-workers reported hexameric dendritic assemblies, held together by hydrogen bonds in non-competitive solvents.[18] The dendritic shell was essential, providing steric buttressing and preventing the formation of linear oligomers. Dendrophanes such as compound **3** provided a point of entry into water stable assemblies.[19] Water soluble dendrophanes similar to **2** but G-0, G-1 and G-2 were hydrophobically threaded onto the steroidal termini of suitably designed molecular rods (Fig. 4). Two molecular rods (**4a** and **4b**, Fig. 5) were synthesised; they consist of rigid oligo(phenylacetylene) spacers with testosterone termini suitable for dendrophane complexation. With their different lengths, these rods were expected to accommodate dendrophanes of different diameters.

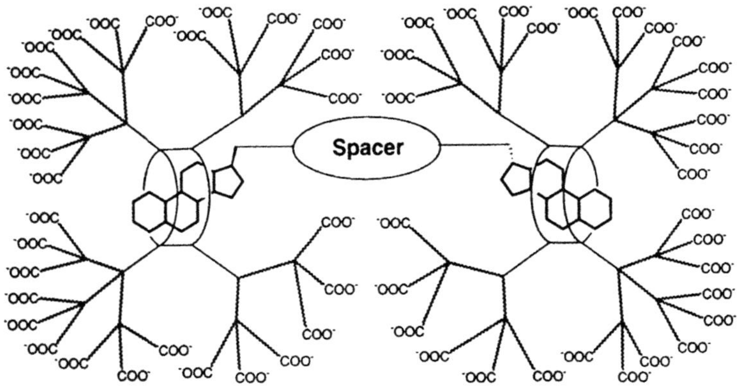

Figure 4. Two dendrophane receptors are threaded onto the testosterone termini of a rigid rod to yield high mass, water-stable assemblies.

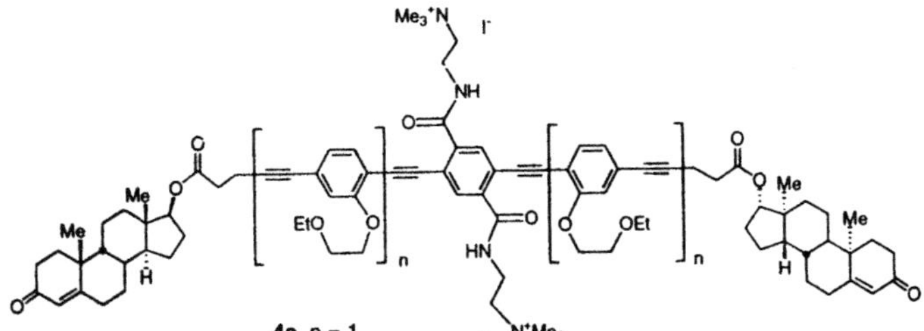

4a n = 1
4b n = 2

Figure 5. Rigid rods with testosterone termini used for dendrophane assembly.

The binding affinities of the dendrophanes for the steroid functionalised rods were higher than those obtained for testosterone alone, presumably due to the additional ion-pairing interactions between the ammonium cations on the rod and the surface carboxylates of the dendrophanes. As planned, complexes of both 2:1 and 1:1 stoichiometry were observed (Table 3). Their relative stabilities were dependent both on the length of the rod and the generation of the dendrophane. With the smaller rod **4a** the G-0 core cyclophane threaded smoothly onto both terminal testosterones. For the G-1 dendrophane, however, the second threading was thermodynamically disfavoured (by 8.8 $kJmol^{-1}$), whilst for the G-2 dendrophane, the 2:1 complex was not observed at all. This is presumably due to repulsion between the negatively charged bulky dendritic shells. For the longer rod **4b**, however, the G-1 dendrophane readily formed both 1:1 and 2:1 (MW= 6642 Da) complexes. The 2:1 complex with the G-2 dendrophane with the remarkable molecular weight of 14714 Da could also be observed. This is due to the greater ability of **4b** to penetrate more deeply into the dendritic shell, as a result of its extra length. As a consequence, the two dendrophanes threaded onto the steroidal termini are positioned a greater distance apart, preventing the electrostatic repulsion that caused the exclusive formation of a 1:1 complex with the shorter rod, **4a**. The dendritic shell therefore controls the stoichiometry and the stability of the assembled structures. This research is currently being extended to the self assembly of larger architectures, and structures with different geometries and functions.

Dendrophane	Rod	ΔG_{11} (kJmol^{-1})	ΔG_{21} (kJmol^{-1})
G-0	**4a**	-25.5	-24.7
G-1	**4a**	-28.1	-19.3
G-2	**4a**	-23.9	not obs.
G-1	**4b**	-25.5	-25.5
G-2	**4b**	-28.5	-20.5

Table 3. Binding constants (kJmol^{-1}) from ^{1}H NMR and/or fluorescence titrations for the assembly of nanoscale architectures in water/methanol (1/1, 0.15 M phosphate buffer pH=8.7, 298K) using dendrophanes and steroid functionalised rods.

5. Synthesis of a New Dendritic Wedge

To achieve a better understanding of how a dendritic shell modulates recognition and other properties, it rapidly became clear to us, that using different branching motifs would provide further information and allow new systems to be investigated. We were particularly interested in the role of hydrogen bonding at a dendritic core, and consequently targeted the synthesis of flexible and soluble ester-linked dendritic branches, such as **5** (Fig. 6) which lacked the amide protons present in the Newkome-type system we had previously used. Multiple amide protons not only confuse investigations of hydrogen bonding; they also directly compete for H-bonding acceptor sites of the guest.

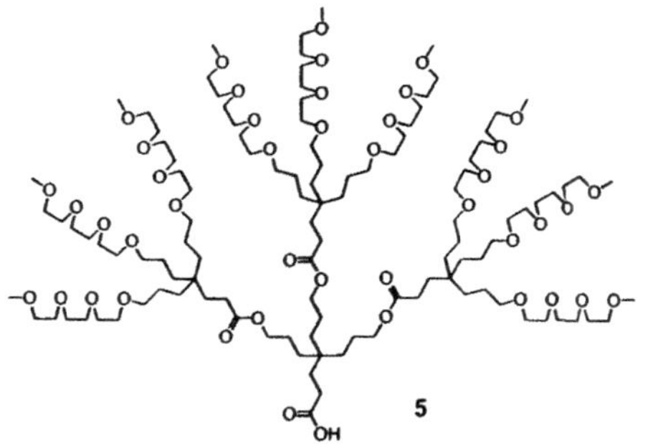

5

Figure 6. New second generation dendritic wedge designed specifically for H-bonding receptors.

The synthetic strategy was as convergent as possible, with parallel routes being used to prepare the key fragments **10** and **15** (Scheme 1).[20] Firstly, nitro-methanetrispropanol **6** was converted to trimesylate **7** using MsCl and NEt_3 in THF. The product was purified using column chromatography (SiO_2, EtOAc/hexane 65:35).

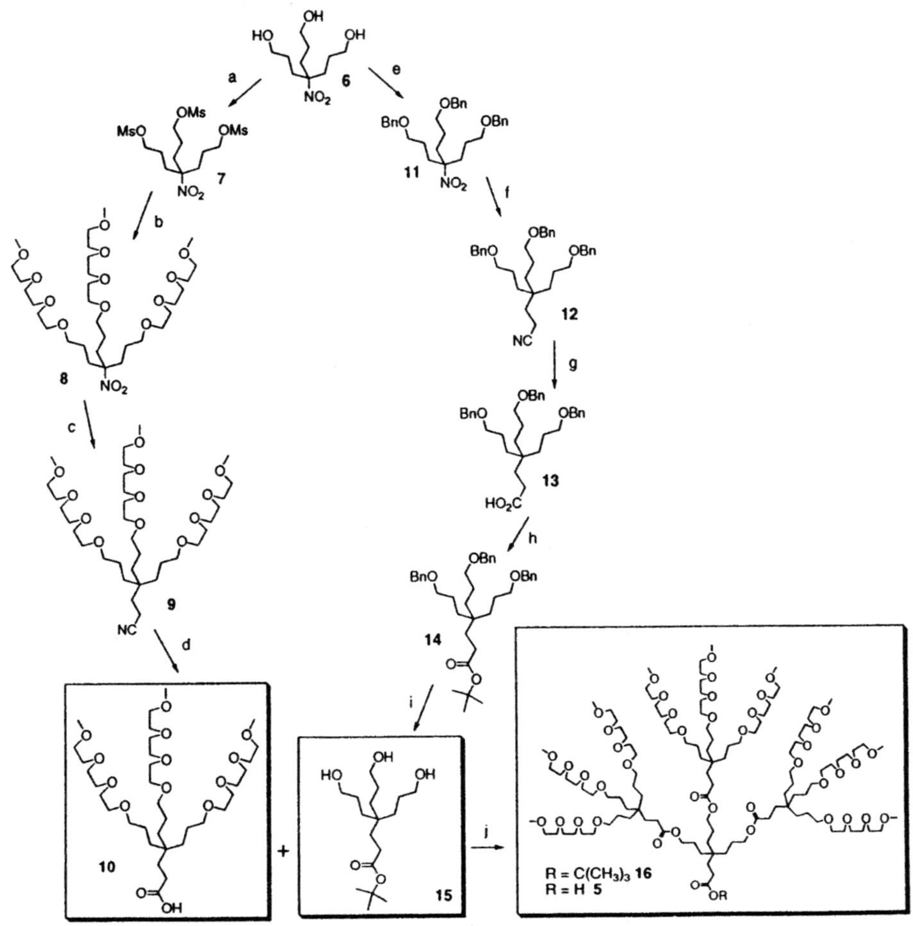

Scheme 1. Synthesis of new 'designer' dendritic branches. a) MsCl, NEt_3, THF, 75%; b) $CH_3O(CH_2)_2O(CH_2)_2O(CH_2)_2OH$, NaH, THF, 45%; c) Acrylonitrile, Bu_3SnH, AIBN, toluene, 65%; d) H_2O_2, KOH, EtOH, H_2O, 80%; e) BnCl, KOH, Me_2SO, 72%; f) see c, 48%; g) see d, 80%; h) 2,4,6-Trichlorobenzoyl chloride, NEt_3, THF; then *t*-butanol, DMAP, C_6H_6, 64%; i) Pd/C, H_2, EtOH, 94%; j) **10**, $(COCl)_2$, CH_2Cl_2; then **15**, THF; then CF_3COOH, CH_2Cl_2.

The solubilising oligo-ether groups were then attached using a Williamson type synthesis. Triethyleneglycol monomethyl ether was deprotonated using NaH in THF, and trimesylate **7** was added dropwise to the solution, which was subsequently heated at 60°C for 4 hours. Product **8** was isolated by column chromatography (SiO_2, CH_2Cl_2/MeOH/NEt_3 94:5:1). The full carbon framework of the dendritic wedge in **9** was formed by a radical addition to acrylonitrile, after loss of the nitro group in **8**. This approach was first used for dendrimer synthesis by Newkome and co-workers.[21] Product **9** was partially purified by column chromatography (SiO_2, CH_2Cl_2/MeOH/NEt_3 94:5:1) and the crude material then hydrolysed by refluxing with H_2O_2 and KOH in aqueous EtOH for 24 hours. Isolation of **10** was achieved by acidification, extraction with CH_2Cl_2 and column chromatography (SiO_2, CH_2Cl_2/MeOH/CF_3COOH 90:9:1). In spite of the excellent water solubility of **8-10**, these compounds could always be extracted from the water into halogenated solvents, greatly facilitating their handling.

The general methodology for the synthesis of **15** is similar to that for **10**. Compound **6** was benzylated by BnCl and KOH in Me_2SO. Purification of the protected product **11** was achieved by column chromatography (SiO_2, hexane/EtOAc gradient elution). The radical reaction was performed as above, with purification of **12** by chromatography (SiO_2, hexane/EtOAc 80:20). The yield was only moderate (48%), but the reaction was highly reproducible. Hydrolysis with H_2O_2 and KOH was, in this case, quantitative, with purification of **13** by extraction from acidic aqueous solution into CH_2Cl_2. The free acid was then protected as a *tert*-butyl ester via formation of an anhydride intermediate (Yamaguchi conditions).[22] Chromatography was used for the isolation of **14** (SiO_2, hexane/EtOAc 80:20). Finally, the benzyl protecting groups were removed by hydrogenation in EtOH using Pd/C (10%) as catalyst. Filtration through Celite yielded pure **15** in excellent yield.

The second generation dendritic branch **5** was synthesised by activating **10** as an acyl chloride (using oxalyl chloride in CH_2Cl_2), and subsequently reacting 3.5 equivalents with 1 equivalent of **15** in the presence of base (NEt_3 and catalytic amounts of DMAP). The resulting protected dendritic wedge **16** was purified by gel permeation chromatography (*Biobeads*, *SX1*, CH_2Cl_2). The *tert*-butyl ester group was then quantitatively removed using CF_3COOH in CH_2Cl_2 and the free wedge **5** isolated by partitioning between HCl_{aq} (0.1M) and CH_2Cl_2. The overall yield for these three steps was very good. All new products have been fully characterised (^1H and ^{13}C NMR, IR, FAB or MALDI MS, elemental analysis, and analytical GPC).

6. Hydrogen Bond Receptors: Dendritic Modulation of Selectivity

We used the dendritic wedges described above to make new dendritic clefts (dendroclefts) for the molecular recognition of monosaccharides through the formation of neutral hydrogen bonds.[23] Hydrogen bond recognition at a dendritic core has also been investigated by others but no specific favourable effect of the branching on binding strength or selectivity has been reported.[24] We targeted enantiomerically pure G-2 dendrocleft (–)-**17** (Fig. 7) by the attachment of the dendritic wedge **5** to a rigid (*R*)-configured 9,9'-spirobi[9*H*-fluorene] core with 2,6-diamidopyridine functionalisation in the 2,2' positions. This dendrimer possesses a buried cleft suitable for binding carbohydrates. The periphery is solubilised with neutral polyether groups.

Recognition studies were performed with octyl α- or β-glucopyranosides (L and D) (**18**, **19** and **20**) by ^1H NMR titrations in dry $CDCl_3$ at 298K. Association constants K_a [M^{-1}] and binding free energies ΔG [kJ mol^{-1}] were evaluated for the formed 1:1 complexes and a number of conclusions drawn.

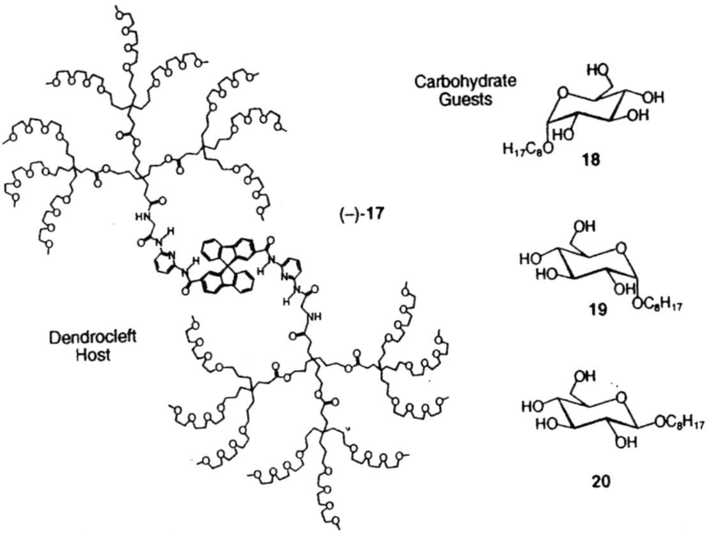

Figure 7. Dendrocleft (–)-**16** for the molecular recognition of monosaccharide guests.

i) The complexes formed by the G-1 and G-2 dendroclefts as well as the G-0 core are of similar strength (K_a between 100 and 600 M^{-1}). Large complexation induced downfield shifts of the diamidopyridine N-H resonances (up to 1.2 ppm at saturation) gave evidence for hydrogen bonds between the sugar O-atoms and the amide protons in the cleft. The dendritic shell still allows access to the core and remarkably does not, in general terms, weaken the binding (in spite of possessing a relatively high density of potentially competitive donor oxygen atoms).

ii) In the complexation of enantiomeric α-glucopyranosides **18** and **19**, the G-0 core exhibited a relatively high enantioselectivity in favour of the D-enantiomer ($\Delta\Delta G$= 3.6 $kJmol^{-1}$). Dendroclefts of first and second generation, however, only showed small enantioselectivities ($\Delta\Delta G$= 0.8 and 0.5 $kJmol^{-1}$, respectively).

iii) On the other hand, for diastereomeric guests **19** and **20**, which differ in their configuration at the anomeric position, the core only showed a small diastereoselectivity ($\Delta\Delta G$= 0.5 $kJmol^{-1}$), whilst the G-1 and G-2 dendroclefts showed larger selectivities ($\Delta\Delta G$= 1.4 and 2.3 $kJmol^{-1}$, respectively).

Not only is this the first report of chiral recognition at the core of a dendrimer, but the results indicate that the dendritic shell is controlling the binding selectivity; an unprecedented result. There are two plausible reasons for this currently under investigation. Firstly, the steric demands of the branching may disfavour certain complexes. Secondly, the oxygen donor atoms in the dendritic shell could participate in the formation of a hydrogen bonding network with the guest, changing the binding selectivity. This type of receptor provides an interesting model system for many sugar binding proteins, which bind their substrates selectively within a deeply buried hydrophobic cleft in the protein superstructure.[6]

The use of such dendrimers as chiroptical sensors is also currently under active investigation. Profound changes in the CD spectra are observed on addition of glucopyranoside guests, the response being selective for different sugars and different for the core compared with the dendroclefts. The dendritic receptor can also be easily recycled owing to the large size difference between host and guest. Simple filtration through a plug of Sephadex gel (LH-20, MeOH) provides quantitative recovery of pure (–)-**17** from host-guest solutions. Recyclability is an attractive feature in sensor design.

7. Summary and Conclusions

Molecular recognition within a controlled environment is a research field very much in its infancy, but as these results clearly illustrate, a dendrimer is certainly able to provide a unique environment for host-guest interactions. As time progresses, we are gaining greater insight into the role played by a dendritic shell in modulating binding strength and selectivity. In the future, it may be that the dendritic shell will exert even greater control over what happens at the core, so that recognition becomes essentially divorced from the solvent in which the chemist chooses to work.

As well as providing information of a fundamental nature about supramolecular dendrimer chemistry, these new dendritic receptors model the behaviour of many enzymes, which possess buried recognition sites. They also provide an entry into new molecular architectures, and the future could see the development of new supramolecular catalysts, switches and other devices based on recognition within the dendritic interior.

8. Acknowledgements

This work would not have been possible without the efforts of a number of talented chemists. They are gratefully acknowledged, and their names appear in the references. For financial support we thank the ETH research council, F. Hoffman-La Roche Ltd. (Basel), the Swiss National Science Foundation and the Royal Society (fellowship to D.K.S.).

9. References

1. a) Lehn, J.-M. (1995), *Supramolecular Chemistry: Concepts and Perspectives*, VCH, Weinheim.
2. For some recent examples see: various authors (1997), *Curr. Opin. Chem. Biol.* **1**, 447-521.
3. Smithrud, D.B. and Diederich, F. (1990) Strength of molecular complexation of apolar solutes in water and organic solvents is predictable by linear free energy relationships: A general model for solvation effects on apolar binding, *J. Am. Chem. Soc.* **112**, 339-343.
4. Kato, Y., Conn, M.M. and Rebek, J. (1995) Hydrogen bonding in water using synthetic receptors, *Proc. Natl. Acad. Sci.* **92**, 1208-1212.
5. a) Bonar-Law, R.P. and Sanders, J.K.M. (1995) Polyol recognition by a steroid capped porphyrin. Enhancement and modulation of misfit guest binding by added water or methanol, *J. Am. Chem.*

Soc. **117**, 259-271; b) Ladbury, J.E. (1996) Just add water! The effect of water on the specificity of protein-ligand binding sites and potential applications to drug design, *Chem. Biol.* **3**, 973-980.

6. Quiocho, F.A. (1989) Protein-carbohydrate interactions: Basic molecular features, *Pure Appl. Chem.* **61**, 1293-1306.
7. Newkome, G.R., Moorefield, C.N. and Vögtle, F. (1996) *Dendritic Molecules: Concepts, Syntheses, Perspectives*, VCH, Weinheim.
8. a) Issberner, J., Moors, R. and Vögtle, F. (1994) Dendrimers: From generations to functional groups and functions, *Angew. Chem., Int. Ed. Engl.* **33**, 2413-2420; b) Smith, D.K. and Diederich, F. (1998) Functional dendrimers: Unique biological mimics, *Chem. Eur. J.* **4**, 1353-1361.
9. Hawker, C.J., Wooley, K.L. and Fréchet, J.M.J. (1993) Solvatochromism as a probe of the microenvironment in dendritic polyethers: Transition from an extended to a globular structure, *J. Am. Chem. Soc.* **115**, 4375-4376.
10. a) Dandliker, P.J., Diederich, F., Gisselbrecht, J.-P., Louati, A. and Gross, M. (1995) Water soluble dendritic iron porphyrins: Synthetic models of globular heme proteins, *Angew. Chem., Int. Ed. Engl.* **34**, 2725-2728; b) Dandliker, P.J., Diederich, F., Zingg, A., Gisselbrecht, J.-P., Gross, M., Louati, A. and Sanford, E. (1997), Dendrimers with porphyrin cores: Synthetic models for globular heme proteins, *Helv. Chim. Acta* **80**, 1773-1801.
11. Zeng, F. and Zimmerman, S.C. (1997) Dendrimers in supramolecular chemistry: From molecular recognition to self-assembly, *Chem. Rev.* **97**, 1681-1712 and references therein.
12. Collman, J.P., Fu, L., Zingg, A. and Diederich, F. (1997) Dioxygen and carbon monoxide binding in dendritic iron(II) porphyrins, *Chem. Commun.* 193-194.
13. Jiang, D.-L. and Aida, T. (1996) A dendritic iron porphyrin as a novel haemoprotein mimic: Effects of the dendrimer cage on dioxygen binding affinity, *Chem. Commun.* 1523-1524.
14. Newkome, G.R. and Lin, X. (1991) Symmetrical, four-directional poly(ether-amide) cascade polymers, *Macromolecules* **24**, 1443-1444.
15. Collman, J.P., Brauman, J.I., Iverson, B.L., Sessler, J.L., Morris, R.M. and Gibson, Q.H. (1983) O2 and CO binding to iron(II) porphyrins: A comparison of the 'picket fence' and 'pocket' porphyrins, *J. Am. Chem. Soc.* **105**, 3052-3064.
16. Mattei, S., Seiler, P., Diederich, F. and Gramlich, V. (1995) Dendrophanes: Water soluble dendritic receptors, *Helv. Chim. Acta* **78**, 1904-1912.
17. a) Wallimann, P., Seiler, P. and Diederich, F. (1996) Dendrophanes: Novel steroid-recognizing dendritic receptors, *Helv. Chim. Acta* **79**, 779-788; b) Mattei, S., Wallimann, P., Kenda, B., Amrein, W. and Diederich, F. (1998) Dendrophanes: Water soluble dendritic receptors as models for buried recognition sites in globular proteins, *Helv. Chim. Acta* **80**, 2391-2417.
18. Zimmerman, S.C., Zeng, F., Reichert, D.E.C. and Kolotuchin, S.V. (1996) Self-assembling dendrimers, *Science* **271**, 1095-1098.
19. Kenda, B. and Diederich, F. (in press) Supramolecular aggregates of dendritic cyclophanes (dendrophanes) threaded on molecular rods with steroidal termini, *Angew. Chem., Int. Ed. Engl.*
20. The full experimental details will be reported in Smith, D.K., Diederich, F. and Zingg, A. (manuscript in preparation), *J. Chem. Soc., Perkin Trans. 2.*
21. Newkome, G.R., Arai, S., Fronczek, F.R., Moorefield, C.N., Lin, X. and Weis, C.D. (1993) Synthesis of functionalised cascade cores: Tetrakis(ω-bromoalkyl)methanes, *J. Org. Chem.* **58**, 898-903.
22. Inanaga, J., Hiroko, K., Saeki, H., Katsuki, T. and Yamaguchi, M. (1979), A rapid esterification by means of mixed anhydride and its application to large ring lactonisation, *Bull. Chem. Soc. Jpn.* **52**, 1989-1993.
23. Smith, D.K. and Diederich, F. (submitted) Dendritic hydrogen bonding receptors: Enantiomerically pure dendroclefts for the selective recognition of monosaccharides, *Chem. Commun.*
24. a) Zimmerman, S.C., Wang, Y., Bharathi, P. and Moore, J.S. (1998) Analysis of amidinium guest complexation by comparison of two classes of dendrimer hosts containing a hydrogen bonding unit at the core, *J. Am. Chem. Soc.* **120**; 2172-2173, b) Newkome, G.R., Woosley, B.D., He, E., Moorefield, C.N., Güther, R., Baker, G.R., Escamilla, G.H., Merrill, J. and Luftmann, H. (1996), Supramolecular chemistry of flexible, dendritic-based structures employing molecular recognition, *Chem. Commun.* 2737-2738.

SUPRAMOLECULAR CATALYSIS IN TRANSITION

JEREMY K. M. SANDERS

University Chemical Laboratory, Lensfield Road,
Cambridge, CB2 1EW, UK

1. Introduction

Supramolecular chemistry is poised at a fascinating moment in its history. Advances in synthesis, structural techniques and computing allow us to devise and prepare complex systems at will, study their structures and dynamics in exquisite detail, and rationalise the observations afterwards. So why, amongst the myriad of new supramolecular building blocks and arrays, do we see so few effective catalysts? How is it that we can apparently understand so much and yet fail the practical test of producing even rudimentary catalysis in any reliable way?

There are several motivations for attempting supramolecular catalysis: the most profound is probably that the essence of chemistry is to tame the molecule and so demonstrate our mastery over matter. It is only when we can predict behaviour and then demonstrate in the laboratory the accuracy of our prediction, that we can truly claim to have understanding. In supramolecular chemistry, success and failure depend on the delicate — and still unpredictable — balance between weak and opposing non-covalent interactions; in the absence of reliable prediction, we usually fail. Yet we know it can be done: enzymes represent the highest expression of chemical catalysis and therefore they are a key source of inspiration for supramolecular chemists. They achieve astonishing selectivities and catalytic efficiencies by deploying intermolecular forces to guide captured substrates very precisely along a reaction pathway towards the transition state and beyond. Additional binding interactions in the transition state ensure that the activated complex of the reaction is stabilised to a larger extent than the enzyme–substrate complex itself, so enzymes can be thought of as complementary in structure to the transition state of the reaction they catalyse. These binding properties, coupled with catalytic functionalities strategically placed within the enzyme active site, decrease the activation energy for reaction. If enzymes can achieve all this, surely so

R. Ungaro and E. Dalcanale (eds.), Supramolecular Science: Where It Is and Where It Is Going, 273–286.

can we? This article, which is a modified version of a recent review,[1] attempts to highlight some recent successes, explore why success is so elusive, and suggest some likely directions for future exploration. In particular I suggest that *in the search for effective catalysis* the fear of entropy in flexible systems has taken us too far in the direction of rigidity and pre-organisation.

2. Successful design approaches

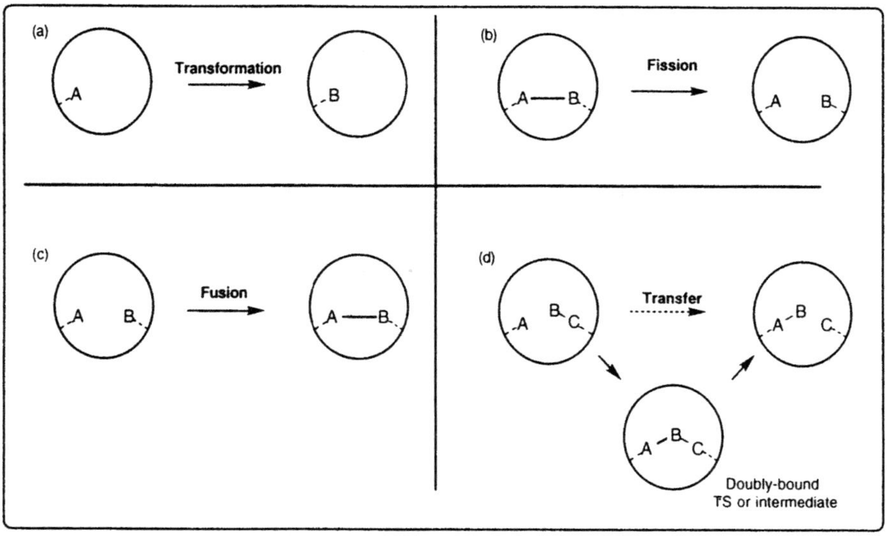

Figure 1. Schematic views of host-catalyzed reactions: (a) simple chemical transformation; (b) fission of a single substrate; (c) fusion of two substrates; (d) group transfer from one substrate to another *via* doubly-bound transition state or intermediate.

Figure 1 summarises the types of reactions one might hope to catalyse or influence. The simplest, and therefore the most common, are those that operate on a single substrate, catalyzing a chemical transformation such as an oxidation or a ring opening or closing reaction Figure 1(a). In such reactions, product binding is not likely to be stronger than substrate binding and catalytic turnover should be achievable except where the host molecule becomes covalently modified in the process. In practice it is proving much easier to achieve regio- or stereoselectivity in such reactions, by directing the substrate along one pathway rather than another, than it is to achieve catalytic turnover. For example, ring opening of cyclic phosphodiesters has been controlled in a specific direction (Fig. 2) by Breslow[2] using modified cyclodextrins and by Hamilton using Cu(bipy) complexes,[3] but in each case the "catalyst" is used in excess; turnover has not been demonstrated, presumably because product binding is comparable with, or even stronger than, substrate binding.

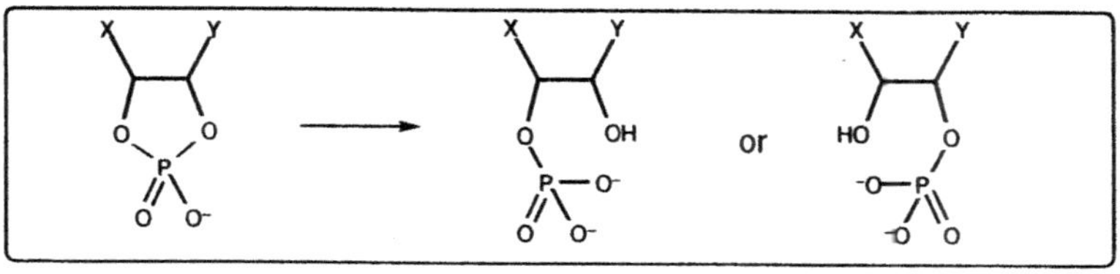

Figure 2 Ring-opening of asymmetric phosphodiesters can give two different products. Suitably-designed hosts can control the regiochemical outcome.[2,3]

Nevertheless, there are now some oxidation systems that do show catalytic turnover. For example, Breslow[4] has reported a manganese porphyrin equipped with four hydrophobic binding sites in the form of cyclodextrins; this mimic of cytochromep450 selectively oxidises unactivated carbon centres that are placed near the metal centre by virtue of the binding geometry (Fig. 3). Catalytic turnover is observed, but oxidative self-destruction of the catalyst competes with the desired chemistry; use of a more robust, halogenated, porphyrin has recently solve this problem.[5]

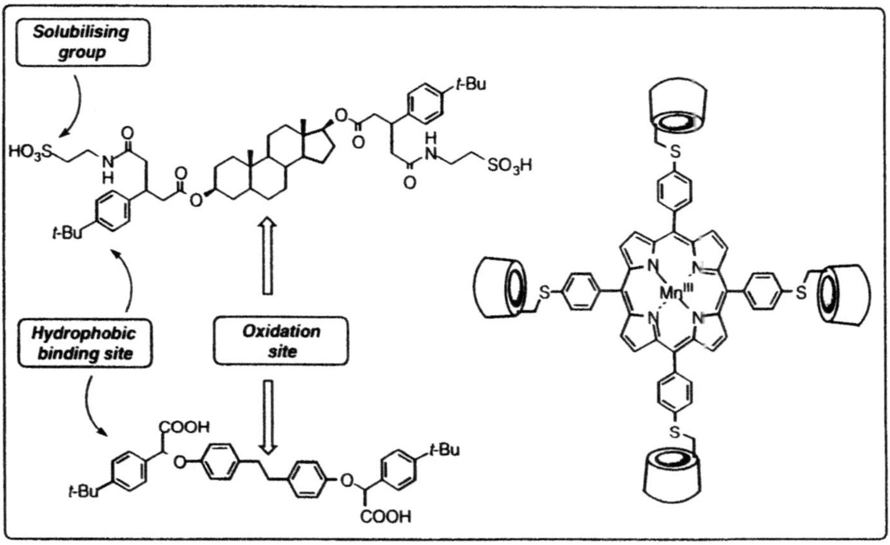

Figure 3 Catalytic oxidation by a cytochrome p450 mimic.[4] Cyclodextrins provide hydrophobic binding sites while the metal centre oxidises nearby unactivated carbons.

Even more impressive is Diederich's pyruvate oxidase mimic based on a cyclophane (Fig. 4):[6] in methanol this host can bind an aromatic aldehyde within its cavity, create a covalent intermediate by reaction with its attached thiazolium group, oxidise this tethered intermediate by intramolecular transfer of a "hydride-equivalent" to the appended flavin, and then release the methyl ester product by solvolysis. Catalytic turnover is achieved by electrochemical regeneration of the flavin. This system achieves a turnover number of around 100, and — unlike most supramolecular systems described to date — can operate on a genuinely preparative scale. The systems in Figs 3 and 4 are both excellent examples of supramolecular catalysis whereby several building blocks, each with its own specific function, are brought together to function in a coordinated and synergistic way.

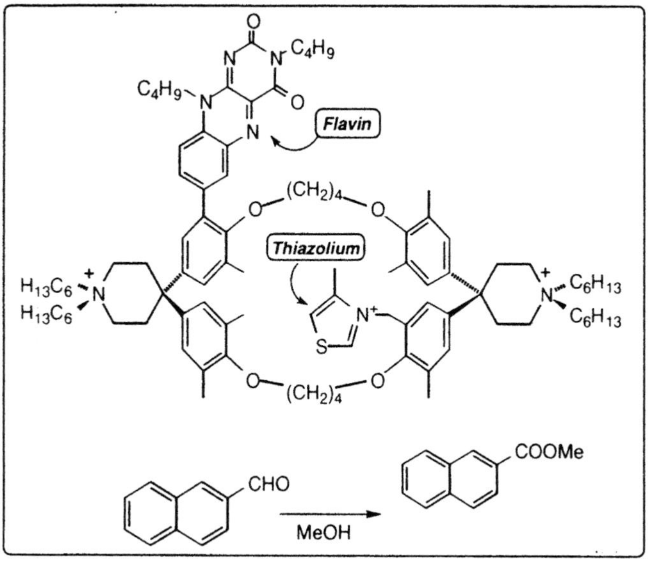

Figure 4 Diederich's model pyruvate oxidase system.[6]

Fission processes as illustrated in Fig. 1(b) should be a relatively easy type of reaction to catalyse with turnover as the products will necessarily be less well bound than the starting material. Once more Breslow has presented an effective system based on hydrophobic binding to cyclodextrins (Fig. 5);[7] a functional metal ion is again the engine of chemical change, this time activating a bound water by enhancing its nucleophilicity. The substrate is a highly activated *p*-nitrophenyl ester.

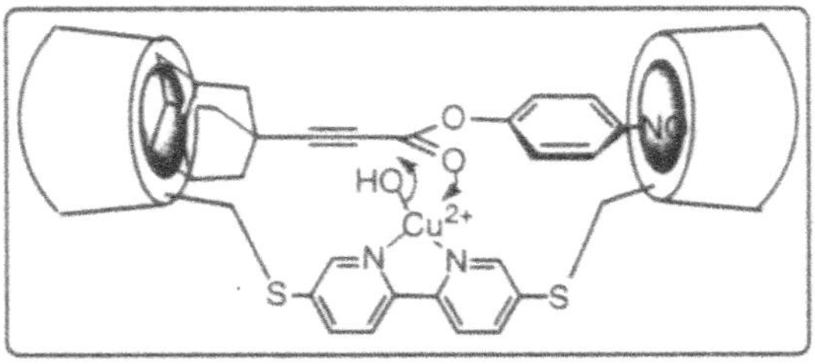

Figure 5. Breslow's hydrolysis system based on hydrophobic binding within the cyclodextrins and nucleophilic attack by a metal-bound, activated water molecule.[7]

Baltzer's group has recently described a fully-synthetic protein that is also capable of hydrolysing *p*-nitrophenyl esters:[8] the polypeptide, which contains 42 amino acids, was designed to fold into a hairpin helix–loop–helix motif that dimerises into a four-helix bundle. The dimer is predicted to present on its surface a shallow reactive site containing several histidine residues. The spectroscopic properties of the peptide are consistent with the predicted folded structure, and the molecule does indeed catalyse ester hydrolysis (and transesterification) more effectively than 4-methylimidazole does. However, there is little substrate selectivity, and not much turnover. The histidine array does not seem to act via general acid-base catalysis, but rather to bind and stabilise ester oxygens in the transition state. We return to this molecule below.

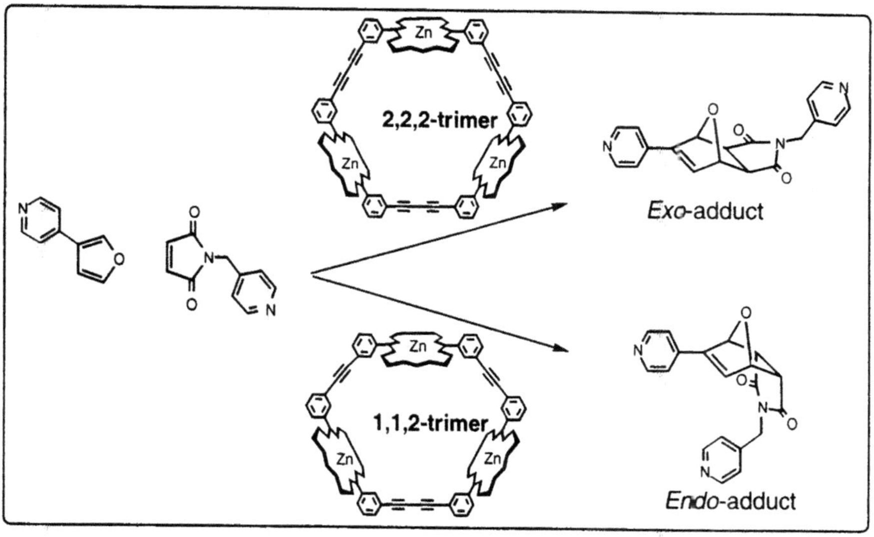

Figure 6 Redirection of a Diels–Alder reaction using the geometrical constraints of a host cavity.[9]

Catalysis of bimolecular reactions is difficult to achieve because two substrates need to be recognised and correctly oriented at the same time. In fusion reactions leading to net bond formation Figure 1(c) there is the additional problem that turnover is likely to be inhibited by strongly-bound product unless a geometry-modifying event can be engineered after the host-catalysed step. However, one can still accelerate the reaction and influence regio- or stereochemistry, using stoichiometric amounts of host to direct the outcome of the reaction. For example, subtle changes in host structure can lead to dramatic changes in the stereochemical outcome of a Diels–Alder reaction where two finely-balanced pathways compete (Fig. 6).[9] The reversal of stereoselectivity between the two cyclic trimers at 30° C is the result of two separate effects, one predicted and one not: the large (500-fold) *endo*-acceleration induced by the smaller 1,1,2-trimer was expected but the fact that this smaller trimer is ineffective at 30° C at accelerating the *exo*-reaction and binding the *exo*-adduct was a surprise. The key difference appears to lie in the greater flexibility of the larger 2,2,2-system: neither of the trimer hosts has the ideal equilibrium geometry to bind the *exo*-transition state or adduct, but at 30° C the larger trimer is more flexible and so is better able to respond to the geometrical demands of the *exo*-pathway. One might claim that the stereochemical reversal at 30° is a major success for the design approach, but it is important to note that at 60° C the small trimer becomes more flexible and loses its stereoselectivity. As in templated synthesis — which depends crucially on concentration effects — there is a narrow line between spectacular success and dismal failure.[10] The even more difficult tasks of changing regiochemistry and engineering catalytic turnover remain a challenge to be addressed.

Self-replicating systems in which two components bind to a template and then react to yield a further molecule of template represent a particularly interesting set of fusion reactions. Wang and Sutherland have recently reported a fully-synthetic system using a Diels–Alder reaction,[11] and there have also been notable successes with peptides:[12,13] through the use of hydrophobic and complementary electrostatic interactions, short α–helical synthetic peptides catalyse the condensation of two shorter peptide fragments. A cysteine at the end of one of the shorter peptide fragments participates in an initial transthiolesterification with a thioester of the other fragment; this is then followed by a rearrangement to give the final native peptide. Incorporation of ionisable groups into the sequence allows pH control over assembly of the correct shape for templated replication,[12] while Ghadiri discovered unexpectedly that a mixture of complementary peptides can lead to a cross-catalytic, symbiotic "hypercycle" in which each long peptide acts as template for the synthesis of the other.[13] Rate enhancements are very modest, but this beautiful result surely points one way forward for supramolecular catalysis, and may well have relevance for origin-of-life discussions.[14]

Finally in this survey of "success" we come to transfer reactions of the type shown in Figure 1(d); these should be ideal for demonstrating catalysis and turnover as the products should be only weakly bound and also because the intermediates or transition states are stabilised by being doubly-bound. Oxidation and ester hydrolysis reactions of the type described above could be considered in this category; otherwise there are few recent successes, probably because most transfer reactions also require the involvement of a catalyst or reagent. We reported some time ago[15] an acyl transfer reaction that is catalysed by the 2,2,2-porphyrin trimer shown in Fig 6, and we presented preliminary evidence consistent with the idea that the reaction proceeds through a tightly-bound intermediate. More detailed kinetic study has revealed that the observed catalysis and turnover owe as much to the action of subtle cavity effects on concentration and reactivity as to the stabilisation of putative intermediates.[16]

3. Why aren't design approaches more successful?

With rare exceptions,[4] synthetic systems achieve rate accelerations that are tiny by comparison with enzymes; furthermore, anecdotal evidence suggests that so far most synthetic systems have turned out to be inactive, however carefully they have been designed. It may be argued that it is too early in the development of the field to expect any better: enzymes have had millions of years in which to optimise their activities, while we have been trying for less than 50. It must of course be true that future work using a design approach with large building blocks will lead to major improvements, but collectively we have perhaps placed too much faith in our current powers of design. We may also have become too wedded to large building blocks by our fear of entropy: efficient binding requires maximising the enthalpic benefit of non-bonded interactions between host and guest while at the same time minimising the entropic cost. Fear of this escalating entropic cost has led many supramolecular chemists to regard host pre-organisation via structural rigidity as the top priority. As a result we have tended to create highly-engineered structures from large conformationally-restricted building blocks that are often decorated with buttresses and braces in order to inhibit any flexibility or responsiveness. A number of receptors have been described recently that respond geometrically to the demands of their guests, but none has been designed to possess useful catalytic properties.[17–19]

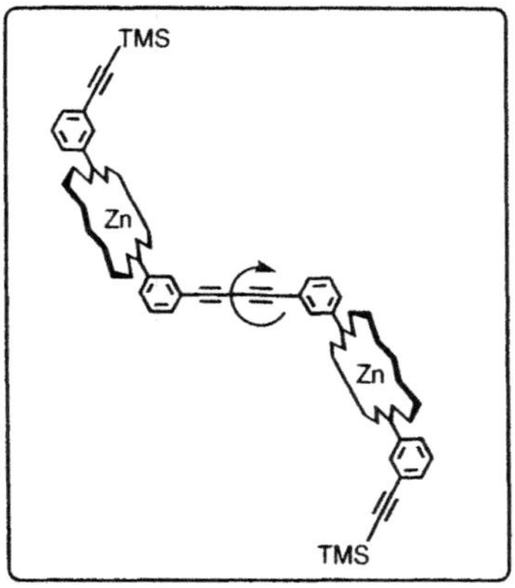

Figure 7 A rotationally-free linear porphyrin dimer that effectively and *exo*-selectively accelerates the Diels–Alder reaction in Fig. 6.[9]

However, we saw above that host flexibility is the key to *exo*-selectivity in the Diels–Alder reaction of Fig 6. Indeed the linear porphyrin dimer in Fig. 7 retains all the stereoselectivity and much of the accelerating power of its cyclic analogue even though the porphyrin units can sweep out great circles in conformational space by rotation about the alkyne linkers.[9] Therefore, I suggest that, in the search for catalysis, the pursuit of rigidity through large building blocks is, in part at least, a mistaken strategy:

(i) a genuinely rigid structure with a slight mismatch to the transition state will not be an effective catalyst;

(ii) the desired rigidity is rarely achieved anyway, but the flexibility is often in a direction that was not anticipated and that frustrates the designer;

(iii) the synthetic effort required to improve these elaborate structures is often so large that we prefer, or are forced, to study poor catalysts rather than to improve them;

(iv) it is difficult to achieve sub-Ångstrom adjustments using components that are large and rigid.

This is not to argue that we should abandon large building blocks in supramolecular chemistry: they will continue to provide us with beautiful structures and unexpected new insights. For example Chris Hunter's influential model of π–π interactions is the result of porphyrin dimers "behaving badly".[20] But I do believe that in the search for catalysis there are certainly two, and perhaps three, lessons we should learn from looking at real enzymes. Firstly, effective pre-organisation can be achieved in a single molecule by using a large number of small flexible building blocks with many competing weak non-covalent interactions; secondly, such a structure can be flexible yet still bind without excessive entropic cost.[21–23] The promising results that have been obtained with designed peptide helices[8,12,13] and with other repetitive oligomers[24–27] show that indeed some chemists are learning these lessons, although it will be a long hard road from repetitive motifs to sophisticated three-dimensional active sites. Our own experience with a variety of systems[28,29] suggests that the optimum synthetic strategy for macrocycles might be to combine relatively rigid building blocks with linkers to give systems that are sufficiently flexible to allow responsiveness but not so flexible as to unleash anarchy.

The strategy of many placing small components in a linear string (Fig. 8), while moving away from amino acid building blocks is an even bigger challenge for design and modelling. It also brings us to the third lesson from nature. There are so many possible structures to explore, and we are so poor at prediction, that selection of a good structure from a combinatorial mixture, or evolution under selective pressure, may prove a far more effective strategy.

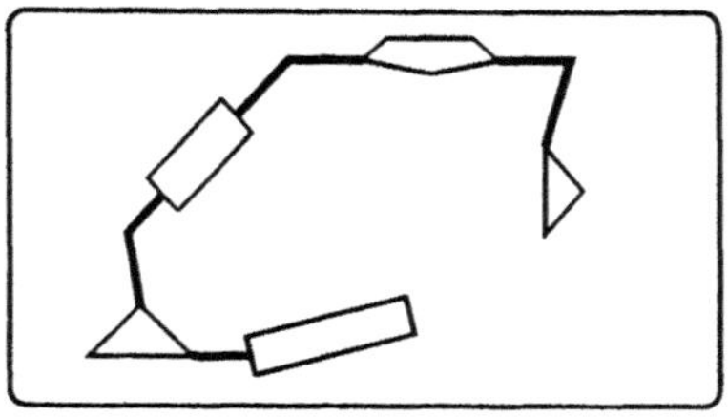

Figure 8 Cartoon of a general linear molecule that is capable of recognition and catalysis by virtue of a three-dimensional structure that is governed by non-covalent interactions.

4. Evolutionary and selection approaches

Given that enzymic catalysis implies selective recognition of the transition state, a suitable transition state analogue (TSA) should be able to elicit or select good catalysts from a mixture of many different molecules. Current approaches sharing this idea include molecular biology, organic and inorganic synthesis, and polymer and solid-state chemistry. The literature to early-1997 has been summarised elsewhere[30,31] so the remainder of this review is mainly restricted to very recent advances and some concluding thoughts.

Molecular biological approaches include catalytic antibodies, ribozymes based on RNA, and "DNAzymes". In each case a large pool of structures is generated chemically or biologically, and the most active is selected by some kind of TSA. All have the enormous advantage over chemical approaches that generic replication and mutation methods are available, so that tiny quantities of material can be identified, amplified and then subjected to further selection pressures. Very recent examples include DNAzymes that metallate porphyrins[32] and a catalytic antibody that promotes regio- and enantioselective Robinson annelations.[33] Biological approaches tend to lack the structural diversity that the synthetic chemist is so good at providing, although there are reports of antibody–metalloporphyrin complexes that are capable of peroxidase-like activity.[34] There also appear to be thermodynamic limits on what can be achieved by a biological system.[35]

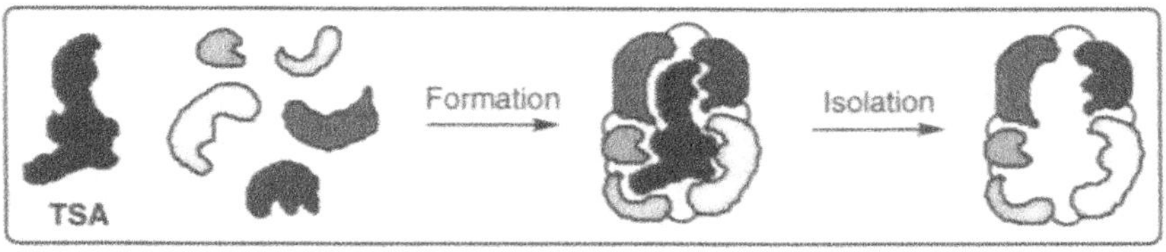

Figure 9 A general selection scheme for generating receptors or polymer cavities that are capable of recognition and catalysis.

A conceptually-related scheme that appears to be suitable for synthetic supramolecular chemistry is shown in Fig. 9: a TSA is used as a template for the optimum receptor or binding site. Several groups have been pursuing this type of approach through molecularly-imprinted polymers: Wulff has recently shown that a mixture of amidine-containing monomers (for functional group binding and catalysis), ethylene dimethylacrylate (as a cross-linker) and a phosphonic acid monoester TSA gave, after TSA-removal, an imprinted polymer that catalyses an unactivated ester cleavage by around 100-fold and demonstrates Michaelis–Menten kinetics.[36]

However, it will always be difficult to obtain homogeneous binding sites or make systematic rational changes using such a polymerisation approach. We have therefore been pursuing a solution-state equivalent in which the bond-formation step is covalent but reversible, and the TSA is tethered to a bead to allow recovery of good receptors.[28,29] The most effective receptors will be stabilised by binding to the TSA-template, while other species in solution will be "proof-read" and consumed. We believe that in this way it will be possible to isolate good TSA-binders (and therefore catalysts) from complex combinatorial mixtures, but it is too early to predict if this approach will prove successful: major technical difficulties still remain to be overcome. Still has described a conceptually similar approach.[37]

These selection approaches acknowledge that (a) truly effective design is currently beyond us, and (b) even if we were better designers, there is insufficient time to explore the multi-dimensional conceptual space that contains all possible catalytic receptors. Selection approaches provide the means to explore that space most efficiently.[14] However, selection schemes using TSAs are prisoners of the approach, in that the real transition state is inevitably different from the TSA. The correlation between TSA-bonding and catalytic activity is not as clear-cut as one would like,[38] and ways also need to be found for inducing catalysts for multi-step reactions. For the moment, therefore, the results of this type of selection experiment should be seen as providing a lead compound that can then be optimised by synthetic modification.

An alternative approach is to use the reaction itself as the tool for selection. Thus a catalytic RNA for the Diels-Alder reaction has been isolated by allowing it to covalently attach itself to a support via the Diels–Alder reaction; [39] similarly an RNA that can make peptide bonds,[40] a DNA that can cleave RNA[41] and a catalytic antibody have been selected by covalent trapping with their own reaction products.[42]

5. Conclusions

This review has highlighted recent progress in supramolecular catalysis, and has indicated new directions in which the field could move. In particular I have suggested that the fear of entropy has taken supramolecular chemists too far in the direction of rigidity and pre-organisation, and that the future may lie in more flexible systems that rely on non-covalent interactions to impose order on three-dimensional structure. I have also suggested that selection approaches may in the long run provide a better route than design for exploring the almost infinite number of possible catalyst structures, even given that modelling will inevitably become more reliable.

However, we the scientists, and our paymasters in government and industry, are impatient for results. For a many different reasons, some more justifiable than others, we want to demonstrate success, value for money, and a rôle in society for our kind of science. Supramolecular catalysis has not yet produced new industrial catalytic processes and it has made little impact in the world of synthetic methodology, so by these short term — and short-sighted — criteria it is a failure.

We must resist this way of thinking. Every supramolecular system has the potential to teach us something new and fundamental about the behaviour of molecules. The great paradox is that although we aspire to successful prediction, achieving such success would leave us with nothing to learn. So, we should not be afraid — or ashamed — of unpredicted failure: we must have the wit to recognise when a failure is more instructive than success, for it is only through the unexpected discovery that we can break truly new ground.

6. Acknowledgements

I am grateful to my many talented co-workers for contributing so much to my understanding of this problem, and to the EPSRC for generously supporting our research in this area over many years.

7. References

1 Sanders, J. K. M. (1998) Supramolecular catalysis in transition, *Chem. Eur. J.*, **4**, 1378–1383.

2 Breslow, R. and Schmuck, C. (1996) Goodness-of-fit in complexes between substrates and ribonuclease mimics — effects on binding, catalytic rate constants, and regiochemistry, *J. Am. Chem. Soc.* **118**, 6601–6605.

3 Liu, S., Luo, Z. and Hamilton, A. D. (1997) Rapid and highly selective cleavage of ribonucleoside 2',3'-cyclic monophosphates by dinuclear Cu-II complexes *Angew. Chem. Int. Ed. Engl*, **36**, 2678–2680.

4 Breslow, R., Huang, Y., Zhang, X. and Yang, J. (1997) An artificial cytochrome P450 that hydroxylates unactivated carbons with regio- and stereoselectivity and useful catalytic turnovers *Proc. Natl. Acad. Sci. U.S.A.*, **94**, 11156–11158.

5 Breslow, R., Gabriele, B. and Yang, J. (1998) Geometrically directed selective steroid hydroxylation with high turnover by a fluorinated artificial cytochrome P-450, *Tetrahedron Letters*, **39**, 2887–2890.

6 Mattei, P. and Diederich, F. (1997) Catalytic cyclophanes .11. A flavo-thiazolio-cyclophane as a biomimetic catalyst for the preparative-scale electro-oxidation of aromatic aldehydes to methyl esters, *Helv. Chim. Acta,* **80**, 1555–1588.

7 Breslow, R. and Zhang, B. L. (1997) Ester hydrolysis by a catalytic cyclodextrin dimer enzyme mimic with a metallobipyridyl linking group, *J. Am. Chem. Soc.*, **119**, 1676–1681.

8 Broo, K. S., Brive, L., Ahlberg, P. & Baltzer, L. (1997) Catalysis of hydrolysis and transesterification reactions of p-nitrophenyl esters by a designed helix-loop-helix dimer, *J. Am. Chem. Soc.*, **119**, 11362–11372.

9 Clyde–Watson, Z., Vidal-Ferran, A., Twyman, L. J., Walter, C. J., McCallien, D. W. J., Fanni, S., Bampos, N., Wylie, R. S. & Sanders, J. K. M. (1998) Reversing the stereochemistry of a Diels–Alder reaction: use of metalloporphyrin oligomers to control transition state stability, *New J. Chem.*, **122**, 493–502.

10 Sanders, J. K. M. (1996) Templated chemistry of porphyrin oligomers in *Comprehensive Supramolecular Chemistry*, *Vol. 9* (Eds.: Atwood, J. L. , Davies, J. E. D., Macnicol, D. D. & Vögtle, F.) Elsevier, Oxford (1996), pp. 131–164.

11 Wang, B. & Sutherland, I. O. (1997) Self-replication in a Diels-Alder reaction, *Chem. Commun.*, 1495–1496.

12 Yao, S., Ghosh, I., Zutshi, R. & Chmielewski, J. (1997) A pH-modulated, self-replicating peptide, *J. Am. Chem. Soc.*, **119**, 10559–10560.

13 Lee, D. H., Severin, K., Yokobayashi, Y. & Ghadiri, M. R. (1997) Emergence of symbiosis in peptide self-replication through a hypercyclic network, *Nature*, **390,** 591–594.

14 Kauffman, S. A. (1993) *The origins of order*, Oxford University Press, New York.

15 Mackay, L. G., Wylie, R. S. & Sanders, J. K. M. (1994) Catalytic acyl transfer by a cyclic porphyrin trimer — efficient turnover without product inhibition, *J. Am. Chem. Soc.*, **116**, 3141–3142.

16 Levy, E. G. (1997) Ph. D. thesis, University of Cambridge

17 Eliseev, A. V. & Nelen, M. I. (1998) Use of molecular recognition to drive chemical evolution, Part 2. Mechanisms of an automated genetic algorithm implementation, *Chem. Eur. J.*, **4**, 825–834.

18 Walliman, P., Mattei, S., Seiler, P. & Diederich, F. (1997) New cyclophanes as initiator cores for the construction of dendritic receptors: Host-guest complexation in aqueous solutions and structures of solid-state inclusion compounds , *Helv. Chim. Acta*, **80**, 2368–2390.

19 Bampos, N., Marvaud, V. & Sanders, J. K. M. (1998) Metalloporphyrin oligomers with collapsible cavities: characterisation and recognition properties of individual atropisomers, *Chem. Eur. J.,*. **4**, 335–343.

20 Hunter, C. A., & Sanders, J. K. M. (1990) The nature of π–π interactions, *J. Am. Chem. Soc.*, **112,** 5525–5534.

21 Zimmerman, S. C., Mrksich, M. and Baloga, M. (1989) Highly efficient complexation of a π-acceptor by a molecular tweezer containing two π-donors: the role of preorganization, *J. Am. Chem. Soc.*, **111**, 8528–8530.

22 Eblinger, F. & Schneider, H. J. (1998) Stabilities of hydrogen-bonded supramolecular complexes with various numbers of single bonds: Attempts to quantify a dogma in host-guest chemistry, *Angew. Chemie-Intl Edn*, **37**, 826–829.

23 Bundle, D. R., Alibés, R., Nilar, S., Otter, A., Warwas, M. and Zhang, P. (1998) Thermodynamic and conformational implications of glycosidic rotamers preorganized for binding,. *J. Am. Chem. Soc.*, **112**, 5317–5318.

24. Gellman, S. H. (1998) Foldamers : a manifesto, *Accounts of Chem.Res.*, **31**, 173–180.

25 Moore, J. S. (1997) Shape-persistent molecular architectures of nanoscale dimension, *Accounts of Chem.Res.*, **30**, 402–413.

26 Lokey, R. S. & Iverson, B. L. (1995) Synthetic molecules that fold into a pleated secondary structure in solution, *Nature* , **375**, 303–305.

27 Hintermann, T., Gademann, K., Jaun, B. & Seebach, D. (1998) gamma-peptides forming more stable secondary structures than alpha-peptides: Synthesis and helical NMR-solution structure of the gamma-hexapeptide analog of H-(Val-Ala-Leu)$_2$-OH, *Helv. Chim. Acta*, **81**, 983–1002.

28 Rowan, S. J. & Sanders, J. K. M. (1997) Building thermodynamic combinatorial libraries of quinine macrocycles, *Chem. Commun.*, 1407–1408.

29 Rowan, S. J., Lukeman, P. S., Reynolds, D. & Sanders, J. K. M. (1998) Engineering diversity into dynamic combinatorial libraries by use of a small flexible building block, *New. J. Chem.*, in press.

30 Brady, P. A. & Sanders, J. K. M. (1997) Selection approaches to catalytic systems, *Chem. Soc. Rev.*, **26**, 327–336.

31 Rowan, S. J. & Sanders, J. K. M. (1997) Enzyme models: design and selection, *Current Opinion in Chemical Biology* , **1**, 483–490.

32 Li, Y. F. & Sen, D. (1997) Toward an efficient DNAzyme, *Biochemistry*, **36**, 5589–5599.

33 Zhong, G., Hoffmann, T., Lerner, R. A., Danishefsky, S. & Barbas III, C. F. (1997) Antibody-catalyzed enantioselective Robinson annulation, *J. Am. Chem. Soc.*, **119**, 8131–8132.

34 Harada, A., Fukushima, H., Shiotsuki, K., Yamaguchi, H., Oka, F. & Kamachi, M. (1997) Peroxidation of pyrogallol by antibody-metalloporphyrin complexes, *Inorg. Chem.* , **36,** 6099–6102.

35 Mader, M. M. & Bartlett, P. A. (1997) Binding energy and catalysis: The implications for transition-state analogs and catalytic antibodies, *Chemical Reviews* , **97**, 1281–1301.

36 Wulff, G., Gross, T. & Schöfeld, R. (1997) Enzyme models based on molecularly imprinted polymers with strong esterase activity, *Angew. Chem. Intl. Ed. Engl.*, **36**, 1962–1964.

37 Hioki, H., Still, W. C. (1998) Chemical evolution: A model system that selects and amplifies a receptor for the tripeptide (D)Pro(L)Val(D)Val, *J. Org. Chem.*, **63**, 904–905.

38 Baca, M., Scanlan, T. S., Stephenson, R. C. &Wells, J. A. (1997) Phage display of a catalytic antibody to optimize affinity for transition-state analog binding, *Proc. Natl. Acad. Sci. U.S.A.*, **94**, 10063–10068.

39 Tarasow, T. M., Tarasow, S. L. & Eaton, B. E. (1997) RNA-catalysed carbon-carbon bond formation, *Nature* , **389**, 54–57.

40 Zhang, B. & Cech, T. R. (1997) Peptide bond formation by in vitro selected ribozymes, *Nature*, **390**, 96–99.

41 Geyer, C. R. & Sen, D. (1997) Evidence for the metal-cofactor independence of an RNA phosphodiester-cleaving DNA enzyme, *Chemistry and Biology*, **4**, 579–593.

42 Janda, K. D., Lo, L.-C., Lo, C.-H. L., Sim, M.-M., Wang, R., Wong, C.-H. & Lerner, R. A. (1997) Chemical selection for catalysis in combinatorial antibody libraries, *Science*, **275**, 945–948.

SUPRAMOLECULAR CHEMISTRY / SCIENCE
Some CONJECTURES and PERSPECTIVES

J.-M. LEHN
Laboratoire de Chimie Supramoléculaire, ISIS
Université Louis Pasteur, CNRS ESA 7006
4, rue Blaise Pascal, 67000 Strasbourg, France

1. Introduction

Supramolecular chemistry has over the last quarter of a century grown into a major field of chemistry and has fueled numerous developments in the related areas of biology and physics, thus giving rise to the emergence and establishment of *supramolecular science* (and *technology*), as a broad multi- and inter-disciplinary domain providing a highly fertile ground for the creativity of scientists from all origins. Its impact has been analyzed in a recent commentary on the theme of the present meeting [1], and illustrated by the perceptive presentation of five selected topics of supramolecular chemistry by some of the major players in the field [2-6]. The chapters in this book provide further extensive coverage and markedly widen the scope.

Rather than adding another facet to this already breathtaking panorama of the field, it might be appropriate here to emphasize "where it is going". The present essay will therefore not be extensively documented but rather present some conjectures for the future, mainly based on and illustrated by work performed in the author's laboratories. An outlook towards the perspectives of supramolecular chemistry, and more generally of supramolecular science, has been presented a few years ago [7]. These perspectives still hold and will be borrowed in part from this earlier text. Considerations on some selected areas will be given along three main themes:

R. Ungaro and E. Dalcanale (eds.), Supramolecular Science: Where It Is and Where It Is Going, 287–304.

- information and programmability;
- dynamics and reversibility;
- combinatorial features and adaptability.

2. From Condensed Matter to Organized Matter. The Information Paradigm

One of the major lines of development of chemical science in the coming century resides in the ever clearer perception, deeper analysis and more deliberate application of information features in the elaboration and transformation of matter, thus tracing the path from merely condensed matter to more and more highly organized matter towards systems of increasing complexity.

A most basic and far-reaching contribution of supramolecular chemistry is the introduction and implementation in chemical science of the concept of molecular information and its corollaries, instructions and programmed chemical systems, with the aim of gaining progressive control over the organization of matter.

In chemistry, like in other areas, the language of information is extending that of constitution and structure as the field develops towards more and more complex architectures and behaviors. Supramolecular chemistry has paved the way towards apprehending chemistry as an *information science*. This change in paradigm will profoundly influence our perception of chemistry, how we think about it, how we perform it. Instructed chemistry extends from the selective synthesis and reactivity of molecular structures to the organisation and function of complex supramolecular entities. It will consider sets of instructed constituents rather than pure substances, performing reactions in complex environments that will lead to the desired substances and properties by the action of built-in self-processes.

Supramolecular chemistry has started and developed as defined by its basic object, the chemistry of the entities generated via intermolecular non-covalent interactions [7]. Through the appropriate manipulation of these interactions, it became progressively the chemistry of molecular information, involving the storage of information at the molecular level, in the structural features, and its retrieval, transfer, processing at the supramolecular level, through molecular recognition processes operating via specific interactional algorithms.

Recent years have witnessed a great increase in activity towards the explicit application of molecular recognition as a means of controlling the evolution of supramolecular species and devices as they build up from their components and operate through *self-processes* (self-assembly, self-organization, self-recognition...). Systems presenting such features have been termed *programmed supramolecular systems* (see [7] and references therein).

Molecular programming involves the incorporation into molecular components of suitable instructions for the generation by self-assembly of a well-defined supramolecular entity. Depending on the design of the interaction patterns between the components, more or less strict programming of the output species will be achieved. The programme is molecular, the information being contained in the covalent structural framework; its operation through non-covalent recognition algorithms is supramolecular. The processing of molecular information via molecular recognition events implies the passage from the molecular to the supramolecular level.

Three stages of information input may be distinguished in self-assembly: 1) molecular *recognition* for the *selective binding* of the basic components; 2) *growth* through binding of the components in the correct relative disposition; 3) *termination* of the process, requiring a built-in feature, a *stop signal*, that specifies the end point and signifies that the process has reached completion.

In addition, *temporal information* is involved since self-assembly is of course a time-dependent process and may display kinetic control, generating kinetic products before reaching the thermodynamic ones. This is the case in the initial assembly of a triple helical complex in a (linear ligand/metal ion) system that progressively evolves towards a circular helicate [8]. Such a process may either be sequential, if the kinetic triple helix is an intermediate located on the pathway towards the final product, or it may be bifurcated, non-sequential, if this is not so (Figure 1). In the sequential case, the progressive build-up of the final superstructure occurs through a *defined sequence* of molecular instructions and algorithms, a given component or recognition event coming into play at a well-defined stage in the overall process. Then, the generation of a given intermediate structure will depend on the previous one and sets the stage for the next one amounting to *hierarchical self-assembly*.

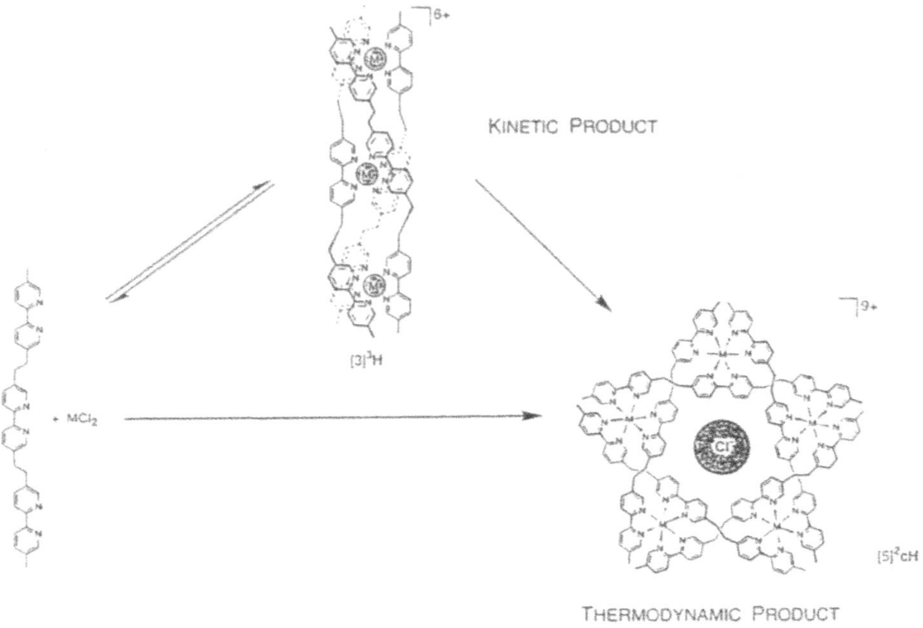

Figure 1. Kinetic and thermodynamic control in self-assembly illustrated by the generation of a triple helix as kinetic product and a circular helicate as thermodynamic product from a linear ligand and metal ions of octahedral coordination geometry (e. g. Fe(II), Ni(II)) [8].

This is observed in the formation of discotic liquid crystals by the assembly of phthalhydrazide derivatives (of "sector" shape) into disks, which thereafter build-up into discotic columns, as schematically represented in Figure 2 [9]. Such sequential self-assembly represents a further step in the design of artificial systems presenting higher levels of complexity.

An important feature of programmed systems is their *robustness*. In a *robust system* the instructions are strong enough for ensuring the stability of the process, i.e. the self-assembly is resistant, stable towards modifications of parameters such as concentrations and stoichiometries of the components, presence of foreign species, etc. Conversely, when the assembly only occurs in a narrow range of parameter values, the system is unstable and presents a singularity; it may also display a bifurcation or a switching point between different assemblies.

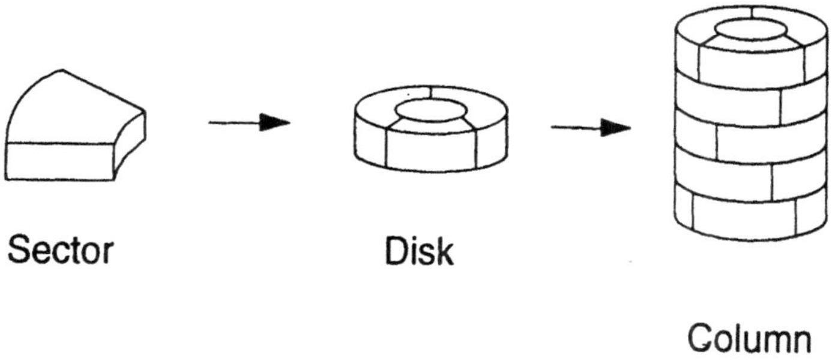

Figure 2. Hierarchical self-assembly. Self-assembly of sector components into a disk are a prerequisite for the subsequent self-assembly of the disks into a discotic columnar architecture [9].

The *self-recognition* observed in the self-assembly of helicates of different lengths indicates that the structural instructions are sufficiently strong [10], whereas this is not the case when there are only small differences in ligand structure so that mixed-ligand helicates are formed [11]. It may be noted that robustness is a key feature in the behavior of biochemical networks [12]. A basic factor in self-recognition is *commensurability* of the components.

In a broad perspective, one may point out [7, 10] the emergence of a new outlook involving a change in paradigm, from "pure compounds" to "instructed mixtures", from "unicity" (pure substance) to "multiplicity + information" (mixture of instructed components + program). Rather than pursuing mere chemical purity of a compound or a material, one seeks the design of instructed components which, as mixtures, lead through self-processes to the spontaneous and selective formation of the desired functional superstructures. This may recall the build-up of complex species (displaying highly integrated functions) that takes place side-by-side in the self-assembly and operation of the machinery of the living cell.

One may venture to predict that this *instructed mixture paradigm* will gain more and more weight in the years to come: the spontaneous but controlled built-up of structurally organized and functionally integrated supramolecular systems from a preexisting "soup" of instructed components following well-defined programs and interactional algorithms. It is also implicit in the implementation of combinatorial chemistry, a major area of present research activity (see also below).

Whereas the assembly of complex architectures usually involves several components of only one type, extension to *multiple components* of different structures and shapes represents an important step forward. Such is the case in the formation of single and multicompartimental inorganic superstructures from a set of ligands of two types [13]. It should be possible to gain access to even more complex entities by using several suitably instructed and functional ligands.

Structural features that code for the self-assembly of a given superstructure make up a specific assembly program. A further logical step is to explore how several such single-code programs, that independently govern the formation of different structures, can be combined as subroutines of an overall program. Specifically, the system may behave either as a linear combination of the subroutines with each yielding its predetermined substructure, or as a more complex combination with interference and crossover between the subprograms. This has been achieved with a ligand containing two different subunits coding respectively for the assembly of a helicate and of a [2x2] grid-type complex. It yields the architecture (Figure 3) that results from a crossover process [14a]; it was however not fully controlled since two outputs (with or without crossover) could be expected. Full control over the outcome of the process has been realized by means of a ligand which generates a single predetermined assembly depending on the type of metal ions used [14b]. This case also illustrates how the operation of different reading/coordination algorithms on the same program/ligand may lead to a different output/superstructure (see also Figure 4 below).

One may expect that it will be possible to devise molecular programs of more and more complex architecture, capable of producing a variety of outputs under the strict control of the interactional recognition algorithms. As is already implicit in the self-recognition and double-subroutine cases mentioned above, such developments lead to the introduction of *parallel processing* into programmed chemical systems, extending eventually to massively parallel systems where numerous self-assembly processes would operate in parallel towards the generation of a given complex entity.

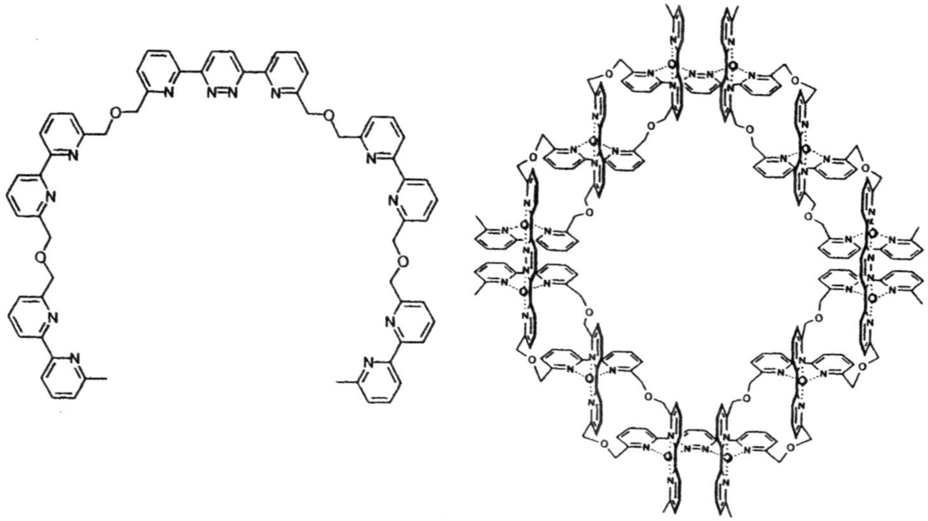

Figure 3. Double-subroutine ligand containing two different subunits coding respectively for the self-assembly of a double helicate and a [2x2] grid, and the crossover coordination architecture formed with Cu(I) ions [13].

The combination of different recognition/instruction features in a molecular program also opens a door to the design of self-assembling systems capable of performing *molecular computation.* Recent studies described the use of DNA-based protocols to solve combinatorial problems [15, 16]. There is no reason why an approach making use of specifically designed non-natural components should not be feasible, with possibly higher diversity, better resistance to fatigue and more compact/smaller size. Such potential may already be perceived in the coordination-controlled and combinatorial assembly of artificial double helical complexes from ligand strands containing defined sequences of different binding subunits by means of specific metal ions, as represented schematically in Figure 4 [17]. And there is no dearth in types of recognition units and interaction types, be they of inorganic or organic nature!

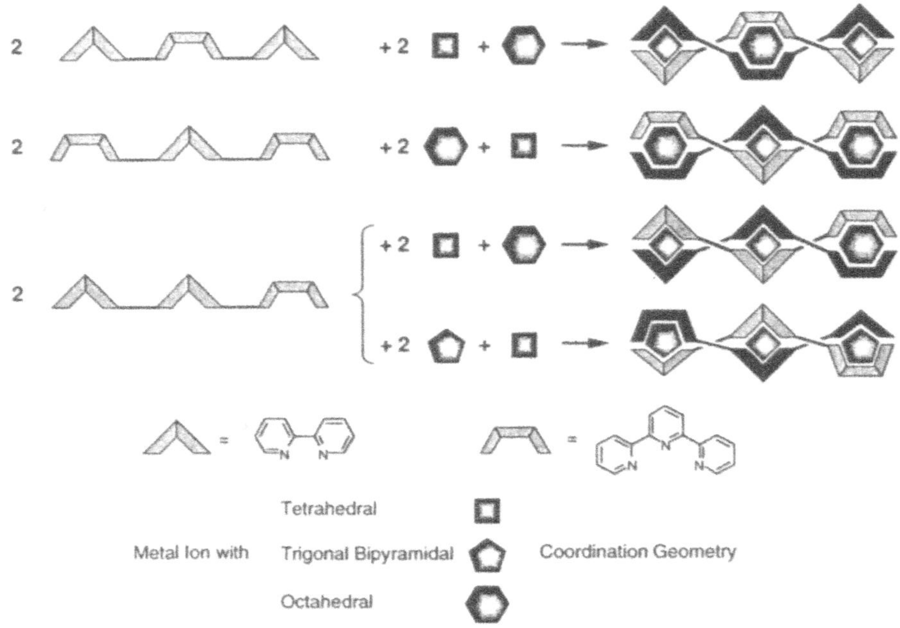

Figure 4. Self-assembly of linear ligands containing a defined sequence of different binding subunits with specific metal ions to yield predetermined double helicates. The third example (from top) represents a case where the reading of the same instructions (ligand subunits) by different interaction algorithms (metal ion coordination geometry) yields different outputs (double helicates).

3. Dynamic Chemistry. Supramolecular Materials: Programmable, Dynamic, Combinatorial Materials

Dynamic Chemistry

Supramolecular chemistry is by nature a *dynamic chemistry* in view of the lability of the interactions connecting the molecular components of a supramolecular entity. Most significantly, the reversibility of the associations allows a continuous reorganization by both modification of the connections between the constituents and incorporation or extrusion of constituents by exchange with the surroundings, thus conferring *combinatorial* features to the system.

Reversibility is a basic and crucial feature of supramolecular systems. It introduces the possibility of *adaptation* by association/growth/dissociation sequences.

The dynamic generation of molecular and supramolecular diversity from the reversible combination of covalently or non-covalently linked building blocks, defines a *dynamic combinatorial chemistry* (DCC) which gives access to dynamic combinatorial libraries, of either virtual (VCL) or real nature depending on the system and the conditions [18-20].

We were led to the DCC/VCL concept by our studies on self-recognition in the formation of correctly paired helicates from mixtures of different ligands and metal ions [10] and on the formation of circular helicates [21]. In the latter case it was found that a set of basic ligand molecules gave different multicomponent assemblies depending on their interaction with a target species, thus expressing a given constituent of the VCL of all possible combinations. The approach was applied to the selective amplification of an enzyme inhibitor from an interconverting set of constituents [18].

Supramolecular Materials. Dynamic Materials

The properties of a material depend both on the nature of the constituents and on the interactions between them. Supramolecular chemistry may thus be expected to have a strong impact on materials science via the explicit manipulation of the non-covalent forces that hold the constituents together. These interactions and the recognition processes that they underlie, allow the design of materials and the control of their build-up from suitable units by self-assembly.

Supramolecular materials are by nature *dynamic materials*, which we define as materials whose constituents are linked through *reversible* connections and undergo spontaneous and continuous assembly/deassembly processes in a given set of conditions. They are in fact either of molecular or of supramolecular nature depending on whether the links between the components are reversible covalent connections or non-covalent ones.

Because of their intrinsic ability to exchange their constituents, dynamic materials have also combinatorial character so that they may be considered as *Dynamic Combinatorial Materials* (DCMs). Supramolecular materials thus are *instructed*, *dynamic* and *combinatorial*; they may in principle select their constituents in response to external stimuli or environmental factors and thus behave as *adaptive materials* (Figure 5).

SUPRAMOLECULAR MATERIALS

★ INSTRUCTED — programmed, recognition-directed self-assembly from components

★ DYNAMIC — reversible incorporation of components
— adaptive, responsive to environmental factors

★ COMBINATORIAL — property-driven selection of components and amplification of preferred combination

⇨ ADAPTIVE MATERIALS

Figure 5. Basic features of supramolecular materials

Combinatorial materials have been considered for instance in the exploration of the electrical or optical properties of solids [22]. A significant further step is brought about by the introduction of reversibility, which confers the dimension of dynamic diversity to the generation of novel materials.

Supramolecular chemistry opens up new perspectives in materials chemistry towards an area of *supramolecular engineering*, which gives access to the molecular-information-controlled generation of well-defined polymolecular architectures and patterns in molecular assemblies, layers, films, membranes, micelles, gels, mesophases and solids as well as in large inorganic entities, polymetallic coordination architectures and coordination polymers.

Supramolecular Polymer Chemistry

A rich domain emerges from the combination of polymer chemistry with supramolecular chemistry, defining a supramolecular polymer chemistry [23, 24]. It involves the designed manipulation of molecular interactions (hydrogen bonding, donor-acceptor effects, etc.) and recognition processes to generate main-chain (or side-chain) supramolecular polymers by the self-assembly of complementary monomeric components (or by binding to lateral groups). In view of the lability of these associations, such entities present features of "living" polymers capable of growing or shortening, of rearranging their interaction patterns, of exchanging components; of undergoing annealing, healing and adaptation processes. Supramolecular versions of the various species and procedures of molecular polymer chemistry may be imagined and implemented [7, 23] providing a wide field of future investigation that may offer a wealth of novel entities and functionalities. *Growth control* may be achieved by the introduction of end-capping recognition units.

Supramolecular polymer chemistry is thus both dynamic and combinatorial and supramolecular polymers are thus DCMs. The recognition directed association of heterocomplementary monomers may yield supramolecular homopolymers or copolymers. The latter become combinatorial if a mixture of monomers containing different core groups R = X, Y, Z etc. is used (Figure 6). The components effectively incorporated into the polyassociation may depend on the nature of the R groups and on the interactions with the environment.

Similar considerations apply to the generation of *supramolecular liquid crystals* [23, 24].

Nanochemistry - Nanomaterials

Nanoscience and *nanotechnology* have become and will remain very active areas of investigation, in view of both their basic interest and their potential applications. Here again, supramolecular chemistry may have a deep impact. Indeed, the spontaneous but controlled generation of well-defined, functional supramolecular architectures of nanometric size through self-organization offers a very powerful alternative to nanofabrication an to nanomanipulation, providing a chemical approach to nanoscience and technology [7]. One may surmise that rather than having to stepwise construct or to top-down prefabricate nanostructures, methodologies resorting to self-organization from instructed components will be developped. The reversibility and dynamic features of such supramolecular architectures confers to them the potential to undergo healing and adaptation, processes of great value for the development of *"smart" nanomaterials*.

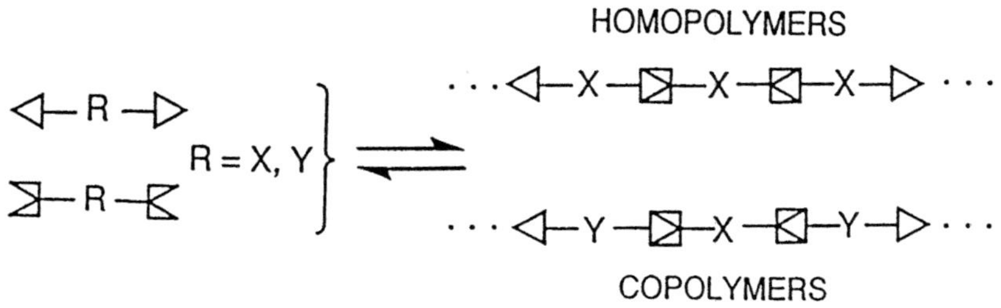

Figure 6. Formation of supramolecular main-chain polymers by polyassociation of complementary components.

Dynamic Devices

The area of molecular and supramolecular devices would desserve full treatment (see [25]) in view of the important developments one may predict for this field in the future. However we shall limit ourselves here to a few considerations concerning dynamic changes in mechano-devices which are the seat of motional processes.

Increasing attention has been given in recent years to the generation and control of molecular motions for instance in rotaxanes and catenanes [3, 26]. One should note in this respect, that early precursors to these processes are the displacements of metal centers about a ligand found in the extensively studied fluxional metal complexes and metallo-organic species [27]. They present a number of interesting features and offer a wide variety of systems; numerous developments may be imagined involving directed, long-range and triggered motions [28]. Examples of related processes are the displacement of metal ions inside ditopic cryptates [29] and the reversible, electrochemically triggered shape-switching between a double helical dinuclear Cu(I) complex and two single strand Cu(II) complexes (Figure 7) [30]. A number of photochromic systems undergo light-triggered interconversions involving molecular motions and shape change. Dynamic devices may formally represent prototypes of machines or motors if externally triggered motions become oriented for instance by attachment to a surface.

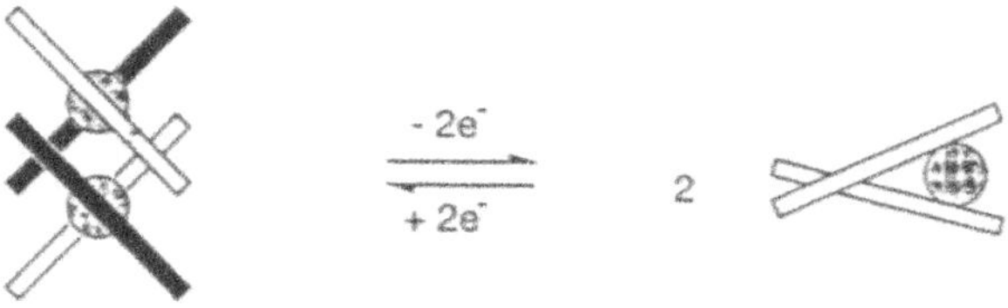

Figure 7. Electroswitching of molecular shape in the double helix-single strand interconversion between a double helical dinuclear Cu(I) complex and two single-strand Cu(II) complexes.

Further developments involve in particular the induction of directional motion, generating work, which requires the input of energy in order not to violate the second law of thermodynamics (for recent reports see [31-33]). On the other hand, devices based on optical, electrical or chemical stimuli and changes would seem to be preferable to mechanical ones, as is electronic or photonic computing with respect to mechanical computing. Much future activity is to be expected in these areas.

4. Towards ADAPTIVE CHEMISTRY

The combination of the features of supramolecular systems: - information and programmability, - dynamics and reversibility, - combinatorics and structural diversity, leads towards the emergence of the global concept of ADAPTIVE CHEMISTRY, which lies higher up the axis of complexity. It implies combinatorial assembly, selection and growth under *time reversibility*.

Beyond programmed systems the next step in complexity consists in the design of chemical *"learning" systems*, systems that are not just instructed but can be trained, that possess *self-modification* ability and *adaptability* in response to external stimuli (Figure 8). Adaptation can be *passive* or *active*. The former is illustrated by the reversible interconversion of pentameric and hexameric circular helicates as a function of the anions present in the medium [21]. The latter implies a learning process.

When *time irreversibility* is incorporated, *evolutionary chemical systems* are generated, representing the induction of *chemical evolution*. This opens perspectives towards systems that would undergo progressive change of internal structure under the pressure of environmental factors. It implies also the passage from closed systems to *open systems* that are connected spatially and temporally to their surroundings.

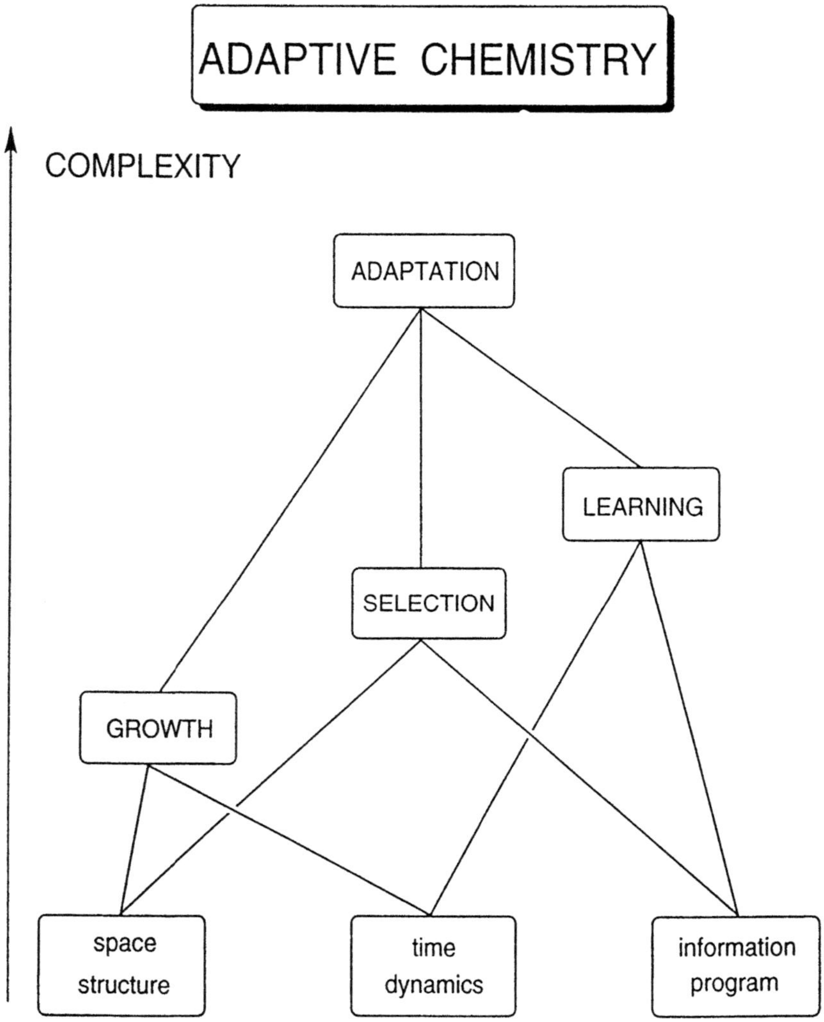

Figure 8. Towards adaptive chemistry, the supramolecular way.

5. Conclusion. A Science of Informed Matter

A major line of development of chemical science is towards complex systems, spanning the widest outlook from divided to condensed matter then to organized and adaptive matter, on to living matter and thinking matter, up the ladder of complexity.

Complexity implies and results from multiple components *and* interactions between them with integration *i.e.* long range correlation, coupling and feedback. The species and properties defining a given level of complexity result from and may be explained on the basis of the species belonging to the level below and of their multibody interaction, e.g. supramolecular entities in terms of molecules, cells in terms of supramolecular entities, tissues in terms of cells, organisms in terms of tissues and so on, up to the behavior of societies and ecosystems along a hierarchy of levels defining the architecture of complexity.

At each level of increasing complexity *novel features emerge* that do not exist at lower levels, that are *deducible* from but *not reducible* to those of lower levels [7].

Supramolecular chemistry, together with the corresponding fields in physics and biology, constitutes a supramolecular science whose already remarkable achievements, point to the challenges that lie ahead. They lead towards a science of *informed matter*, of *organized, complex, adaptive matter*, from inanimate to animate matter and beyond (Figure 9). The perspectives that open up and become wider and wider as progress is being made, will constitute horizons of the future for quite some time!

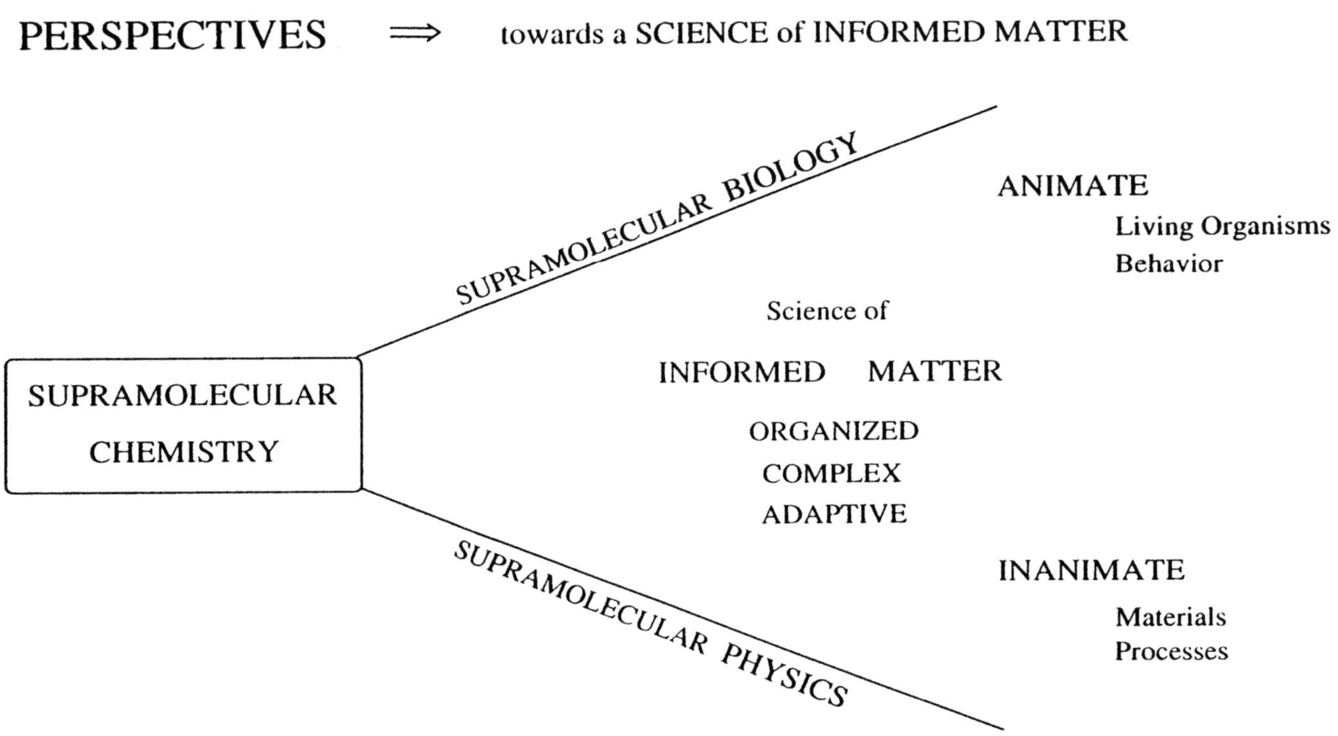

Figure 9. Supramolecular science as the science of informed matter.

6. References

1. Reinhoudt, D. N., Stoddart, J. F. and Ungaro, R. (1998) *Chem. Eur. J.* **4**, 1349.
2. Smith, D. K. and Diederich, F. (1998) *Chem. Eur. J.* **4**, 1353.
3. Chambron, J.-C. and Sauvage, J.-P. (1998) *Chem. Eur. J.* **4**, 1362.
4. Hartgerink, J., Clark, T. D. and Ghadiri, M. R. (1998) *Chem. Eur. J.* **4**, 1367.
5. de Mendoza, J. (1998) *Chem. Eur. J.* **4**, 1373.
6. Sanders, J. K. M. (1998) *Chem. Eur. J.* **4**, 1378.
7. Lehn, J.-M. (1995) *Supramolecular Chemistry - Concepts and Perspectives*, VCH, Weinheim, see especially chapters 9 and 10.
8. Hasenknopf, B., Lehn, J.-M., Boumediene, N., Leize E. and Van Dorsselaer, A. (1998) *Angew. Chem.*, in press.
9. Suárez, M., Lehn, J.-M., Zimmerman, S. C., Skoulios, A. and Heinrich, B. (1998) *J. Am. Chem. Soc.*, **120**, 9526; see also references therein.
10. Krämer, R., Lehn, J.-M. and Marquis, A. (1993) *Proc. Natl. Acad. Sci. USA* **90**, 5394.
11. Funeriu, D.-P., He, Y.-B., Bister, H.-J. and Lehn, J.-M. (1996) *Bull. Soc. Chim. Fr.* **133**, 673.
12. Barkai, N. and Leibler, S. (1997) *Nature* **387**, 913.
13. Baxter, P. N. W., Lehn, J.-M., DeCian, A. and Fischer, J. (1993) *Angew. Chem. int. Ed. Engl.* **32**, 69. Baxter, P. N. W., Lehn, J.-M., Kneisel, B., Baum, G. and Fenske, D. *Chem. Eur. J.*, in press.
14. a) Funeriu, D. P., Lehn, J.-M., Baum, G. and Fenske, D. (1997) *Chem. Eur. J.* **3**, 99; b) Funeriu, D. P. and Lehn, J.-M., to be published.
15. Adleman, L. M. (1994) *Science* **266**, 1021; Lipton, R. J. (1995) *Science* **268**, 542.
16. Guarnieri, F., Fliss, M. and Bancroft, B. (1996) *Science* **273**, 220.
17. Smith, V. and Lehn, J.-M. (1996) *Chem. Commun.* 2733.
18. Huc, I. and Lehn, J.-M. (1997) *Proc. Natl. Acad. Sci. USA* **94**, 2106; *ibid.* **94**, 8272.
19. Eliseev, A. V. and Lehn, J.-M., *Current Top. Microbiol. Immuno.*, in press; Ganesan, A. (1998) *Angew. Chem. Int. Ed.* **37**, 2828.
20. Brady, P. A. and Sanders, J. K. M. (1997) *Chem. Soc. Rev.* **26**, 327 and references therein.

21. Hasenknopf, B. Lehn, J.-M., Baum, G., Kneisel, B. O. and Fenske, D. (1996) *Angew. Chem. Int. Ed. Engl.* **35**, 1838; Hasenknopf, B., Lehn, J.-M., Boumediene, N., Dupont-Gervais, A., Van Dorsselaer, A., Kneisel, B. O. and Fenske, D. (1997) *J. Am. Chem. Soc.* **119**, 10956.

22. Gabriel, B., Chang, H., Sun, X. and Schultz, P. G. (1995) *Science* **270**, 273; Sun, X.-D., Wang, K.-A., Yoo, Y., Wallace-Freedman, W. G. Gao, C., Xiang, X.-D. and Schultz, P. G. (1997) *Adv. Mater.* **9**, 1046.

23. Lehn, J.-M. (1993) *Makromol. Chem., Macromol. Symp.* **69**, 1.

24. Antonietti, M. and Heinz, S. (1992) *Nachr. Chem. Techn. Lab.* **40**, 308; Paleos, C.M. and Tsiourvas, D. (1995) *Angew. Chem. Int. Ed. Engl.* **34**, 1696.

25. See Chapter 8 in ref. [7] for a broad presentation of the area; see also the chapter by Balzani, Credi and Venturi in the present volume.

26. Balzani, V. Gómez-López, M. and Stoddart, J. F. (1998) *Acc. Chem. Res.* **31**, 405; Sauvage, J.-P. (1998) *ibid.* **31**, 611.

27. Cotton, F. A. (1968) *Acc. Chem. Res.* **1**, 257. See also in (1994) *Encyclopedia of Inorganic Chemistry*, Wiley, Chichester, King, R. B., Ed. in chief, vol. 5, pp. 2581.

28. Fernandez, D., Pfammater, M., Khoury, R. and Lehn, J.-M., work in progress.

29. Lehn, J.-M. and Stubbs, M. E. (1974) *J. Am. Chem. Soc.* **96**, 4011; see for instance also: Zelikovich, L., Libman, J. and Shanzer, A. (1995) *Nature* **374**, 790.

30. Gisselbrecht, J.-P., Gross, M., Lehn, J.-M., Sauvage, J.-P., Ziessel, R., Piccinni-Leopardi, C., Arrieta, J. M., Germain, G. and Van Meerssche, M. (1984) *Nouv. J. Chim.* **8**, 661.

31. Faucheux, L. P., Bourdieu, L. S., Kaplan, P. D. and Libchaber, A. J. (1995) *Phys. Rev. Lett.* **74**, 1504.

32. Kelly, T. R., Tellitu, I. and Sostelo, J. P. (1997) *Angew. Chem. Int. Ed. Engl. 36*, 1866; Davis, A. P. (1998) *ibid.* **37**, 909.

33. Gimzewski, J. K., Schlitter, R. R., Langlais, V., Joachim, C., Tang, H. and Johannsen, I. (1998) *Science* **298**, 531; Sebastian, K. L. (1998) *Current Science* **75**, 536.

Author Index

Subject Index

GPSR Compliance

The European Union's (EU) General Product Safety Regulation (GPSR) is a set of rules that requires consumer products to be safe and our obligations to ensure this.

If you have any concerns about our products, you can contact us on ProductSafety@springernature.com

In case Publisher is established outside the EU, the EU authorized representative is:

Springer Nature Customer Service Center GmbH
Europaplatz 3
69115 Heidelberg, Germany

Batch number: 10755284

Printed by Printforce, the Netherlands